Analogue Electronic Circuit Design

Analogue Electronic Circuit Design

Jan Davidse

Professor, Faculty of Electrical Engineering
Delft University of Technology

Prentice Hall
New York London Toronto Sydney Tokyo Singapore

First published 1991 by
Prentice Hall International (UK) Ltd
66 Wood Lane End, Hemel Hempstead
Hertfordshire HP2 4RG
A division of
Simon & Schuster International Group

© Prentice Hall International (UK) Ltd, 1991

All rights reserved. No part of this publication may be
reproduced, stored in a retrieval system, or transmitted,
in any form, or by any means, electronic, mechanical,
photocopying, recording or otherwise, without prior
permission, in writing, from the publisher.
For permission within the United States of America
contact Prentice Hall Inc., Englewood Cliffs, NJ 07632.

Typeset in 10 on 12 point Times
by MCS Typesetters, Salisbury, Wiltshire, England

Printed and bound in Great Britain
by Dotesios Limited, Trowbridge

Library of Congress Cataloging-in-Publication Data

Davidse, J., 1929–
    Analogue electronic circuit design/by Jan Davidse.
      p.   cm.
    Includes bibliographical references and index.
    ISBN 0-13-035346-9 (pbk.): $45.00
    1. Linear integrated circuits – Design and construction.
2. Electronic circuit design.   3. Signal processing. I. Title.
TK7874.D37 1991   91-8200
621.381$'$5 – dc20   CIP

British Library Cataloguing in Publication Data

Davidse, Jan 1929–
    Analogue electronic circuit design.
    1. Analogue circuits. Design
    I. Title
    621.3815

ISBN 0-13-035346-9

---

1 2 3 4 5   95 94 93 92 91

Contents

Preface ix

1 Signals and signal handling 1
1.1 Types of electrical signals 1
1.2 When and where analogue signal processing is needed 3
1.3 Time domain and frequency domain 5
1.4 Symbols and sign conventions 6

2 Active devices and integrated circuits: a survey 9
2.1 Introduction: passive and active devices 9
2.2 Semiconductors 10
2.3 *PN*-junction 13
2.4 The bipolar transistor 18
2.5 Second-order effects in the bipolar transistor 25
2.6 Small-signal transistor modelling 29
2.7 The junction field-effect transistor 35
2.8 The MOS-transistor 39
2.9 Small-signal model for FETs 43
2.10 Body-effect 46
2.11 Subthreshold behaviour of the MOST 48
2.12 Monolithic technology 51
2.13 Monolithic bipolar devices 54
2.14 MOS-ICs 61
2.15 BICMOS-technology 64
2.16 Design styles 64

3 Signal transfer in linear signal-processing circuits 68
3.1 Introduction 68
3.2 Poles and zeros of linear transfer functions 68
3.3 Transient response of networks with given poles and zeros 74
3.4 Poles and zeros of feedback systems: root loci 81

4 Low-pass amplifier techniques 84
4.1 Amplification is the most important electronic signal-processing function 84
4.2 The problem of accurate signal transfer: definition of quality factors 85
4.3 Methods for achieving accuracy 90

4.4 Practical implementation of feedback amplifiers 94
4.5 Consequences of the imperfection of the active block; asymptotic-gain model of feedback amplifiers 101
4.6 A single-stage feedback amplifier 105
4.7 A voltage amplifier with feedback over two stages 107
4.8 Some general remarks concerning the application of negative feedback 110
4.9 Error-feedforward techniques 112
4.10 Compensation techniques 113

5 Basic amplifying circuit configurations 116

5.1 Introduction 116
5.2 CE- and CS-amplifier stages 116
5.3 Stages with local feedback 118
5.4 High-frequency properties of single-transistor stages 120
5.5 The CB- and CG- configurations 129
5.6 Emitter follower and source follower 132
5.7 Long-tailed pair 134
5.8 Cascode circuit 135
5.9 Series stage 136
5.10 Shunt stage 138
5.11 Compound active components 139
5.12 Biasing of active devices 142
5.13 Biasing circuits 143
5.14 Interstage level shifts 149
5.15 Drift effects 152

6 Differential amplifiers and operational amplifiers 154

6.1 Introduction: differential-mode and common-mode signals 154
6.2 Current sources 159
6.3 Signal transfer of current mirrors 167
6.4 Reference voltage and current sources 170
6.5 MOS reference sources 178
6.6 Active loads 181
6.7 Active loads in MOS-technology 186
6.8 Operational amplifiers 190
6.9 Offset and offset-compensation 191
6.10 Slew rate 195
6.11 Dedicated OPAMPS 198
6.12 OPAMP circuits: the inverting amplifier 199
6.13 Integrating amplifier 203
6.14 Non-inverting amplifier 204
6.15 Effects of offset-voltages 206
6.16 Logarithmic amplifier 208

7 Power amplifiers 210

7.1 Introduction 210
7.2 Driving considerations 210
7.3 Classes of operation 217
7.4 Implementation of class-AB output stages 221
7.5 MOST-output stages 226

8 Noise in electronic circuits 230

8.1 Introduction 230
8.2 Description of noise in electronic systems 230
8.3 Autocorrelation and spectral power density 231

8.4 Physical aspects of noise: thermal noise 233
8.5 Shot noise 235
8.6 Other types of noise 236
8.7 Noise in bipolar transistors 237
8.8 Noise in field-effect transistors 238
8.9 A general model for noise in electronic circuits 240
8.10 Application of the generalized model for noisy twoports to the modelling of noise in active devices 243
8.11 Qualification of noisy twoports 245
8.12 Optimization of the noise behaviour of a twoport 248
8.13 Noise matching in feedback amplifiers 253
8.14 Noise behaviour of basic amplifier configurations 257
8.15 Differential pair 259
8.16 Current sources 261

9 Low-pass wideband amplifiers 266
9.1 The basic problem of wideband amplification 266
9.2 Improving the bandwidth of single stages 267
9.3 Optimum pole–zero patterns for multistage wideband amplifiers 273
9.4 Complex poles by judicious design of a second-order loop transfer 276
9.5 A design example of a wideband feedback amplifier 285

9.6 Use of local-feedback stages 289
9.7 Distributed amplifiers 291

10 Oscillators 295
10.1 Introduction 295
10.2 Specification of oscillator properties 295
10.3 Harmonic oscillators 296
10.4 Practical implementation of harmonic oscillators 302
10.5 Amplitude stabilization 306
10.6 Realization of the negative resistance for undamping the resonator 311
10.7 Crystal oscillators 317
10.8 First-order or relaxation oscillators 321
10.9 Two-integrator relaxation oscillator 323

11 Bandpass amplifiers and active filters 331
11.1 Introduction 331
11.2 Methods for achieving bandpass characteristics 331
11.3 Bandpass filtering by resonators in interstage couplings 334
11.4 Use of coupled resonant circuits 337
11.5 Selective amplifier circuits 344
11.6 Neutralization 350
11.7 Methods for on-chip filtering 351
11.8 Sampled-data methods 358
11.9 Switched-capacitor filters 363

12 Circuits for modulation, demodulation and frequency conversion 368

12.1 Introduction 368
12.2 Amplitude modulation 369
12.3 Receiver architectures 371
12.4 Consequences of non-linearity in non-selective parts of the receiving system 375
12.5 General principles of circuits for modulation and frequency conversion 379
12.6 Switching modulators 383
12.7 Single-sideband modulators 389
12.8 Demodulation of AM-signals 391
12.9 Angle modulation 392
12.10 Circuits for FM-modulation 395
12.11 Demodulation of FM-signals 403

Index 413

Preface

Analogue electronics is the oldest branch in the field of electronics. Although much signal processing is done digitally at the present time, analogue electronics remains an indispensable discipline. The field has constantly been in a process of rejuvenation and, with monolithic technology as the driving force, this process has more gained speed during the last decade, than in any preceding period.

At present, analogue electronics is a wide field. Full coverage within the bounds of a single volume is illusory and any author of a textbook has to take decisions with regard to the selection of the subjects to be included and to the depth of their treatment. In writing this book I have adopted two guiding principles. First, emphasis should be on the pursuit of high performance in integrated implementations of analogue signal-processing circuits. Hence, subjects such as the judicious application of negative feedback and the mastering of noise in circuit design have obtained more attention than they get in most textbooks. Low-noise design is nearly always of the essence in high-performance signal processing, but experience reveals that suboptimal designs are not uncommon. Second, topics of a specialist nature have been omitted or only touched upon. As criteria for declaring a topic to be of a specialist nature, the availability of modern specialized books devoted to the particular subject has been taken. Examples in this category are A-to-D and D-to-A converters, signal processing with charge-coupled devices, switched-capacitor filters and phaselock loops.

The book has grown from course material which I taught for many years in graduate and upper undergraduate courses. Hence, most of the material has been tried out extensively on generations of students. The reader is assumed to possess basic knowledge of elementary network and signal theory and of electronic devices and circuits. In order to support readers whose knowledge of these topics is lacking to some extent, the first three chapters survey the prerequisites for fruitful study of the subsequent chapters.

A collection of problems, together with elaborated solutions is available from the publisher in a separate booklet. This material has been especially devised for the benefit of readers who use the book for self-study and who lack everyday guidance by a teacher. I believe that much of the material included in this book is valuable for analogue circuit designers who are willing to extend or to update their knowledge.

I headed the electronics laboratory of the Department of Electrical Engineering of the Delft University of Technology for many years. Acknowledgements are due to all members of the staff of this laboratory. Special thanks go to Leo de Jong, Gerard Meijer, Kees Wissenburgh and Albert van der Woerd for reading and commenting on the manuscript. They have saved me from many pitfalls, and their comments have contributed considerably to the final shape of the text. I am very grateful to Mrs Olfien Lefèbre-van den Broeke who undertook the enormous amount of typing work involved in the preparation and finalization of the text, and to Rob Janse, who prepared all the drawings using his self-developed drawing program 'tekplot'. Finally, I wish to thank the many generations of students who have continually challenged my didactic qualities and who, in a process of mutual interaction, in one sense were also my teachers.

Jan Davidse
Delft, September 1990

■ Signals and signal handling

1.1 Types of electrical signals

Electronics is the art of handling information-carrying electrical signals. The electrical quantity constituting the signal can be electrical charge, magnetic flux, current or voltage. Nearly always, the primary information is not available in the form of an electrical quantity but is available, for example, in the form of an optical, acoustic, thermal or chemical signal. To render these signals manageable by electronic devices and circuits they must be converted into electrical signals. The devices serving this purpose are called *input transducers* or *sensors*. After being transported and/or processed in an electronic system, the output signals of these systems must, in most cases, be converted again into the physical quantities that are appropriate to the final representation of the information. This is accomplished by devices called *output transducers* or *actuators*.

Figure 1.1 depicts the structure of a generalized system for the transport and/or processing of information. The link between the input transducer and the electronic processing system is called the 'front interface' of the system. Similarly, the corresponding link at the output side is called the 'rear interface'. A television system can serve as an example of an electronic system for the transport and processing of information. The primary information source is an optical image.

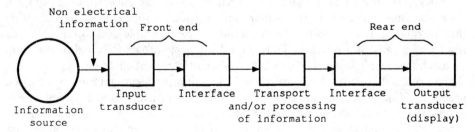

Fig. 1.1 *Generalized electronic system for the transport and/or processing of information.*

The signal obtained from the camera is amplified and modified in such a way that it can be transmitted by a television transmitter. At the reception site, the signal is picked up, amplified and processed for appropriate driving of the display system that reconverts the signal into an optical image.

The question may arise as to why the detour is needed for the employment of electronic means for the manipulation of the information. The reason is the unparalleled flexibility of electronic means. Electronic signal handling can be extremely fast. In addition, it can be achieved at the cost of very little power and material consumption. Microwatts and nanowatts are current power measures in electronics. The material of greatest interest in electronics is silicon, which is the second most abundant chemical element on earth. Even if the preferred material were not so ubiquitous, this would not be a problem since in modern electronics most circuits are realized in the form of chips, tiny pieces of material whose volume rarely exceeds 1 mm^3. The physical basis of the extreme versatility of electronic means is the very large specific charge (ratio of charge and mass) of the electron, viz. 1.76×10^{11} C kg^{-1}. It is therefore possible to exert large forces on the electron with very small fields and, hence, at the expense of very little power consumption.

The signals generated by the sources of primary information are almost always continuous-time signals, and the same holds for the electrical signal produced by the input transducer. The signal, as a function of time, corresponds to the time-dependent physical phenomenon providing the primary information. Such an electrical signal is an analogue of the primary information stream; it is, therefore, called an *analogue signal*. For example: the signal generated by a microphone is an electric analogue of the sound signal picked up by the device. It is characteristic of a true analogue signal that its value is relevant at any moment. However, the possible rate of change dS/dt of a signal $S(t)$ is always finite. In spectral terms this is expressed by the finite nature of the spectral bandwidth of the signal. Information theory states that a signal confined within a spectral bandwidth B is fully determined by $2B$ samples per second. Hence, without loss of information, a continuous-time signal can be converted into a discrete-time sampled data signal, provided the samples are taken at equal time intervals and at a rate exceeding $2B$ per second. For instance, a video signal confined to 5 MHz bandwidth can be fully specified by 10^7 samples per second. Figure 1.2 sketches both types of signals.

Though a sampled-data signal is discrete in time, the succession of sampled values is still an analogue of the primary information. A further step in discretizing the information is to measure the magnitude of each sample and to represent the results by means of a number. Thus the information stream is coded in the form of a sequence of numbers. This way of formatting is called 'digitizing' and the signal obtained is called a 'digital signal'. The number representation most often used is the binary representation, which uses only two digits, 0 and 1. A convenient electrical representation is in the form of pulses where, for example, no pulse denotes 0 and a pulse denotes 1. In present-day electronics, digital signal processing plays a dominant role. It has many attractive features, the most obvious being its robustness. Only two signal values can exist and, hence, noise and interference have

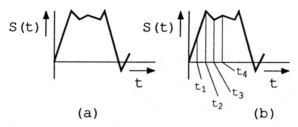

Fig. 1.2(a) *Example of an analogue signal.* (b) *Sampling of the signal at discrete time intervals.*

little grip on the signal. Moreover, circuits for handling digital signals are excellently suited to integrated-circuit (IC) implementation and to computer-aided circuit design (CACD) methods. A digital signal-processing system can, in principle, be entirely built up out of one type of elementary circuit, the logic decision circuit, and there is always a direct relationship between the function to be performed and the structure of the circuit implementing the function. The circuits can be built using transistors only, sometimes augmented with resistors. The transistors operate as switches and, in essence, only their switching behaviour is involved. In the days of discrete device electronics, the vast number of transistors needed to implement the desired functions was considered to be a drawback, but in the era of IC-electronics this is no problem at all, since hundreds of thousands of transistors can be accommodated within the confines of a single chip. There is no doubt that digital signal processing will continue to prevail wherever feasible. However, in certain areas analogue signal processing remains indispensable or desirable. These areas will be outlined in the following section.

1.2 When and where analogue signal processing is needed

The outside world, as it presents itself to a human observer, is essentially of an analogue nature. Our senses are organized for communicating with this analogue world. The signals delivered by the sensors at the front end of an electronic signal-processing system are almost always in analogue format. Customarily, these signals are very weak and therefore prone to noise and interference. 'Noise' is the term generally used to indicate the stochastic variations that fundamentally accompany all physical processes. These stochastical phenomena are a direct consequence of the second law of thermodynamics, which implies that they cannot be avoided. Hence, any primary signal is contaminated with noise and any signal handling adds noise to the signal. The ratio of signal power to noise power is called the 'signal-to-noise ratio' (S/N). This quantity is a measure of the accuracy with which the information content of a signal can be recovered. The reduction of S/N that accompanies any signal processing is insignificant when signal power is large in comparison with the noise added by the signal handling. Hence, the weak primary signal delivered by the transducer has to be amplified at the earliest possible stage of the signal handling. Accurate and low-noise amplification is an important task of the interface electronics (Figure 1.1). This interface has to cope with analogue signals. Since sensor properties vary widely, depending on the type of transduction process involved, interface electronics circuitry must always be designed to measure. Standard solutions are rarely applicable. From the foregoing, it will be clear that low-noise design is at a premium at this stage. In fact it is of great importance in electronic design in general, and for this reason much attention is given to the topic in this book.

Having obtained robust signals, one can choose between digital or analogue processing, whichever is most practical in the relevant situation. At the rear end of the chain the signal is applied to an actuator or a display system. Frequently, the final presentation of the information will be in analogue format again, for instance a sound signal or an optical image. At this rear end interface, noise will rarely be a problem because signal power can be made large. Here, the main requirements are usually the accuracy and power efficiency of the information transfer. In summary, it can be stated that analogue signal processing is nearly always indispensable in the front end and the rear end sections of the signal-processing system. Moreover, these analogue parts have to comply with critical and application-specific requirements.

A further area in which analogue signal processing is and remains of the essence is in systems where spectral bandwidth is a fundamentally scarce commodity, as is the case in non-directed communications, such as broadcasting. Given the amount of information to be handled, a digital signal has a much larger spectral width than an analogue signal. According to information theory, the information capacity C of a communication channel is given by

$$C = B \cdot {}^2\log\left(1 + \frac{S}{N}\right) \text{ bit s}^{-1} \quad (1.1)$$

where B is the bandwidth of the channel and S/N the ratio of signal power to noise power under the assumption of Gaussian noise. Conversion of an analogue signal to a digital signal implies a large increase in spectral width. As a consequence, the required capacity is obtained with low S/N. The analogue format requires minimum spectral bandwidth but, according to eq. (1.1), this implies that S/N should be high. Obviously, bandwidth and power can be exchanged within certain bounds and analogue signal processing minimizes bandwidth.

An important quantity related to signals and systems is the *dynamic range* of the signal involved. This is the ratio of the maximum and minimum signal levels that can occur. The maximum level that can be handled by a system is usually determined by its power handling capability; the minimum is determined by the noise with which the signal is contaminated. In certain systems the dynamic range encompassed by the signals to be processed can be quite large. An example is the signal delivered by an antenna for the medium wave broadcast band (bandwidth ≈ 1.5 MHz). The dynamic range in the relevant spectrum encompasses about 130 dB. The minimum signal level is on the microvolt level. Conversion into a digital signal would involve about 23 bits per sample, with the least significant bit at the level of, say, 10^{-7} V, which is far beyond present capabilities. A similar conclusion applies to the analogue-to-digital conversion of signals of extremely large spectral width. Bit rates in excess of some hundreds of Mbit s^{-1} are hardly feasible, given the present state of the art. The feasibility limit will increase in the course of time, but there will continue to be a limit.

Finally, note that in certain situations where, in principle, both digital and analogue signal processing are viable options, the analogue approach leads to much simpler solutions. The choice for the digital option implies the requirement of analogue-to-digital conversion (A/D-conversion), and vice versa. An accurate converter is a rather complex subsystem and its cost effectiveness can be disputed, particularly in cases where a modest amount of signal processing is required. Whether the analogue or the digital approach provides maximum cost effectiveness depends, of course, on the state of the art both in digital and in analogue techniques, and is, therefore, subject to change.

It is common practice to use the terms 'analogue electronics' and 'digital electronics'. Strictly speaking, digital electronics is a nonsense term in contrast to the term 'digital signal processing'. The latter applies to the system level where it is an appropriate indication of the signal format involved. Electronics is the art of manipulating electrons. On fundamental grounds, it is impossible to change the operation of an electronic device or circuit between two discrete states within an infinitely short time interval. Any change from the 0-state to the 1-state involves the passing of an infinitely large amount of intermediate states. In other words: during the transient from 0 to 1, the signal is a continuous-time or analogue signal. For the proper control of the transient situation, knowledge of analogue signal behaviour is indispensable. Admittedly, trying to banish the term 'digital electronics' would be an exercise in futility but, on the other hand, it must be stated that the designer of 'digital circuits' should be aware of the intricacies of transient behaviour, since these can to a large extent determine the ultimate performance of the designer's circuits.

1.3 Time domain and frequency domain

Any signal, whether analogue, sampled analogue or digital, is a function of time. In many cases it can and should be treated as such. Alternatively, a function of time can be described in an indirect way by applying a mathematical transform. One that has found widespread use is the Fourier transform. The Fourier transform $F(\omega)$ of a time-dependent signal $S(t)$ is defined by

$$F(\omega) = \int_{-\infty}^{+\infty} S(t) \exp(-j\omega t) \, dt \tag{1.2}$$

with the inverse transform

$$S(t) = \frac{1}{2\pi} \int_{-\infty}^{+\infty} F(\omega) \exp(j\omega t) \, d\omega \tag{1.3}$$

Equation (1.3) can be interpreted by stating that $S(t)$ is the superposition of an infinite number of sinusoidal signals with different frequencies. Thus the signal is represented by a spectral function. As has already been stated, the rate of change of a signal is always limited. In spectral terms this implies that the spectral bandwidth is limited. The relation between the two representations will be taken up again in Chapter 3.

The two ways of specifying signals have given rise to the terms *time domain representation* and *spectral domain* or *frequency domain representation*. One advantage of using the frequency domain approach is that, usually, the response of a circuit to a sinusoidal signal as a function of frequency is easily found. This applies to the majority of circuits of analogue electronics, and particularly to filter networks, modulators and frequency converters. The signal-processing operations that these last-mentioned circuits perform are essentially spectral manipulations. Describing their behaviour in terms of time domain signal processing usually leads to cumbersome calculations and formulas.

In frequency domain considerations, the quantity commonly used to specify frequency is the angular frequency ω. The relation between frequency f and angular frequency ω is $\omega = 2\pi f$. Strictly speaking, where ω is referred to, the term 'angular frequency' should be used. We will not stick to this tribute to accuracy in terminology. Instead, we will loosely use the term 'frequency' for either of the quantities f and ω, unless confusion would possibly be invoked by this mild form of sloppiness.

Readers who are not familiar with the various transforms used in signal theory are advised to consult one of the many textbooks dealing with this topic. For a quick refresher the compact survey in Gregorian and Themes (1986) may be useful.

1.4 Symbols and sign conventions

The conventions concerning the use of symbols in textbooks and papers are fairly well standardized, but in practice some variability can be observed. For the sake of completeness, a survey is given here on the conventions used in this book. Figure 1.3 shows the most important symbols used. For resistors, impedances and their reciprocals (conductor and admittance) the same symbol (Figure 1.3a) is used. Which meaning is applicable follows from the context. For the symbols used for field-effect transistors reference is made to Chapter 2. Wherever the type of field-effect transistor involved is irrelevant in principle, the general symbols of Figure 1.3g will be used. The symbols for voltage sources and for current sources are in agreement with international standardization, although many books and papers use alternative symbols. The symbols used here are logical and self-explanatory and they should therefore be preferred.

As to sign conventions: the plus- and minus-signs at the terminals of voltage sources or electronic circuits indicate which polarity is considered to be positive or negative. In order to prevent any confusion: plus does not mean 'this terminal is

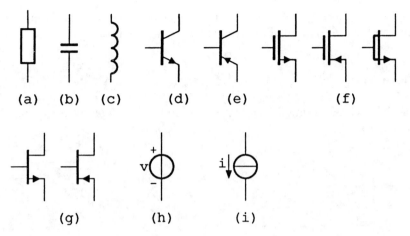

Fig. 1.3 *Symbols for circuit elements and electronic devices:* (a) *resistance or impedance;* (b) *capacitance;* (c) *inductance;* (d) *npn-bipolar transistor;* (e) *pnp-bipolar transistor;* (f) *various symbols for MOS field-effect transistors;* (g) *symbols for junction field-effect transistors and general symbols for field-effect transistors;* (h) *voltage source;* (i) *current source.*

positive with respect to the terminal carrying the minus sign', but rather 'if this terminal is positive with respect to the terminal with the minus sign, the voltage is labelled positive'. In like manner for currents: the direction of the arrow does not mean 'the current flows in the direction of the arrow', but rather 'if the current flows in the direction of the arrow it is labelled positive'. Further, in agreement with the conventions usually followed in treatments on circuit theory, positive current direction will be chosen to correspond with the direction *into* the network. Admittedly, this convention can lead to formulas that at first sight lack logic. Figure 1.4 gives a simple example. For this network,

$$\frac{i_o}{i_i} = -\frac{R_3}{R_2 + R_L + R_3}$$

hence, if $R_3 \to \infty$, $i_o/i_i = -1$, though it is obvious that the physical output current

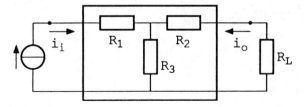

Fig. 1.4 *Current transfer in a simple resistive network.*

flows in the same direction as the input current. However, deviating from the established use would possibly provoke misinterpretations.

References

Gregorian, R. and Temes, G. C. (1986) *Analog MOS Integrated Circuits*, Wiley, New York.

2 Active devices and integrated circuits: a survey

2.1 Introduction: passive and active devices

Electronic circuits are composed of electronic components or devices. These components can be distinguished in several ways. An important distinction is that between *passive* and *active* components. *Passive* components can only dissipate or store energy. A resistor is a dissipating component, whereas capacitors and inductors, in their ideal form, are capable of storing energy without dissipation. Strictly speaking, electronic components must be distinguished from the corresponding elements defined in circuit theory. The latter are defined by mathematical relations and only by these. They lack physical properties such as weight, size and colour, whereas electronic devices do have these properties. Formally, resistors, capacitors and inductors or coils should not be indicated as resistances, capacitances and inductances. Though the use of the latter terms should properly be reserved to denote network elements, the distinction is usually only loosely maintained and we will adopt this practice here.

Active devices are devices that are capable of amplification of signal power. Figure 2.1 depicts the general principle underlying amplifying devices. The device contains at least one electrode that is capable of emitting charge carriers, very often

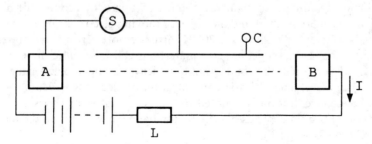

Fig. 2.1 *The general principle of operation of an active device; S is the signal source.*

electrons, in a region where their behaviour can be controlled (electrode A in Figure 2.1). There is also at least one electrode where the charge carriers are collected (electrode B). And there is at least one electrode (C) that is driven by the signal-carrying quantity S (voltage, current, charge), which is capable of controlling the process of emission or transport of charge carriers between A and B. Several names are used to denote these electrodes. The emitting electrode is called the *cathode* in electron tubes, the *emitter* in bipolar transistors and the *source* in field-effect transistors. The collecting electrode is denoted by the names *anode, collector* and *drain* respectively. For the control electrode the names *grid, base* and *gate* are used. Under control of the signal S, the current I in the external circuit is modified. With proper dimensioning of the device and its environment, the signal power P_o developed in the load L can exceed the input signal power P_i which implies power amplification. Of course, the law of conservation of energy is not violated; in fact the amplifying device converts supply energy into signal energy.

The first electronic amplifying device was the vacuum triode. By general consensus it is agreed that the invention of this device by Lee de Forest in 1906 marks the birth of electronics.

Electron tubes are no longer used as amplifying devices, except in very special cases, for instance in applications where very high power is involved. In essence, tubes are field-effect devices; they behave much like field-effect transistors. In this book, vacuum tubes will not be used as building elements of electronic circuits. It is not entirely impossible that the old device will enjoy a comeback in quite a new fashion. Attempts have been made to construct integrated circuits wherein field-effect devices are formed which are essentially vacuum tubes and behave as such. For the time being this is speculation. For the rest it can be stated that in the realm of the primary subject of this book, electronic circuit design, device properties are second to circuit configurations.

2.2 Semiconductors

Transistors are made from semiconducting materials. Solid-state physics classifies a crystalline material as a semiconductor when the two energy bands denoted as valence band and conduction band are about 1 eV apart. In a conductor both bands overlap so that there is an abundance of free moving electrons. In an insulator the bands are so far apart that at $T = 300$ K virtually no electrons are present in the conducting band. Pure semiconductor material is called *intrinsic* material. In such material each electron in the conduction band leaves an open place, called a 'hole', in the valence band. The intrinsic concentration of electrons is denoted by n_i; the intrinsic hole concentration is denoted by p_i. As a matter of course, $p_i = n_i$. The intrinsic concentration depends heavily on the bandgap W_G and on the temperature T. The relation is given by

$$n_i^2 = cT^3 \exp(-W_G/kT) \qquad (2.1)$$

where C is a proportionality constant.

In silicon, which is the dominating semiconductor in electronics by far, at $T = 300$ K, $W_G \approx 1.16$ eV and $n_i^2 = 2.25 \times 10^{20}$ cm^{-6}.

Silicon is a tetravalent material. If it is doped with a pentavalent material, such as phosphorus or arsenic, each doping atom contributes an electron that is free – in terms of the energy band model it is in the conduction band – at $T = 300$ K. Silicon doped in this way is called n-type silicon; the dopant is called a donor material. By virtue of the equilibrium between free charge-carrier generation and recombination, the product of the electron concentration n and the hole concentration p is constant, independent of the doping level. Consequently, $pn = n_i^2$. The electron concentration is very much dominated by the doping, hence $n = N_D$, where N_D denotes the concentration of the dopant atoms. Since n_i^2 is fixed, according to eq. (2.1), and $n = N_D$ holds, $p = n_i^2/N_D$. Practical doping levels range from 10^{13} to 10^{19} cm^{-3}. Hence for any practical value of N_D, $n \gg p$ holds. The electrons are called *majority* carriers, the holes *minority* carriers.

EXAMPLE
Let N_D be 10^{18} cm^{-3}, a common doping level in practice, then $p = 2.25 \times 10^{20}/10^{18} = 225$ cm^{-3}, so that the qualification of minority carriers is amply justified.

Conversely, if silicon is doped with a trivalent material, such as boron, each boron atom provides a hole. Since, again, $np = n_i^2$, in this type of silicon, called p-type silicon, the electrons are minorities, whereas the holes are majorities. The dopant is called an acceptor material and the acceptor concentration is indicated by N_A.

Since n_i^2 depends heavily on the temperature T, and N_A or N_D do not depend on T, the majority concentration is nearly independent of T, whereas the minority concentration is strongly temperature-dependent.

If, for some reason or other, the minority concentration is brought above the equilibrium level, recombination temporarily exceeds generation. As a consequence, the excess concentration fades out gradually. Let n_o be the equilibrium concentration and Δn the excess concentration at time $t = 0$, then

$$n = n_o + \Delta n \cdot \exp(-t/\tau) \tag{2.2}$$

where t is called the 'lifetime' of the minority carriers. This quantity is an important measure for the perfection of the crystal lattice. Since the working mechanism of bipolar transistors depends on the possibility of temporarily maintaining an excess minority concentration, bipolar transistors must be made of high-quality monocrystalline silicon.

Charge transport in semiconductors can take place due to two mechanisms. If in a homogeneously doped crystal the carrier concentrations differ at different locations, transport by *diffusion* of carriers occurs. The flow of carriers is proportional to the concentration gradient, hence

$$I_n = qD_n \frac{dn}{dx} \tag{2.3a}$$

where q denotes electron charge, I_n is the electron current density and D_n is a constant called the 'diffusion constant for electrons'. Similarly,

$$I_p = -qD_p \frac{dp}{dx} \tag{2.3b}$$

The minus sign expresses that the 'charge of a hole' is $-q$. In general, the diffusion constants for holes and electrons in a given material are different. In silicon D_n is almost three times D_p.

The second transport mechanism is carrier flow due to the presence of an electric field E. The current density is proportional to E, hence for electrons

$$I_n = qn\mu_n E \tag{2.4a}$$

and for holes

$$I_p = qp\mu_p E \tag{2.4b}$$

where μ_n and μ_p are called the electron mobility and the hole mobility. The quantities D and μ are interrelated by $\mu/D = q/kT$, where k is Boltzmann's constant.

If both mechanisms are active simultaneously, we have

$$I_n = q\mu_n nE + qD_n \frac{dn}{dx} \tag{2.5a}$$

and

$$I_p = q\mu_p pE - qD_p \frac{dp}{dx} \tag{2.5b}$$

Table 2.1 *Some important physical constants and data*

Constant	Symbol	Value
Charge of electron	q	1.60×10^{-19} C
Mass of electron	m	9.11×10^{-31} kg
Specific charge of electron	q/m	1.76×10^{11} C kg^{-1}
Boltzmann's constant	k	1.38×10^{-23} J K^{-1}
Band-gap energy of silicon	W_G	1.16 eV
Electron concentration in intrinsic silicon ($T = 300$ K)	n_i	1.5×10^{10} cm^{-3}
Electron mobility in intrinsic silicon ($T = 300$ K)	μ_n	1350 cm^2 V$^{-1} \times$ s^{-1}
Hole mobility in intrinsic silicon ($T = 300$ K)	μ_p	480 cm^2 V$^{-1} \times$ s^{-1}
Number of atoms in silicon per cm^3		5×10^{22} cm^{-3}

Table 2.1 gives the numerical values of the relevant quantities. By the way, note that electron currents and hole currents are independently in equilibrium, implying that both currents cannot mutually compensate for obtaining overall charge neutrality. Such a mechanism would violate the second law of thermodynamics, which requires that generation and recombination should be locally in equilibrium. This is called the 'principle of detailed balance'.

2.3 PN-junction

At the boundary of an n-type region and a p-type region, electrons tend to diffuse into the p-region, while holes tend to diffuse into the n-region. As a consequence, a space-charge layer builds up. According to the laws of electrostatics, the presence of space charge ρ gives rise to an electric field E and a potential distribution V. The relation between these quantities is determined by Poisson's equation. If we consider a junction between homogeneously doped regions, the distribution of space charge is unidimensional. Let the direction perpendicular to the junction be denoted by x, then Poisson's equation reads

$$-\frac{d^2V}{dx^2} = \frac{dE}{dx} = \frac{\rho}{\varepsilon\varepsilon_0} \tag{2.6}$$

where $\varepsilon\varepsilon_0$ is the permittivity of the material. Figure 2.2 sketches space charge ρ, electric field E and potential distribution V for an abrupt junction. As the diagram shows, a potential barrier is established. This barrier counteracts the diffusion of charge carriers. On the other hand, the electric field sweeps electrons generated in the p-region to the n-region and holes generated in the n-region to the p-region. In the steady state the diffusion currents and the field driven currents are in equilibrium. The height of the potential barrier is determined by the laws of Boltzmann statistics:

$$V_d = \frac{kT}{q} \ln \frac{n_n}{n_p} = \frac{kT}{q} \ln \frac{N_A N_D}{n_i^2} \tag{2.7}$$

where n_n and n_p denote electron concentrations in the n-region and the p-region respectively.

EXAMPLE
A junction is formed between an n-region with $N_D = 10^{19}$ cm^{-3} and a p-region with $N_A = 10^{16}$ cm^{-3}. What is the height of the potential barrier? With $kT/q = 26$ mV at $T = 300$ K:

$$V_d = 26 \ln \frac{10^{35}}{2.25 \times 10^{20}} \text{ mV} \approx 875 \text{ mV}$$

If an external voltage V is applied to the junction, the equilibrium is disturbed. If

14 ANALOGUE ELECTRONIC CIRCUIT DESIGN

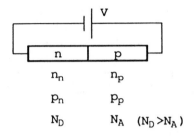

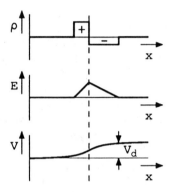

Fig. 2.2 *Sketch of space charge ρ, electric field strength E and electron potential distribution V in an abrupt junction.*

the external voltage is such that it increases the potential barrier, diffusion stops altogether, so that only the current due to thermal generation of minorities remains. The junction is said to be reverse-biased and the resulting current is called the 'saturation current' I_s. If the biasing is such that the potential barrier is lowered, the diffusion current increases. The junction is now said to be forward-biased. The current-to-voltage characteristic obeys the equation

$$I = I_s \ [\exp(qV/kT) - 1] \qquad (2.8a)$$

where

$$I_s = Aq \left[\frac{p_n D_p}{L_p} + \frac{n_p D_n}{L_n} \right] \qquad (2.8b)$$

In this equation A denotes junction area, while L_p and L_n denote diffusion lengths for holes and electrons. These quantities represent the average distance over which the minorities travel before recombining with the abundant majorities. Diffusion length is directly related to minority lifetime. The relation is given by $L = (D\tau)^{1/2}$.

In practical junctions, as used in semiconductor technology, it is customary to have one of the regions doped much more strongly than the other one. According to eq. (2.8b), then, doping of the lightly doped region determines the properties of

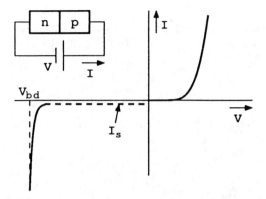

Fig. 2.3 *I–V characteristic of a pn-junction.*

the junction. If $N_D \gg N_A$, so that $p_n \ll n_p$, eq. (2.8b) simplifies into

$$I_s = Aqn_i^2 D_n / N_A L_n \qquad (2.9)$$

The factor n_i^2 reveals the origin of the current under reverse bias, namely thermal generation of minorities. And, as is apparent from eq. (2.9), the magnitude of I_s is in inverse proportion to the doping of the lightly doped region.

Figure 2.3 shows the current-to-voltage characteristic of the pn-junction. As is apparent from this diagram, the device constituted by such a junction has rectifying properties. The value of I_s as given by eq. (2.9) is extremely small, of the order of 10^{-15} A. In practice, the value of the current under reverse-bias is larger than I_s because of leakage due to imperfections of the semiconductor surface. As shown in Figure 2.3, at a certain value of the reverse-bias voltage a sudden rise in the current occurs. This is due to 'junction breakdown'. Two breakdown mechanisms can be distinguished:

(a) Avalanche breakdown
This is the mechanism that is responsible for the breakdown in most practical cases. It occurs by acceleration of charge carriers in the space-charge layer – the so-called depletion layer – to the extent that they can liberate secondary charge carriers by impact ionization. The required breakdown field strength E_{bd} in Si is about 300 kV cm^{-1}.

(b) Zener breakdown
This type of breakdown can occur in very narrow junctions. The width of the space-charge region depends on the doping level, as is apparent from Figure 2.2. If the doping level is very high, the depletion layer can be so narrow that the length of the path travelled by the charge carriers is too small to provoke avalanche breakdown. At very high field strength (of the order of 1 MV cm^{-1}) breakdown can occur by the disruption of covalent bonds in the crystal lattice.

By the way, note that both breakdown mechanisms have opposite temperature coefficients. Therefore, the temperature dependence can be small in junctions where both mechanisms occur simultaneously. For an abrupt junction, applying Poisson's equation yields, for the peak value of the electric field,

$$E_{max} = \left(\frac{2qV}{\varepsilon\varepsilon_0} \cdot \frac{N_A N_D}{N_A + N_D} \right)^{1/2}$$

where $\varepsilon\varepsilon_0$ denotes the permittivity of the material. Hence, the breakdown voltage V_{bd} follows from

$$V_{bd} = \frac{\varepsilon\varepsilon_0}{2q} \frac{N_A + N_D}{N_A N_D} E_{bd}^2 \qquad (2.10)$$

revealing that V_{bd} depends strongly on the doping level of the lightly doped region. A relatively heavy doping of this region yields a low breakdown voltage. Consequently, if a high breakdown voltage is desired, light doping is essential. As is seen from eq. (2.9), a high breakdown voltage goes along with a relatively large value of I_s.

EXAMPLE
Calculate V_{bd} for a pn-junction with $N_D = 0.9 \times 10^{19}$ cm^{-3} and $N_A = 5 \times 10^{16}$ cm^{-3}. For silicon $\varepsilon\varepsilon_0 = 1.04 \times 10^{-12}$ F cm^{-1} and $E_{bd} = 300$ kV cm^{-1}. These doping data are typical for a base–emitter junction of an integrated transistor. Substitution of the data in eq. (2.10) yields $V_{bd} = 5.85$ V.

Reverse-biased junctions are often used as voltage references because of the steep breakdown characteristic. Used in this mode, a pn-diode is commonly called a Zener diode. As to the reproducibility of the breakdown voltage it should be noted that, as follows from eq. (2.10), this voltage depends on the doping levels. Henceforth, the accuracy with which the voltage can be defined is determined by the attainable doping accuracy.

If $V \gg kT/q$, the characteristic of the forward-biased junction is expressed by

$$I = I_s \exp(qV/kT) \qquad (2.11)$$

or, alternatively,

$$V = \frac{kT}{q} \ln \frac{I}{I_s} \qquad (2.12)$$

The quantity kT/q is often called the 'thermal voltage', and for brevity is written V_T. From here on we will often use this notation. Where $T = 300$ K, $V_T = 26$ mV.

As is assumed to be known by the reader, in analogue electronics it is customary to bias electronic devices at a certain point of their characteristics. This is called the 'biasing point' or the 'quiescent point' of the relevant characteristic. When the signal quantity (voltage or current) applied to the device is small with respect to the associated bias quantity, the non-linear characteristic can be approximated by its tangent in the quiescent point. The voltage-to-current transfer for the signal quantity is then given by $g = dI/dV$. Using eq. (2.11), we obtain for the so-called *small signal conductance* g_d of the pn-diode

$$g_d = \frac{dI}{dV} = \frac{I}{V_T} \tag{2.13}$$

The reciprocal value is given by

$$r_d = \frac{V_T}{I} = \frac{26}{I} \quad (I \text{ in mA}) \tag{2.14}$$

For ease of calculation V_T is often set at 25 mV.

Note that g_d depends solely on the bias current I, irrespective of doping levels and of junction geometry. Even the material that the junction is made of is irrelevant, hence eq. (2.13) is equally valid for, e.g. a germanium junction.

In practical applications, the range of currents occurring in a given circuit will rarely encompass more than two decades. From eq (2.12) it is seen that the voltage over the junction varies only slightly within this range of currents. With the doping levels commonly encountered in integrated circuits, and in the current range of, say, 10 μA to 1 mA, the voltage across the junction is about 0.7 V. With every doubling of the current, the junction voltage increases by $\Delta V = V_T \ln 2 \approx 18$ mV. And for a tenfold increase in current, $\Delta V = V_T \ln 10 \approx 60$ mV.

As we have seen, around the metallurgical junction a space charge builds up. The region where the space charge resides is called a 'depletion layer', since it is depleted of free carriers. The space charge consists of ionized atoms. The lighter the doping of the relevant area, the larger the width of the ionized area must be in order to accommodate the space charge. Since the depletion layer occupies a certain space, it possesses a certain capacitance C_j. The amount of space charge and, consequently, the width d of the depletion layer, depends on the voltage across it. Hence, the capacitance it represents is also voltage-dependent. For an abrupt junction, the capacitor formed by the depletion layer can be modelled as a simple plate capacitor. Then,

$$C_j = \frac{dQ}{dV} = \frac{A\varepsilon\varepsilon_0}{d} \tag{2.15}$$

where A denotes junction area and d the width of the depletion layer.

Using Poisson's equation for determining the width of the depletion layer, one finds for an abrupt junction:

$$C_j = A \left[\frac{q\varepsilon\varepsilon_0}{2V_{\text{tot}}(1/N_A + 1/N_D)} \right]^{1/2} \tag{2.16}$$

where V_{tot} denotes total voltage across the junction, i.e. the sum of the built-in voltage V_d, according to eq. (2.7), and the external reverse-bias voltage V. If

$N_D \gg N_A$; then eq. (2.16) simplifies to

$$C_j = A\sqrt{\frac{\varepsilon\varepsilon_0 q N_A}{2V_{tot}}} \qquad (2.17)$$

revealing that the capacitance of a reverse-biased abrupt junction is proportional to $\sqrt{N_A}$ and inversely proportional to $\sqrt{V}$. A reverse-biased pn-junction can therefore be used as a voltage-controlled capacitance. If the junction is not abrupt, but gradual, the voltage dependence remains, but the relation between C_j and V is modified.

EXAMPLE
Calculate the capacitance per μm^2 of an abrupt junction with $N_D \gg N_A$ and $N_A = 10^{16}$ cm^{-3}, while $V_{tot} = 2.5$ V reverse bias. The permittivity $\varepsilon\varepsilon_0$ of silicon is 104×10^{-14} F cm^{-1}. Using expression (2.17) we find that for $A = 1\ \mu m^2$, $C_j \approx 18.2 \times 10^{-5}$ pF.

2.4 The bipolar transistor

In a bipolar transistor three regions can be distinguished: emitter, base and collector. Emitter and collector have the same type of doping, whereas the base doping is of the opposite type. Hence, there are two types of transistors: npn- and pnp-transistors. There are two pn-junctions, called emitter–base junction and collector–base junction or, for short, emitter junction and collector junction. Both junctions can be in forward or reverse bias. If both junctions are reverse-biased no current can flow, except for the very small leakage currents occurring under reverse bias. This region of operation is called the *off-region*. If both junctions are in forward bias, the transistor is said to operate in its *saturation region*. To a first approximation, the transistor behaves as a short circuit. A more accurate description of the behaviour will be given later. Anyway, in analogue electronics this region is of little significance. If the emitter junction is in forward bias, whereas the collector junction is in reverse bias, the transistor operates in its *active region*. This region is the normal region of operation in analogue electronics. By being forward-biased, the emitter injects minority carriers in the base region (electrons in an npn-transistor), which are subsequently collected by the collector, thus giving rise to the collector current. Finally, the emitter junction can be in reverse bias, whereas the collector junction is forward-biased. The transistor now operates in its *inverse region*. Since in a good transistor the doping level of the emitter is much higher than that of the collector, the properties of the inversely operating transistor differ considerably from those of the transistor in its normal active mode of operation.

As a vehicle for the discussion we will use the npn-transistor, which is the version usually used. All results are equally valid for the pnp-version, provided that all voltage and current polarities are reversed and it is taken into account that electrons

ACTIVE DEVICES AND INTEGRATED CIRCUITS: A SURVEY   19

and holes change roles. For brevity we will often indicate the bipolar transistor by the acronym BJT, which stands for bipolar junction transistor.

We start with a description of the static behaviour which is generally valid, irrespective of whether the BJT is in its active, saturation or inverse region. The essence of transistor operation is that the minority currents injected in the base region by a forward-biased junction travel through the base to the opposite electrode. For prevention of loss of minorities due to recombination in the base, the basewidth must be small compared to the diffusion length. Since some recombination will always take place, it must be taken into account that part of the current injected into the base leaves the transistor via the base contact, thus constituting the base current.

Figure 2.4 shows the sign conventions we will adopt. In accordance with the conventions normally used in circuit theory, all currents going into the device are assumed positive. Figure 2.5 shows a model of the transistor that makes explicit the two primary operating mechanisms, namely the exponential current-versus-voltage relation of a pn-junction and the occurrence of a finite external base current. The junctions are symbolized by diodes, while α denotes the fraction of the junction current that reaches the opposite electrode.

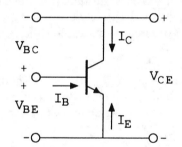

Fig. 2.4 *Sign conventions for a BJT.*

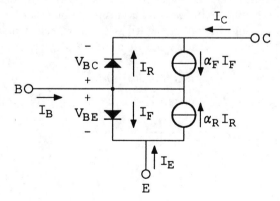

Fig. 2.5 *A simple model for the static behaviour of the BJT (injection model).*

The static behaviour is expressed by the following equations:

$$I_F = I_{ES}[\exp(V_{BE}/V_T) - 1] \tag{2.18}$$

$$I_R = I_{CS}[\exp(V_{BC}/V_T) - 1] \tag{2.19}$$

$$I_C = \alpha_F I_F - I_R \tag{2.20}$$

$$I_E = -I_F + \alpha_R I_R \tag{2.21}$$

$$I_B = -(I_C + I_E) = (1 - \alpha_F)I_F + (1 - \alpha_R)I_R \tag{2.22}$$

A detailed analysis of transistor behaviour reveals that

$$\alpha_F I_{ES} = \alpha_R I_{CS} = I_S \tag{2.23}$$

This relation is called the reciprocity relation; in essence I_S corresponds with the quantity I_s, according to eq. (2.8). Using eq. (2.23) we can write eq. (2.20) in the form

$$I_C = I_S[\exp(V_{BE}/V_T) - 1] - I_R \tag{2.24}$$

When the collector junction is reverse-biased, $I_R \to 0$. Equation (2.24) reveals that I_C accurately depends on V_{BE}, since I_S is solely determined by transistor construction. Since I_{ES} is not a constant quantity, but depends to some extent on the biasing conditions, I_E is not accurately exponentially related to V_{BE}. This fact is important in the design of circuits that exploit the exponential relation, such as logarithmic amplifiers.

The model in Figure 2.5, which is described by eqs. (2.18)–(2.22), is called the *injection model* for the static behaviour of the BJT. Taking eq. (2.23) into account, it is obvious that it is characterized by three quantities, viz. α_F, α_R and I_S. Alternatively, instead of α_F and α_R the quantities β_F and β_R are often used, defined by

$$\beta_F = \frac{\alpha_F}{1 - \alpha_F} \quad \text{and} \quad \beta_R = \frac{\alpha_R}{1 - \alpha_R}$$

Though the injection model is useful in many situations of practical interest, an obvious drawback is that it conceals the accurate exponential relationship between I_C and V_{BE}. By varying the model slightly, this drawback is obviated. Figure 2.6 shows this model, which is now known as the *transport model*. The relevant equations are now

$$I_{CT} = I_S[\exp(V_{BE}/V_T) - 1] \tag{2.25}$$

$$I_{ET} = I_S[\exp(V_{BC}/V_T) - 1] \tag{2.26}$$

$$I_C = I_{CT} - \frac{I_{ET}}{\alpha_R} \tag{2.27}$$

$$I_E = I_{ET} - \frac{I_{CT}}{\alpha_F} \tag{2.28}$$

$$I_B = I_{CT}\left(\frac{1}{\alpha_F} - 1\right) + I_{ET}\left(\frac{1}{\alpha_R} - 1\right) \tag{2.29}$$

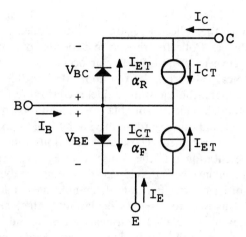

Fig. 2.6 *Alternative representation of BJT static behaviour (transport model).*

These equations are fully equivalent to eqs. (2.18)–(2.22). Note that the transport model is characterized by the same set of parameters as the injection model, namely I_S, α_F and α_R. However, now the exponential relation between the forward junction voltage and the current in the collecting electrode is made explicit directly.

These sets of equations are valid, irrespective of the biasing conditions of the junctions. We can therefore use them to derive simplified equations pertaining to the various modes of operation of the transistor. Most important by far for analogue signal processing is the active region, where $V_{BE} > 0$ and $V_{BC} < 0$. Then $I_{ET} \to 0$ and

$$I_C = I_{CT} = I_S[\exp(V_{BE}/V_T) - 1] \qquad (2.30)$$

$$I_E = -I_C/\alpha_F \qquad (2.31)$$

$$I_B = I_C/\beta_F \qquad (2.32)$$

Similarly, in the reverse region $I_{CT} \to 0$, so that

$$I_E = I_S[\exp(V_{BC}/V_T) - 1] \qquad (2.33)$$

$$I_C = -I_E/\alpha_R \qquad (2.34)$$

$$I_B = I_E/\beta_R \qquad (2.35)$$

As will be explained later, in a good transistor collector doping is much lighter than emitter doping. One consequence of this is that α_R and β_R are much smaller than the corresponding quantities α_F and β_F. A typical value for α_F is 0.995, so that $\beta_F \approx 200$. The values of α_R and β_R depend on the actual transistor construction; even $\beta_R \approx 1$ is not uncommon, particularly in the power devices.

In the saturation region both the emitter and the collector act simultaneously as emitting and collecting regions. Since $I_S \ll I_{CT}$ and $I_S \ll I_{ET}$, from eqs. (2.25) and (2.26) it follows that $V_{CE} = V_{BE} - V_{BC} = V_T \ln(I_{CT}/I_{ET})$. In practice the value of

V_{CE} in saturation is dominated by the voltage drop over the ohmic resistance of the collector region. Since the collector region is lightly doped it exhibits a relatively large specific resistance. Collector resistance depends, of course, on the device structure. Values between 20 and 500 Ω can occur. As a consequence, $V_{CE(sat)}$ depends on I_C. Typical values range from about 0.05 to 0.3 V.

The injection model and the transport model are also referred to as two versions of the Ebers–Moll model, which is suitable for calculations concerning the biasing conditions and for calculations regarding driving by large signals as long as quasi-static operation may be assumed. The model equations can be considered as the mathematical representation of the static transistor characteristics. Various extensions of the model have been developed accounting for dynamic operation and for several second-order effects. Because of the non-linear nature of these models they are not very suitable for 'hand-calculations'; they are mainly used for computer-aided analysis. A detailed treatment can be found in Getreu (1978).

We will now turn our attention to the active region. From now on we will replace I_S by I_s, in accordance with eq. (2.8) for the pn-junction. Hence

$$I_C = I_s \exp(V_{BE}/V_T) \tag{2.36}$$

If the emitter doping is much stronger than the base doping, transistor theory yields

$$I_s = \frac{AqD_n n_i^2}{\int_0^W (N_A - N_D) \, dx} \tag{2.37}$$

where A is the emitter junction area, D_n is the diffusion constant for electrons in the base, W is the width of the base region (see Figure 2.7), and $N_A - N_D$ is the net doping concentration of the base. The quantity $\int_0^W (N_A - N_D) \, dx$ is often referred to as the *Gummel number* of the base region. This is an important parameter, since many transistor parameters are related to it.

The base current, as expressed by the non-zero value of $1 - \alpha_F$, is mainly due to the following three mechanisms.

1. Injection of holes from the base into the emitter gives rise to a base current component

$$I_{B\,inj} = I_{s\,inj} \exp(V_{BE}/V_T) \tag{2.38}$$

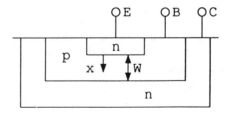

Fig. 2.7 *Schematic drawing of a planar BJT.*

The *emitter efficiency* γ, defined by $\gamma = I_C/(I_C + I_{B\,inj})$ can be shown to be

$$\gamma = 1 - \frac{D_p}{D_n} \frac{W}{L_p} \frac{p_b}{n_e} \qquad (2.39)$$

where D_p is the diffusion constant for holes in the emitter region and D_n the diffusion constant for electrons in the base region; L_p is the diffusion length for the holes in the emitter region, while p_b and n_e denote the equilibrium majority concentrations in base and emitter respectively. These concentrations are equal to the doping concentrations. From eq. (2.39) it is seen that, for good emitter efficiency, W has to be small, while $p_b \ll n_e$. Preferably p_b and n_e should differ by about a factor of 100.

2. Recombination of electrons in the base region causes a base current component

$$I_{B\,rec} = I_{s\,rec} \exp(V_{BE}/V_T) \qquad (2.40)$$

The *base efficiency* or *transport factor* b, defined as $b = I_C/(I_C + I_{B\,rec})$, can be shown to be given approximately by

$$b = 1 - \tfrac{1}{2}\left(\frac{W}{L_n}\right)^{1/2} \qquad (2.41)$$

where L_n is the diffusion length for electrons in the base. Obviously, for a good base efficiency $W \ll L_n$ must hold. For modern transistors, manufactured in planar technology, the imperfect emitter efficiency usually dominates the imperfect base efficiency.

3. Recombination in the emitter–base depletion layer gives rise to a base current contribution

$$I_{B\,rj} = I_{s\,rj} \exp(V_{BE}/2V_T) \qquad (2.42)$$

At very low current levels this fraction of the base current can be non-negligible. Since the exponent in eq. (2.42) is half that in eqs. (2.40) and (2.41), the relative magnitude of $I_{B\,rj}$ decreases with increasing current level. The current given by eq. (2.42) flows in the emitter–base circuit, hence it does not affect the collector current anyway. This implies that the collector current is not contaminated by a component that is not proportional to $\exp(V_{BE}/V_T)$, whereas the emitter current is. This explains why the accurate exponential relationship holds over a much larger range of current values for I_C than for I_E. The exponential relationship for I_C can even hold for a current range encompassing nine decades.

According to eq. (2.36), I_C does depend on V_{BE} only, irrespective of V_{BC}, provided $V_{BC} > 0$. In practice, I_C turns out to be slightly dependent on V_{BC}. This is due to the fact that the width of the depletion layer associated with the collector junction depends on the magnitude of the reverse-bias voltage V_{BC}. As a consequence the effective basewidth W depends on V_{BC} and according to eqs. (2.39) and (2.41) γ and b also do. Hence I_B depends on V_{BC} and therefore also on I_C. This effect is referred to as *basewidth modulation* or *Early-effect*. It can be shown that

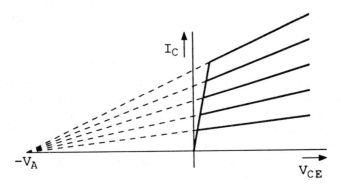

Fig. 2.8 *Determination of the Early-voltage V_A from the I_C-versus-V_{CE} characteristics of the transistor.*

the extrapolated I_C-versus-V_{CE} characteristics approximately have a common intersection, as shown in Figure 2.8. The intersect point is called the *Early-voltage* V_A. Mathematically this effect can be expressed by extending eq. (2.36) by a factor $(1 + V_{CE}/V_A)$, hence

$$I_C = I_s \left(1 + \frac{V_{CE}}{V_A}\right) \exp \frac{V_{BE}}{V_T} \qquad (2.43)$$

The Early-effect is minimized by taking care that the collector-junction depletion layer extends predominantly in the collector region rather than in the base region. This is accomplished by doping the collector region lightly in comparison with the base region. This measure has the additional advantage that the collector-junction capacitance is minimized. A very light doping of the collector region causes the specific resistance of this region to be high. This can give rise to an undesirable high value of the collector series resistance. This effect can be kept within bounds by confining the light doping to that part of the collector region that is adjacent to the junction.

If we now collect all the conditions that should be fulfilled by the doping levels, we find that for a high value of β_F it is required that the emitter doping be much heavier than the base doping. And a high value of V_A requires the collector to be doped lightly in comparison to the base. Figure 2.9(a) (not drawn to scale) surveys these conditions. In practice the doping of the various regions is accomplished through compensation. Starting from a lightly doped n-type collector region the base is formed by introducing acceptor atoms, after which the emitter is formed by overruling the acceptor doping by a heavier doping with donors. Figure 2.9(b) sketches the doping profile in a transistor manufactured in this way.

Does available technology permit us to comply with the stipulated doping requirements? In 1 cm^3 Si there are 5×10^{22} atoms. At a doping level of about 10^{19} cm^{-3}, degeneration occurs; that is, above this doping level the material no longer behaves as doped silicon but as an alloy with different properties. The

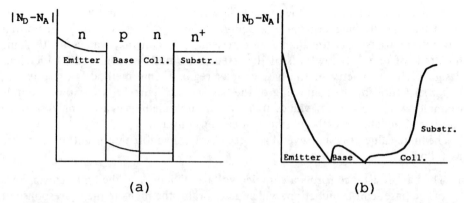

Fig. 2.9 (a) *Idealized doping profile for BJT.* (b) *Sketch of practical profile of epitaxial planar BJT (not to scale).*

physical basis for this behavioural change is that the average distance between doping atoms becomes of the same order of magnitude as the Broglie wavelength of the charge carriers. The doping atoms can then no longer be regarded as isolated atoms, so that mutual interaction occurs. The attainable purity of the undoped starting material obtained from the crystal growing process is about $1:10^{10}$, implying that there are about 5×10^{12} uncontrolled impurity atoms per cm^3. This implies that controlled doping starts at the level of about 5×10^{14} atoms cm^{-3}. Hence, the available doping levels range from about 5×10^{14} to about 10^{19} cm^{-3}. This suffices to comply with all the requirements for adequate transistor performance.

2.5 Second-order effects in the bipolar transistor

In the foregoing section, only first-order aspects of transistor operation in the active region have been considered: injection of charge carriers into the base by the forward-biased emitter junction, the mechanisms of charge transport in the base, with the associated occurrence of a finite base current, and collection of charge carriers in the collector with the associated Early-effect. Several additional effects occur, of which a number cannot pass unmentioned, because of their effect on transistor behaviour in commonly encountered applications.

First, attention must be paid to the effects of the ohmic resistances inherent in the construction of the BJT. The emitter is low-ohmic by virtue of its high doping level. The ohmic resistance of the emitter region can thus usually be neglected. The collector region is lightly doped and henceforth collector resistance cannot always be neglected, as was already pointed out in the context of the occurrence of the finite value of V_{CE} in saturation. Of greater importance is the ohmic resistance of the base region. The base is lightly doped and, what is more, the basewidth W must be small to achieve high β_F. Due to the high specific resistance and the unfavourable geometric layout, base resistance is bound to be high (Figure 2.10a). In normal low-current transistors the ohmic base resistance r_b can take values up to some hundreds of ohms. If low r_b is of the essence this must be achieved by judicious design of the geometry of the base-contact regions. One method is the use of a finger structure for the contacting of the base area. Figure 2.10(b) shows a simple example. When very low r_b-values are required, multiple fingers are applied with the ensuing disadvantage of large silicon area consumption.

The finite value of r_b has several adverse effects, the more important of which are as follows:

1. The base resistance forms a complex voltage divider with the input capacitance of the base–emitter junction, which deteriorates the high-frequency response of the transistor.
2. The base resistance acts as a source of thermal noise.
3. Since the base current flows through r_b, a lateral field is present in the base region. The associated potential drop causes the forward-bias voltage V_{BE} to be

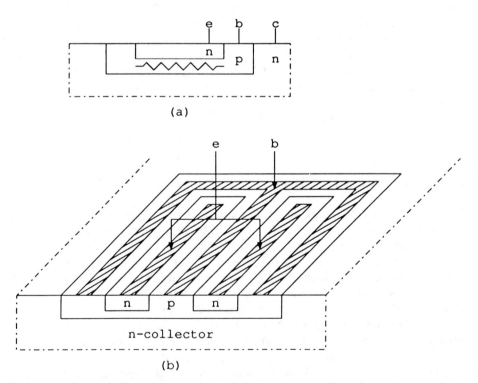

Fig. 2.10 (a) *Symbolic representation of the resistance of the base region below the emitter.* (b) *Low base resistance is obtained by applying a number of long narrow emitters in parallel.*

higher close to the base contact than away from this contact. This causes the injection current to concentrate in a small region near the base contact. This phenomenon is referred to as *emitter crowding*. It causes r_b to be current-dependent; it contributes to current dependency of β_F; and it lowers the safe current range.

At high current levels, so-called high-level injection phenomena take place. Minority concentration in the base can increase so much that at the emitter side of the base region it comes close to the equilibrium majority concentration. As a consequence, emitter and base efficiency, and hence β_F, decrease at high current levels. Figure 2.11 sketches a typical plot of β_F as a function of I_C, showing the decline of β_F at high-current levels. The decline at low-current levels is mainly due to the effects of generation and recombination in the emitter–base depletion layer.

Second in importance are breakdown mechanisms. Under reverse bias the junctions are subject to avalanche breakdown. For the emitter junction of an integrated transistor $V_{bd} \approx 6$ V. In general, the occurrence of avalanche breakdown should be avoided, since it incurs irreversible deterioration of the crystal lattice, owing to overheating, giving rise to lowering of β_F. For the collector junction $V_{bd} \approx 80$ V, due to the much lower doping levels. However, before collector-junction breakdown occurs, another phenomenon can jeopardize transistor integrity. At voltages not too far below V_{bd}, the impact ionization that can ultimately lead to the avalanche effect starts to produce extra charge carriers. This is called 'collector multiplication'. The production of extra carriers can outnumber the loss of carriers in the base, so that α_F can effectively grow beyond unity value. At the point where $\alpha_F = 1$, the value of $\beta_F = \alpha_F/(1 - \alpha_F)$ shows a steep rise. This is a potentially hazardous situation. If no current limiting occurs, the transistor can become irreversibly damaged. Another breakdown mechanism that can occur is due to the small width W of the base region. When V_{CB} increases, the collector depletion layer starts to widen. Ultimately, the entire base region can be depleted,

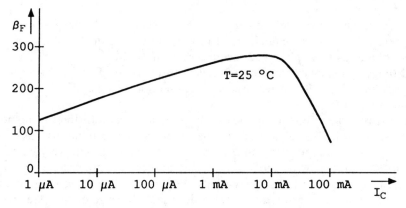

Fig. 2.11 *A typical plot of β_F as a function of I_C.*

so that the emitter and collector depletion regions join. This effect is called 'punch-through'. Manufacturers of transistors and ICs provide data concerning the maximum safe collector-junction voltage. It depends on the mode of operation of the transistor. Two values are commonly given, namely

$V_{CBO \, max}$: the maximum value of V_{CB} with open emitter
$V_{CEO \, max}$: the maximum value of V_{CE} with open base

Typically $V_{CBO \, max}$ is close to the avalanche breakdown voltage of the collector junction, while V_{CEO} is about half that value. The difference is due to the multiplication effect. In the case of an open base there is no outlet for the extra carriers via the base terminal.

Finally, we consider briefly the effects of charge storage. A change in V_{BE} or in I_B does not immediately produce the corresponding response in I_C. This is due to the time needed to build up the charge distribution that conforms with the driving condition. When charge carriers are injected into the base, they need time to reach the collector. The transport mechanism is dominated by diffusion. It can be shown that this transport mechanism closely resembles the transport in a transmission line. However, introducing such a structure into an equivalent circuit is unpractical. To a first approximation, the transport slowness can be attributed to the accumulation of charge in the base region. This can be modelled by introducing an extra capacitance across the emitter junction, called the diffusion capacitance C_d. It stands to reason that this capacitance is proportional to the injected current and that it depends on basewidth. Closer analysis yields, for a homogeneously doped base region,

$$C_d = \frac{I_E}{V_T} \frac{W^2}{2D_n} \qquad (2.44)$$

Obviously, to keep C_d small, W should be minimal. As we have seen, this is also beneficial for obtaining high β_F. However, low r_b and high V_{CEO} are benefited by large basewidth. A compromise must be found between these conflicting requirements.

In the discussion of the properties of the pn-junction, we have found that this structure exhibits a capacitance due to the space charge in the depletion layer. It goes without saying that these capacitances also have to be taken into account when modelling the dynamic behaviour of the transistor. In the next section, dealing with the a.c.-behaviour of the transistor, due attention will be given to the various capacitive effects.

Apart from the effects mentioned in this section, several further phenomena can be listed. Their influence on transistor behaviour is marginal in most cases. Taking into account all minor effects in transistor modelling ends with very complicated, not to say nightmarish, models that can at best be practicable in computer calculations for basic transistor circuits. For calculations on circuits containing more than a few transistors they are hardly usable because of the excessive CPU-time involved in the execution of the computations.

2.6 Small-signal transistor modelling

In processing circuitry for analogue signals the signal levels are often small compared to the bias currents and voltages. In such cases it is possible to greatly simplify calculations by the use of *incremental* or *small-signal* models. In such models the characteristic of interest is replaced by its tangent in the bias point. This comes down to approximating the non-linear characteristic by the first term of its Taylor expansion. As an example, consider the I_C-versus-V_{BE} characteristic of a BJT (Figure 2.12), biased at point $Q(I_C, V_{BE})$. A small signal v_{be} gives rise to a change in i_c in the collector current. The relation is given by $i_c = v_{be} \tan \Theta$, with

$$\tan \Theta = \frac{dI_C}{dV_{BE}} \quad (V_{CE} = \text{const})$$

From $I_C = I_s \exp(V_{BE}/V_T)$

$$\tan \Theta = g_e = I_C/V_T \tag{2.45}$$

follows. The reciprocal value has the dimension of a resistance

$$r_e = 1/g_e = V_T/I_C$$

Since $V_T \approx 25$ mV, $g_e = 40 I_C$ (g_e in mA V^{-1}, I_C in mA) and $r_e = 25/I_C$ Ω (I_C in mA).

Note that g_e and r_e solely depend on I_C, irrespective of any other quantity, even the kind of material the transistor is made of!

Likewise, for the relation between I_C and I_B,

$$\beta_0 = \frac{dI_C}{dI_B} \quad (V_{CE} = \text{const}) \tag{2.46}$$

As long as $I_C = \beta_F I_B$ holds, $\beta_0 = \beta_F$. As we have found in the preceding section β_F is not constant over the range of safely tolerable current values, so in general β_0 is also current-dependent.

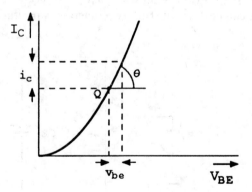

Fig. 2.12 *Sketch of I_C–V_{BE} characteristic with indication of the biasing point and small-signal quantities.*

Similarly,

$$\frac{dI_C}{dI_E}(V_{CE} = \text{const}) = \alpha_0 \quad (2.47)$$

and

$$\beta_0 = \frac{\alpha_0}{1 - \alpha_0} \quad (2.48)$$

Further,

$$\frac{dI_B}{dV_{BE}} = \frac{g_e}{\beta_0} \quad \text{and} \quad \frac{dV_{BE}}{dI_B} = \beta_0 r_e = r_\pi \quad (2.49)$$

Finally, due to the Early-effect I_C depends on V_{CE}. From eq. (2.43) we find

$$\frac{dV_{CE}}{dI_C}(V_{BE} = \text{const}) \approx \frac{V_A}{I_C} = r_{ce} \quad (2.50)$$

We can now draw up a small-signal equivalent circuit for the small-signal operation of the transistor in common-emitter (CE) configuration, as depicted in Figure 2.13. In this diagram, configuration base and emitter constitute the input terminals, collector and emitter the output terminals, hence the emitter is common to input and output circuit. Biasing is schematically indicated with two d.c. voltage sources. In practice, biasing is never accomplished in this way. Practical methods will come to the fore in Chapter 5. According to eq. (2.49), $i_b = v_{be}/r_\pi$, while $i_c = g_e v_{be}$. Further, with V_{BE} constant, hence $v_{be} = 0$, $i_c = v_{ce}/r_{ce}$ holds. Figure 2.14 shows the equivalent circuit representing these relations. The voltage-to-current transfer from v_{be} to i_c is usually labelled the *transconductance*. It is often denoted g_m. In accordance with this custom we will write g_m instead of g_e.

The equivalent circuit of Figure 2.14 is too simple for most practical calculations. For calculations concerning frequency-dependence of signal transfer, the capacitances discussed previously must be taken into account. As we have found, three capacitances can be distinguished, two junction capacitances C_{be} and C_{bc}, plus

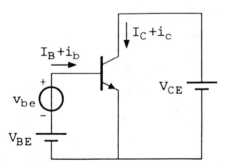

Fig. 2.13 *Schematic representation of CE-circuit.*

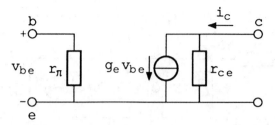

Fig. 2.14 *Elementary equivalent circuit.*

the diffusion capacitance C_d. The junction capacitances are voltage-dependent, while C_d is proportional to current. Both C_{be} and C_d are in parallel with r_π. It is customary to take these capacitances together under the label C_π. However, it should be noted that it is composed of two capacitances with different qualities. The collector-junction capacitance is commonly denoted $C_{b'c}$ or, alternatively, C_μ. The use of b' instead of b is to discriminate between the base of the *intrinsic* transistor and the *extrinsic* transistor. In this terminology 'intrinsic' denotes the 'inner' transistor, consisting of the junctions and the transport mechanism, whereas 'extrinsic' is related to the complete transistor, including its ohmic resistances and its environmental capacitances. Figure 2.15 gives the equivalent circuit of the intrinsic transistor, Figure 2.16 that of the extrinsic transistor. The latter model includes the base resistance r_b, discussed previously, together with the ohmic resistances r_E and r_C and the extrinsic capacitances C_{BE}, C_{BC}, C_{CE}.

In practice r_E and r_C can often be omitted; in integrated transistors r_E is typically 1–2 Ω. The extrinsic capacitances C_{CE} and C_{BE} are usually assumed to be part of the driving and load impedances, while C_{BC} is often taken together with $C_{b'c}$. This is done to improve the useability of the equivalent circuit for quick calculations. For more precise computer calculations the complete equivalent circuit can be used. In practical problems concerning the analysis and design of transistor circuits it is nearly always sensible to work first with the simple equivalent circuit to obtain consistent insight into transistor behaviour and into the way it depends on various quantities. If more precise quantitative data are needed, computer analysis can be called in. Figure 2.17 shows the reduced equivalent circuit that is simple enough to be useful for manual calculations, while still being sufficiently complete for

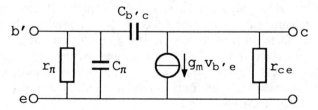

Fig. 2.15 *Equivalent circuit of the intrinsic transistor.*

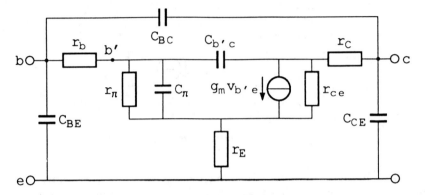

Fig. 2.16 *Equivalent circuit of the extrinsic transistor.*

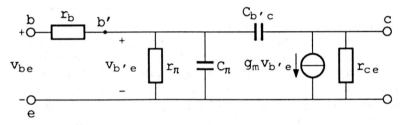

Fig. 2.17 *Hybrid-π equivalent circuit.*

adequate performance. This equivalent circuit is commonly referred to as the *hybrid-π circuit*. It has found general acceptance and is in widespread use.

EXAMPLE
A transistor is biased at $I_C = 0.5$ mA, $C_{b'c} = 0.15$ pF, $C_{be} = 1$ pF, $C_d/I_C = 19$ pF mA^{-1}, $\beta_0 = 200$, $V_A = 150$ V, $r_b = 200$ Ω. Determine the parameters of the hybrid-π circuit.

Since $I_C = 0.5$ mA, $r_e = 50$ Ω and $r_\pi = \beta_0 r_e = 10$ kΩ. Further $r_{ce} = V_A/I_C = 300$ kΩ and $C_d = 0.5 \times 19$ pF $= 9.5$ pF. Hence, $C_\pi = 10.5$ pF. Finally, $g_m = 1/r_e = 20$ mA V·.

Even the model of Figure 2.16 is still incomplete. It can be extended by parameters accounting for non-linear effects such as dependence of β_0 on I_C, high-level injection effects, such as emitter crowding, and so on. Such extended models can be useful for the computer analysis of simple circuits. For circuits containing many transistors, such models, which may contain twenty or more parameters, are too complex to be practical even for computer-assisted calculations to be performed within reasonable CPU-time.

Before embarking on amplifier circuits, we must first lay down some symbol conventions. Bias or d.c.-quantities will be represented by upper-case symbols with an upper-case subscript, such as I_C, V_{CE}. Small-signal quantities will be represented by lower-case symbols with lower-case subscripts, such as i_c, v_{ce}. In cases where the superposition of both quantities is involved, we will use lower-case symbols with upper-case subscripts, such as $i_C = I_C + i_c$. For the complex notation of a.c.-signal quantities we will use upper-case symbols with lower-case subscripts, hence in complex notation i_c is written as I_c. These conventions are similar to those used in most books on electronics.

Let us now consider the simple CE-amplifier configuration shown in Figure 2.18(a). Biasing arrangements are omitted altogether, since our interest is directed towards small-signal response. The input is a voltage source V_g with internal impedance Z_g, the load impedance is denoted Z_L. Figure 2.18(b) shows the equivalent circuit. The capacitance $C_{b'c}$ has been omitted. In a well-constructed transistor it is small and omitting it simplifies the calculation greatly. In Chapter 5 methods will be discussed for coping with this capacitance in calculations concerning signal transfer.

In order to keep notation simple we write Z_π for the parallel combination of r_π and C_π and Z'_L for the parallel connection of Z_L and r_{ce}. By Figure 2.18(b) we write down the relations

$$V_o = -g_m V_{b'e} Z'_L \qquad (2.51)$$

$$V_{b'e} = \frac{Z_\pi}{Z_\pi + r_b + Z_g} V_g \qquad (2.52)$$

Eliminating $V_{b'e}$ from eqs. (2.51) and (2.52), we find for the voltage transfer

$$A_v = \frac{V_o}{V_g} = -\frac{g_m Z'_L Z_\pi}{Z_\pi + r_b + Z_g} \qquad (2.53)$$

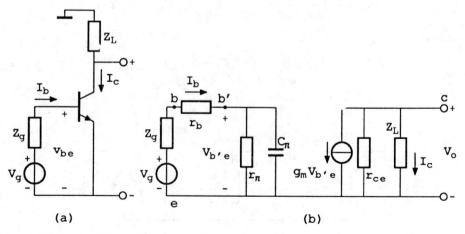

Fig. 2.18 *Basic CE-configuration: (a) small-signal circuit; (b) equivalent circuit.*

For low frequencies, where $Z_\pi = r_\pi$, $Z_L = R_L$ and $Z_g = R_g$:

$$A_v = \frac{V_o}{V_g} = -\frac{g_m R_L' r_\pi}{r_\pi + r_b + R_g} \qquad (2.54)$$

The circuit of Figure 2.18(b) contains three time constants, associated with Z_g, Z_π and Z_L' respectively. Very often the frequency dependence of the transfer is dominated by the time constant due to Z_π. Putting

$$Z_g = R_g, \quad Z_L' = R_L' \quad \text{and} \quad Z_\pi = \frac{r_\pi}{1 + j\omega r_\pi C_\pi}$$

eq. (2.53) can be written

$$A_v = \frac{V_o}{V_g} = -\frac{g_m R_L' r_\pi}{r_\pi + r_b + R_g} \cdot \frac{1}{1 + j\omega \tau_h} \qquad (2.55)$$

where

$$\tau_h = C_\pi \frac{(R_g + r_b) r_\pi}{R_g + r_b + r_\pi}$$

and

$$\frac{(R_g + r_b) r_\pi}{R_g + r_b + r_\pi}$$

represents the parallel connection of $R_g + r_b$ and r_π. The -3 dB cutoff frequency ω_h follows from $\omega_h = 1/\tau_h$.

Of course, other transfer quantities can be found in similar ways. If, for instance, V_g is replaced by a current source, we can find the transimpedance $Z_t = V_o/I_g$. From the equivalent circuit of Figure 2.19 we find

$$V_{b'e} = -I_g \frac{Z_\pi Z_g}{Z_g + Z_\pi + r_b} \quad \text{and} \quad V_o = -g_m Z_L' V_{b'e}$$

so that

$$Z_t = \frac{V_o}{I_g} = \frac{g_m Z_L' Z_\pi R_g}{Z_g + Z_\pi + r_b} \qquad (2.56)$$

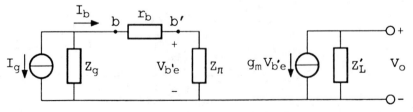

Fig. 2.19 *Equivalent circuit of current-driven CE-circuit.*

If, again, $Z_g = R_g$, $Z_L' = R_L'$ and $Z_\pi = r_\pi/(1 + j\omega r_\pi C_\pi)$, eq. (2.56) yields

$$Z_t = \frac{g_m r_\pi R_L' R_g}{r_\pi + R_g + r_b} \cdot \frac{1}{1 + j\omega \tau_h} \tag{2.57a}$$

By substituting $g_m r_\pi = \beta_0$, eq. (2.57a) can also be written

$$Z_t = \frac{\beta_0 R_L' R_g}{r_\pi + R_g + r_b} \cdot \frac{1}{1 + j\omega \tau_h} \tag{2.57b}$$

EXAMPLE

In the circuit given in Figure 2.18a, let $R_g = 1$ kΩ, $R_L = 2$ kΩ, $V_A = 100$ V, $I_C = 2$ mA, $C_\pi = 50$ pF, $\beta_0 = 200$, $r_b = 100$ Ω. Find the low-frequency voltage gain and its -3 dB bandwidth f_h. From $I_C = 2$ mA we derive $r_e = 12.5$ Ω, $g_m = 80$ mA V^{-1}, $r_\pi = 2.5$ kΩ and $r_{ce} = 50$ kΩ. Substituting these values in eq. (2.55) we obtain

$$A_v = \frac{-200 \times 1.92}{2.5 + 1 + 0.1} \cdot \frac{1}{1 + j\omega \tau_h}$$

with $\tau_h = 50 \times 10^{-12} \times 0.76 \times 10^3$. Hence, $A_v(\omega \to 0) = -107$ and $f_h = 1/2\pi\tau_h = 4.19$ MHz.

In a practical circuit the bandwidth will be smaller because of the extrinsic capacitances and $C_{b'c}$, which have been omitted in the calculation.

2.7 The junction field-effect transistor

Figure 2.20 depicts quite schematically the structure of a field-effect transistor (FET). In essence, it consists of a piece of semiconductor material, called the *channel*, that is provided with two end contacts, called *source* and *drain*. The conductance of the channel can be modified by an electrical field, which can be controlled by means of the potential V_G of a third electrode, the *gate*.

Assume that the channel consists of n-type silicon. The channel conductance depends on the dimensions of the channel and on the doping level. When a voltage is applied between drain (positive) and source (negative) an electron flow is established. By modifying the channel conductance with the aid of V_G, the current is modified. Hence, a small-signal voltage v_g, superimposed on V_G, gives rise to a

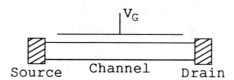

Fig. 2.20 *Schematic representation of the basic structure of an FET.*

36 ANALOGUE ELECTRONIC CIRCUIT DESIGN

small-signal current i_d in the drain circuit. Note that the current consists of majority carriers. This implies that recombination effects do not spoil charge transport, as in the BJT. Therefore, the FET need not be made of monocrystalline material. However, practical FETs almost always are made of monocrystalline material. The reason is that such material can be reliably manufactured with well-defined properties. Polycrystalline FETs are called 'thin film transistors' (TFT). Because of their whimsical properties they are used only as switches in very special applications.

As is apparent from Figure 2.20, two electric fields can be distinguished: a lateral field directed from drain to source and a transverse field, perpendicular to this and under control of the gate. The lateral field is the driving force behind the electron flow. The gate is capacitively coupled with the channel, so there is no galvanic connection and no power is absorbed by the gate circuit.

Two practical embodiments of the structure sketched in Figure 2.20 have been developed, the junction–FET (JFET) and the metal-oxide-semiconductor field-effect transistor (MOSFET or MOST). In the former the gate is separated from the channel by a reverse-biased pn-junction, in the latter by a dielectric layer, mostly SiO_2. As might be expected, both structures behave similarly. In both cases the channel can consist of either n-type or p-type material. The corresponding structures are called n-channel and p-channel FETs.

One advantage of the JFET is that it can be made using bipolar technology. Figure 2.21(a) shows a possible structure for an n-channel JFET. We will use this type as a vehicle for our discussion, the polarities being similar to those of the npn-transistor. Note, though, that in monolithic circuits p-channel FETs are more often used than n-channel types. Monolithic ways to construct FETs will be treated at a later stage.

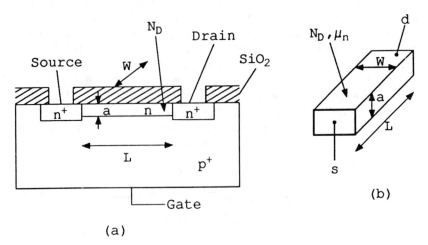

Fig. 2.21 *Junction FET: (a) structure of a planar JFET; (b) channel geometry.*

The source and the drain are made of heavily doped n-type material (commonly denoted as n+). The channel is lightly doped n, whereas the heavily doped p-substrate is used as the gate electrode. The pn-structure is reverse-biased. Figure 2.21(b) details the channel dimensions; L, W and a denote channel length, width and depth respectively. We will now summarize the mechanism of operation. Let the voltage between drain (D) and source (S) be V_{DS}. And let the voltage between gate and source be V_{GS}. For reverse biasing of the pn-junction, V_{GS} must be negative. At the source end of the channel the reverse-bias voltage is V_{GS}; at the drain end it is $V_{GS} - V_{DS}$.

As we have found previously, the depletion layer extends mainly in the lightly doped region, hence in the channel area. Due to the current flowing in the channel there is a gradual increase in the magnitude of the reverse-bias voltage when one goes from source to drain. Henceforth, the depletion layer exhibits a wedge-like shape. At a certain value of the junction voltage the depletion layer occupies the total depth (a) of the channel. When this occurs, the channel conductance in a small part of the channel near the drain becomes very small. This is called *pinch-off* and the junction voltage needed to reach this state is called the *pinch-off voltage* V_p. We have now two relations: first the local width of the (non-conducting) depletion layer as a function of the local value of junction voltage V_j, which can be derived with the help of the theory of the pn-junction; second, Ohm's law which relates V_j and the resistance R_{ch} of the channel section between the source and a selected channel location: $V_j = V_{GS} - R_{ch} \cdot I_D$, where I_D is the drain current. From these two equations the relation between V_{GS}, V_{DS} and I_D can be found, provided no pinch-off occurs (i.e. $V_{DS} < V_{GS} - V_p$). The result is

$$V_p = -\frac{a^2 q N_D}{2\varepsilon\varepsilon_0} \tag{2.58}$$

and

$$I_D \approx \beta V_{DS}(V_{GS} - V_p - \tfrac{1}{2} V_{DS}) \tag{2.59a}$$

where

$$\beta = \frac{4 W \mu_n \varepsilon \varepsilon_0}{3 a L} \tag{2.59b}$$

If V_{DS} increases to the point where pinch-off occurs, the drain current saturates. To a first approximation a further increase of V_{DS} has no effect on I_D. The saturation current follows from eq. (2.59) by putting

$$V_{DS} = V_{GS} - V_p$$

hence

$$I_{D\,sat} \approx \tfrac{1}{2}\beta(V_{GS} - V_p)^2 \tag{2.60}$$

The maximum value of $I_{D\,sat}$ occurs for $V_{GS} = 0$, since with $V_{GS} > 0$ the pn-junction becomes forward-biased. This value is commonly denoted I_{DSS}. From eq. (2.60) $I_{DSS} = \frac{1}{2}\beta V_p^2$ follows, so that eq. (2.60) can also be written in the form

$$I_D = I_{DSS}\left(1 - \frac{V_{GS}}{V_p}\right)^2 \qquad (2.61)$$

EXAMPLE
A JFET is constructed with a homogeneously doped channel with $N_D = 10^{16}$ cm^{-3} while channel depth $a = 0.5$ μm. Further, $W/L = 10$. Calculate V_p, β and I_{DSS}. With $\varepsilon\varepsilon_0 = 104 \times 10^{-14}$ F cm^{-1} we find from eq. (2.58) $V_p = -1.9$ V. From eq. (2.59b) we find with $\mu_n = 1350$ cm^2 V^{-1} s^{-1}, $\beta = 374$ μAV^{-2}. From $I_{DSS} = \frac{1}{2}\beta V_p^2$ we find $I_{DSS} = 675$ μA.

Equation (2.60) is the result of a calculation based on an approximative mathematical modelling of the mechanisms underlying FET operation. The model used conforms to the so-called gradual channel approximation, which assumes that channel conductance is controlled by transverse field only. When a practical FET is operated with $V_{DS} > V_{GS} - V_p$, so that pinch-off occurs, it is observed that V_{DS} continues to have some effect on channel conductance by modulating the effective channel length. As a consequence, I_D is not fully independent of V_{DS}. This effect is comparable with the Early-effect in a BJT. As there, the I_D-versus-V_{DS} characteristics of the FET, to a first approximation, have a common intersection on the V_{DS}-axis. Here, too, the absolute value of the corresponding voltage V_A is often designated as the Early-voltage, though there is no Early-effect whatsoever. In analogy with eq. (2.43), the effect can be accounted for by extending eq. (2.60)

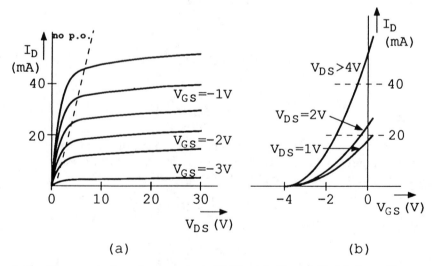

Fig. 2.22 Characteristics of a practical JFET with $V_p = 4$ V: (a) I_D-vs-V_{DS}; (b) I_D-vs-V_{GS}.

appropriately:

$$I_D = \tfrac{1}{2}\beta(V_{GS} - V_p)^2\left(1 + \frac{V_{DS}}{V_A}\right) \quad (2.62)$$

Often, one finds the factor $1 + V_{DS}/V_A$ written in the form $1 + \lambda V_{DS}$, where $\lambda = 1/V_A$.

Figure 2.22(a) shows the I_D-versus-V_{DS} characteristics of the JFET. The portion to the left of the broken line corresponds to the non-saturation region of operation, often designated the 'triode region'; the portion to the right of the broken line corresponds to operation above pinch-off. This region is commonly called the 'saturation region'. Unfortunately, in FET-terminology saturation has a meaning that is essentially different from the meaning of the same term in BJT-terminology. Figure 2.22(b) shows the I_D-versus-V_{GS} characteristics.

2.8 The MOS-transistor

In the MOS-version of the FET, the separation between gate and channel is accomplished by a dielectric layer, mostly SiO_2, sometimes silicon nitride (Si_3N_4), or a combination of these materials. Though MOS originally stands for metal-oxide-semiconductor, the acronym is presently being used for all types of FETs with a dielectrically isolated gate. The gate material is either aluminum or heavily doped polysilicon. Figure 2.23 sketches the structure of a n-type MOST in planar technology. The channel is formed by the area under the gate. The properties of the silicon surface layer under the gate depend on the doping of the silicon substrate and the lateral field under the gate. The magnitude of this field is determined by several effects, as follows:

1. At a site where two different materials are in contact, a contact potential occurs. Its magnitude is determined by the difference of the Fermi levels in the materials involved. The built-in potential difference of a pn-junction is an example of this phenomenon.

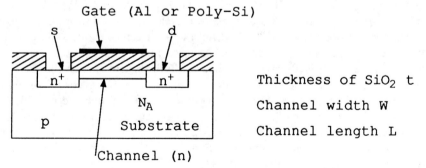

Fig. 2.23 *Sketch of planar n-channel MOST.*

2. The mechanism of oxide formation causes the existence of fixed oxide charges. Acceptors have affinity to the oxide and therefore during the oxydation of p-type silicon the acceptors crowd up in the oxide. Conversely, donors prefer to reside in the silicon.
3. Ionic charges can be present in the oxide due to contamination of the oxide, usually by alkali ions (mostly sodium).
4. At the interface of silicon and silicon oxide, the silicon crystal lattice ends abruptly. This causes defects at the interface, giving rise to so-called charge traps, which can exchange mobile carriers with the semiconductor, acting as donors or acceptors.
5. Externally a field can be established through a voltage at the gate electrode.

The last-mentioned component is the only one that can be manipulated by the circuit designer. The sum of the other causes of electric field formation gives rise to a 'built-in' field and consequently a charge distribution associated therewith.

Due to the existence of the electric field at the interface of Si and SiO_2, in a lightly doped p-type substrate a channel can exist exhibiting conductivity for electrons, even when no external field is applied. In fact, when this occurs, the surface layer behaves as a thin layer with n-type properties. Such a layer is called an *inversion layer*. If the substrate doping is light enough to give rise to this phenomenon, no explicit channel formation by additional doping of the surface layer is necessary for the formation of the device. In a n-type substrate no spontaneous inversion layer builds up, due to the affinity of donors to the silicon rather than the oxide, and due to the different properties of the interface between n-type Si and SiO_2.

The substrate is isolated from the active structure by reverse biasing the junctions between the substrate and the source and drain areas. For the time being we assume that the substrate potential has negligible effect on the properties of the active structure. Let us now assume that a voltage V_{DS} is applied between drain and source. Further, we assume that $V_{GS} = 0$. Whether or not a current can flow depends on whether or not an n-type channel is available. And this depends on the collective effect of the mechanisms responsible for the establishment of the built-in field. If a channel exists when $V_{GS} = 0$, the device is designated as a *normally-on* device or *depletion device*. If an external field is applied by making $V_{GS} > 0$, the channel conductance increases, so that I_D increases. With $V_{GS} < 0$, the channel conductance is lowered, so that I_D decreases. At a certain voltage $V_{GS} = V_t$, the channel ceases to exist and I_D approaches zero value.

As we have seen, it is also possible that the built-in field is such that at $V_{GS} = 0$ no electron conductive channel exists. Such a device is called a *normally-off* or *enhancement device*. Now, at a certain value of $V_{GS} > 0$ the external field opens the channel. By increasing V_{GS}, I_D increases. This can be expressed by saying that V_t is positive. Current flow is possible when $V_{GS} > V_t$. The voltage V_t is generally designated as the *threshold voltage*. Figure 2.24 sketches the I_D-versus-V_{GS} characteristic for the two types of MOSTs.

The substrate is always very lightly doped. With heavy doping, channel

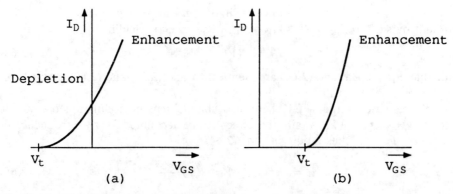

Fig. 2.24 *Characteristics of MOSTs:* (a) *normally-on (depletion type);* (b) *normally-off (enhancement type).*

formation would not occur, i.e. V_t would have an impractically large value. As has already been mentioned, in a lightly doped p-type substrate an n-type surface channel normally occurs without an external field being applied. In other words: if nothing special is done, the n-type MOST is of the 'normally-on' variety. Conversely, in an n-type substrate the formation of a p-conductive channel requires the application of an external field, that is: p-type MOSTs are of the 'normally-off' variety. In some cases it is convenient to have available n-type MOSTs of both the normally-on and the normally-off varieties. For the formation of both types of transistors with different threshold levels, an additional doping step is included in the process sequence.

The *I–V* characteristics of MOSTs are largely similar to those of JFETs. This is not surprising, since the physical mechanisms underlying the operation of the devices are essentially the same. Here, too, channel conductivity as a function of place is controlled by the local value of field strength and this varies along the channel due to the gradual ohmic voltage drop along the channel caused by the current flowing in the channel. Using a first-order modelling approach of the relevant mechanisms ruling the behaviour of the device, the current-vs-voltage characteristics can be derived. As in the case of the JFET, two regions of operation can be distinguished.

If $V_{DS} < V_{GS} - V_t$, the device operates in the *non-saturation region* (also often designated the 'triode region'). In this region:

$$I_D = \beta V_{DS}(V_{GS} - V_t - \tfrac{1}{2}V_{DS}) \qquad (2.63a)$$

where

$$\beta = \frac{\varepsilon_0 \varepsilon_{ox} \mu_n W}{Lt} \qquad (2.63b)$$

wherein $\varepsilon_0 \varepsilon_{ox}$ is the permittivity of the dioxide and t is the thickness of the oxide layer (typically 30–50 nm).

If $V_{DS} > V_{GS} - V_t$, the device operates in the *saturation region*. The I–V relation is found by substituting $V_{DS} = V_{GS} - V_t$ in eq. (2.63a), yielding

$$I_D = \tfrac{1}{2}\beta(V_{GS} - V_t)^2 \tag{2.64}$$

Note that eqs. (2.63a) and (2.64) are identical to eqs. (2.59) and (2.60), apart from the different definitions of β.

As in the case of the JFET, here also channel-length modulation occurs for voltages $V_{DS} > V_{GS} - V_t$. Again, this effect can be accounted for by introducing an 'Early-voltage' V_A, thus modifying eq. (2.64)

$$I_D = \tfrac{1}{2}\beta(V_{GS} - V_t)^2 \left(1 + \frac{V_{DS}}{V_A}\right) \tag{2.65}$$

which is identical to eq. (2.62).

In bipolar transistors V_A is determined by the doping levels of the collector and base regions and by the width of the latter. It is therefore fully determined by the manufacturing process and independent of the geometrical dimensions under the control of the IC designer. Conversely, in MOSTs, V_A depends on channel length L. To a first approximation V_A is proportional to L. Typical values are in the range of 3–5 V μm^{-1}.

The proportionality constant β is an important device parameter. The factor $\varepsilon_0\varepsilon_{ox}/t$ is the capacitance of the gate per unit area; it is often written as C_{ox}. For SiO$_2$, $\varepsilon_{ox} = 3.9$. When calculating $C_{ox}\mu_n$ one should take into account that μ_n in the surface channel is smaller than in the bulk of the silicon. When electrons move parallel to the interface between Si and SiO$_2$ they are pulled toward the oxide by the vertical field. The *surface mobility* can be assumed to be roughly half the value of the mobility in the bulk. The quantity $C_{ox}\mu_n$ is determined by the manufacturing process alone. Apparently, it is the β of a square MOST (i.e. a MOST with $L = W$) and it is often designated by β_{sq}, so that $\beta = \beta_{sq}W/L$. The factor W/L is often referred to as the *shape factor*. It is determined by the geometry of the gate area, which is a free design parameter for the designer of MOS-ICs. This is an interesting feature of MOS-technology since it provides the designer with a versatile design tool that can very often be used to good advantage. The value of β_{sq} in current MOS-processes is 30–40 μA V^{-2}.

As has been shown above, JFETs and MOSTs exhibit quite similar properties. One important difference is that V_{GS} in a JFET is not allowed to become positive, since the pn-junction must be in reverse bias. With the MOST no such constraint occurs since here the gate is isolated by a true isolator. Of course, the gate voltage should not exceed the breakdown voltage of the oxide layer. By their very nature all JFETs are normally-on devices. This makes them less suitable for some special applications. Apart from the obvious differences, MOSTs differ in some more respects from JFETs. These will come to attention at a later stage. However, in many applications JFETs as well as MOSTs behave similarly. Where the choice is irrelevant for the topic under discussion, in this book as a rule the symbol for the JFET will be used. Figure 2.25 shows the symbols we will use.

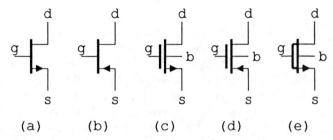

Fig. 2.25 *Symbols used for various types of FETs:* (a) *n-type JFET;* (b) *p-type JFET;* (c) *enhancement-type n-MOST with indication of the substrate b (body);* (d) *idem p-MOST;* (e) *depletion-type n-MOST.*

2.9 Small-signal model for FETs

In this section we will discuss the small-signal behaviour of FETs, using the equation for the large-signal behaviour presented in the preceding sections. At a later stage some additional details of MOST operation will be discussed that can affect circuit properties under certain conditions. Their consequences for small-signal behaviour will give rise to an extension of the model.

In the non-saturation region the FET behaves as a voltage-controlled resistor. From eq. (2.63) we obtain, for the channel conductance,

$$g = \frac{dI_D}{dV_{DS}} = \beta(V_{GS} - V_t - V_{DS}) \tag{2.66}$$

For values of V_{DS} complying with $V_{DS} \ll V_{GS} - V_t$, the conductance is approximately linear and controllable by varying V_{GS}. Several methods have been devised for minimizing non-linearity, some of which will be discussed in the context of an important application of the MOST as a controllable resistor in Chapter 11, Section 11.7.

In the majority of applications the MOST is used in the saturation region. The relevant small-signal parameters are the transconductance g_m and the internal resistance of the drain r_d:

$$g_m = \frac{dI_D}{dV_{GS}} \; (V_{DS} = \text{const}) \text{ and } r_d = \frac{1}{g_d} \text{ with } g_d = \frac{dI_D}{dV_{DS}} \; (V_{GS} = \text{const})$$

From eq. (2.65):

$$g_m = \beta\left(1 + \frac{V_{DS}}{V_A}\right)(V_{GS} - V_t) = \left[2\beta I_D\left(1 + \frac{V_{DS}}{V_A}\right)\right]^{1/2} \tag{2.67}$$

If $V_{DS}/V_A \ll 1$, which is nearly always the case, eq. (2.67) can be written

$$g_m = \beta(V_{GS} - V_t) = \sqrt{2\beta I_D} \tag{2.68}$$

Further,

$$g_d \approx I_D/V_A \text{ or } r_d \approx \frac{V_A}{I_D} \qquad (2.69)$$

Eq. (2.68) reveals important differences between the BJT and the FET. In the BJT, g_m depends solely on the current in the transistor: $g_m = I/V_T$. The linear relationship is the consequence of the exponential I–V characteristic, which is firmly rooted in a basic law of nature. The relation is therefore accurate on fundamental grounds. Conversely, in the FET, g_m is controllable by the device geometry through β; and it is proportional to $I_D^{1/2}$, but this particular relation is not fundamentally accurate. Essentially, the I–V characteristic is determined by the solution of the field equations, governed by the equation of Poisson. The quadratic behaviour arises from adapting the solution of this equation to the boundary conditions, which are set by the device geometry. The quadratic characteristic is therefore an approximation, although it turns out that practical FETs comply rather precisely with the quadratic rule.

Figure 2.26 shows the relevant small-signal equivalent circuit. For low-frequency signals, no gate current occurs because of the purely capacitive nature of the input. Figure 2.27 shows a simple equivalent circuit encompassing the capacitive effects inherent in the operation of the FET. A complete model, taking into account the detailed behaviour of charging effects, is considerably more complex. However, for 'manual calculations' such a model is impractical; its use should be reserved for computer-aided calculations. The various existing software packages for CAD are

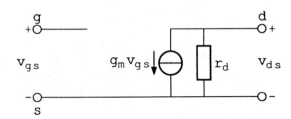

Fig. 2.26 *Basic small-signal equivalent circuit.*

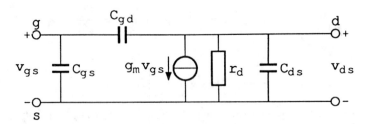

Fig. 2.27 *A simple model accounting for capacitive effects.*

equipped with models of a varying degree of sophistication. In the simple model of Figure 2.27 all charging effects are concentrated in three interterminal capacitances. The input capacitance is by far the largest. It comprises the intrinsic capacitance between gate and channel, together with the extrinsic interelectrode capacitance. The capacitance C_{gd} is much smaller. This is understandable in the light of the fact that the voltage between drain and gate has little effect on the charge distribution because of the pinch-off at the drain side. The extrinsic capacitance can be made small by placing the gate electrode somewhat off the drain. However, this is only possible in depletion-type devices, since in an enhancement-type device a non-controlled portion of the channel would stay below threshold, so that current flow would be blocked. Finally, the capacitance C_{ds} is mainly determined by the capacitance between the drain region and the substrate.

Figure 2.28(a) shows a basic voltage amplifier in the common-source (CS) configuration. For the time being we neglect the influence of C_{gd} on the signal transfer. Then the equivalent circuit of Figure 2.28(b) applies. Using this equivalent circuit, we find

$$A_v = \frac{V_{ds}}{V_g} = -\frac{1/j\omega C_{gs}}{Z_g + 1/j\omega C_{gs}} g_m Z_L' \tag{2.70}$$

where Z_L' is the parallel connection of Z_L, r_d and C_{ds}. The low-frequency gain is $-g_m R_L'$. If $|Z_g| \ll 1/\omega C_{gs}$ and Z_L consists of a resistor R_L and a capacitance C_L in parallel, the 3-dB bandwidth follows from

$$f_h = \frac{1}{2\pi R_L'(C_{ds} + C_L)}$$

It is instructive to compare the voltage-amplification capabilities of the BJT and the MOST. For low-frequency signals where capacitance effects can be neglected, the maximum voltage gain of a BJT in CE amounts to $g_m r_{ce}$ (when $R_L \to \infty) = V_A/V_T$. With $V_A = 150$ V and $V_T = 25$ mV, $A_{v\,max} = 6000$, irrespective of

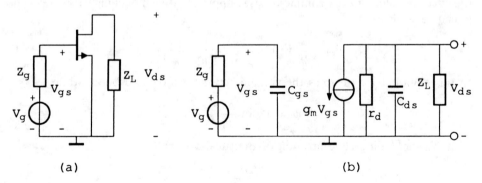

Fig. 2.28(a) *CS-voltage amplifier.* (b) *Equivalent circuit of the configuration of Figure 2.28(a).*

the bias current. Consider now the MOS-variant. Let $\beta_{sq} = 40\ \mu A\ V^{-2}$, $W = 50\ \mu m$ and $L = 10\ \mu m$, so that $W/L = 5$. Then, at $I_D = 1$ mA, $g_m = \sqrt{2\beta I_D} = 0.63$ mA V^{-1}. If V_A is 5 V μm^{-1}, then $V_A \approx 50$ and $r_d = 50$ kΩ, so that $A_v(\max) \approx 32$. Of course it is possible to increase A_v by increasing the gate area. However, this approach is utterly consumptive of silicon real estate. If L is made larger by a factor of 10, then to maintain g_m, W should increase by the same factor. Hence, for a tenfold increase of the maximum voltage gain, silicon area increases by a factor of 100! Clearly, if large amplification is important, the BJT is the active device of choice. Why then ever use FETs? In many practical situations MOS-technology is the natural choice for the main portion of the signal-processing circuitry. This is so, for instance, in digital signal processing or in applications where active filters are required. If it is desirable to include the peripheral analogue circuitry on chip, this circuitry must be implemented in the same technology. Another reason to prefer MOS-technology is that in certain applications MOSTs have some very desirable properties that the BJT is not endowed with. Particularly where very high impedance levels and low leakage currents are of the essence, the MOS-transistor is a very useful device.

2.10 Body-effect

Up to now we have neglected the possible influence of the substrate on the properties of the MOST. In fact, unless the substrate is an ideal insulator, its charge distribution affects the characteristics of the MOST. In certain special MOS manufacturing processes, the active structure is formed on an insulating substrate, mostly alumina (Al_2O_3). Such processes are designated by the term SOI (silicon on insulator). However, in the more common MOS-processes the substrate consists of lightly doped silicon. The substrate is commonly designated by the term *body*, and its influence on MOS-characteristics is called the *body-effect*. A detailed investigation of the effect the substrate has on transistor properties reveals that the charge distribution in the substrate gives rise to two effects. First, it slightly modifies the effective value of β. A commonly used approximative modelling of the relevant phenomena yields

$$\beta_{\text{eff}} = \frac{\beta}{2n} \text{ with } n = 1 + \frac{\gamma}{2\sqrt{(2\phi_B + V_{SB})}} \qquad (2.71)$$

In this equation V_{SB} denotes the voltage between source and body, while

$$\phi_B = V_T \ln \frac{N_B}{n_i} + 3 V_T$$

with N_B the doping concentration of the substrate, and

$$\gamma = \frac{\sqrt{2q\varepsilon_0\varepsilon_{si} N_B}}{C_{ox}}$$

Practical values of N_B are in the range of 10^{14}–10^{15} cm^{-3}, so that $\phi \approx 12\ V_T$ to

14 V_T, hence $2\phi_B \approx 600$ to 700 mV. Assuming $C_{ox} \approx 600$ pF mm^{-2}, $N_B = 10^{15}$ cm^{-3} and $V_{SB} = 0$ we find $n = 1.2$. With larger values of V_{SB}, n decreases. Generally $n = 1.2$ can be assumed to be an upper limit, so that the effect on β is not very significant.

Second, V_{SB} affects the threshold voltage V_t. For large values of V_{SB} this effect can be quite significant. The following approximation proves to be practicable for small values of V_{DS}:

$$V_t = V_{to} + \gamma\{\sqrt{(2\phi_B + V_{SB})} - \sqrt{2\phi_B}\} \tag{2.72}$$

where V_{to} is the value of V_t with $V_{SB} = 0$. For large values of V_{DS}, the threshold shift tends to be smaller.

EXAMPLE

If $N_B = 10^{15}$ cm^{-3}, $\varepsilon_{si} = 11.7$ and $C_{ox} = 600$ pF mm^{-2}, we find $\gamma \approx 0.30$ V$^{1/2}$. If $V_{SB} = 3$ V and $2\phi_B = 0.6$ V, eq. (2.72) yields $\Delta V_t = V_t - V_{to} \approx 0.34$ V.

Taking into account that normal threshold values are on the order of 1 to 2 V, the shift is considerable. If substrate doping is lighter, the shift is smaller. With $N_B = 10^{14}$ cm^{-3} we find $\Delta V_t = 0.11$ V.

In fact the substrate acts as a second gate, commonly referred to as the *back gate*, as opposed to the front gate. Substituting eq. (2.72) into eq. (2.64) we obtain

$$I_D = \tfrac{1}{2}\beta[V_{GS} - V_{to} - \gamma\{\sqrt{(2\phi_B + V_{SB})} - \sqrt{2\phi_B}\}]^2 \tag{2.73}$$

revealing that I_D is a function of both V_{GS} and V_{BS}. To describe small-signal behaviour we can introduce two transconductances, associated with the front gate and the back gate respectively:

$$g_m = \frac{dI_D}{dV_{GS}} \ (V_{BS} = \text{const}) \text{ and } g_{mb} = \frac{dI_D}{dV_{BS}} = -\frac{dI_D}{dV_{SB}} \ (V_{GS} = \text{const})$$

Note that in both definitions the source is assumed to be the reference electrode. Using eq. (2.73) and executing the appropriate differentiations, we obtain

$$g_m = \beta[V_{GS} - V_{to} - \gamma\{\sqrt{(2\phi_B + V_{SB})} - \sqrt{2\phi_B}\}] = \sqrt{2\beta I_D} \tag{2.74}$$

which is identical to eq. (2.68), and

$$g_{mb} = bg_m \text{ with } b = \frac{\gamma}{2\sqrt{(2\phi_B + V_{SB})}} \tag{2.75}$$

This expression is a good approximation for moderate values of V_{DS}. With increasing V_{DS}, b tends to decrease. Note that g_{mb} is non-zero also when $V_{SB} = 0$. However, in general, when $V_{SB} = 0$, this is accomplished by interconnecting source and substrate. In that case no signal v_{bs} can occur and henceforth $g_{mb}v_{bs} = 0$ also.

EXAMPLE

Let, as in the preceding example, $\gamma \approx 0.30$ V$^{1/2}$ and $2\phi_B = 0.6$ V. With $V_{SB} = 3$ V, we find $b = 0.08$. The assumption $\gamma \approx 0.30$ V$^{1/2}$ is rather pessimistic. If we choose $N_B = 10^{14}$ cm^{-3}, we find $\gamma \approx 0.10$ V$^{1/2}$ and $b = 0.025$. It appears that, fortunately, the transconductance of the back gate is usually much smaller than g_m. However, its effect on small-signal behaviour cannot always be neglected.

The best measure to prevent the body-effect is to connect the source to the substrate. This is, however, not always possible. In CMOS circuits the possibilities for taking this measure are more favourable than in NMOS, since in CMOS either the n-type or the p-type MOSTs, or both, are located in wells (see Section 2.14). Analogue circuits are predominantly implemented in CMOS.

If the back gate must be taken into account, the small-signal transfer is given by

$$i_d = g_m v_{gs} + g_{mb} v_{bs} + g_d v_{ds} \tag{2.76}$$

with $g_d = 1/r_d$.

2.11 Subthreshold behaviour of the MOST

According to eq. (2.64), $I_D = 0$ when $V_{GS} = V_t$. This equation is based on the assumption that the inversion layer is formed under the condition of so-called *strong inversion*. For values of V_{GS} approaching V_t and below V_t, this assumption is no longer valid. In strong inversion, the driving mechanism for carrier transport from source to drain is the longitudinal electric field. Diffusion plays an insignificant role. In the region near threshold operation, field-driven transport fades out and diffusion takes over as the remaining driving mechanism. This continues to work when V_{GS} is below threshold. In this subthreshold region the channel is said to be in *weak inversion*. Obviously, the term 'threshold' should not be understood in absolute sense. The threshold voltage is formally defined on the basis of strong inversion theory. For practical purposes it is more convenient to define the threshold voltage as the intersection of the square law portion of the characteristic with the V_{GS}-axis. In the subthreshold mode, where charge transport is by diffusion, the characteristic is exponential, just as in the bipolar transistor. Modelling of the operation mechanism under weak inversion conditions yields the approximate relation

$$I_D = \frac{W}{L} I_{Do} \varepsilon^{V_{GB}/mV_T} (\varepsilon^{-V_{SB}/V_T} - \varepsilon^{-V_{DB}/V_T}) \tag{2.77}$$

where

$$m \approx 1 + \frac{\gamma}{2\sqrt{(1.5\phi_B + V_{SB})}}$$

(Tsividis, 1987, Ch. 4). I_{Do} is a quantity that can be compared with I_s in a bipolar transistor. It is determined by the manufacturing process and it depends slightly on V_{SB}. Typical values can differ widely; commonly they are in the range

10^{-16}–10^{-19} A. Moreover, it appears that even for chips from the same wafer spread can be considerable, up to some 40% (Vittoz and Fellrath, 1977). Typical values for the parameter m range from 1.1 to 1.5 with little spread among chips manufactured in a given process.

If $V_{SB} = 0$, eq. (2.77) simplifies to

$$I_D = \frac{W}{L} I_{Do} \varepsilon^{V_{GS}/mV_T}(1 - \varepsilon^{-V_{DS}/V_T}) \qquad (2.78)$$

For values of $V_{DS} > 2.5 V_T$, $1 - \exp(-V_{DS}/V_T) \approx 1$, so that I_D depends solely on V_{GS}:

$$I_D = \frac{W}{L} I_{Do} \exp(V_{GS}/mV_T) \qquad (2.79)$$

This equation resembles the corresponding equation for the BJT. As there, g_m is proportional to I_D:

$$g_m = \frac{I_D}{mV_T} \qquad (2.80)$$

which is independent of W/L.

The 'specific transconductance' g_m/I_D is smaller by a factor of $1/m$ in comparison with the BJT. With V_{GB} fixed, eq. (2.77) would yield $g_{mb} = dI_D/dV_{BS} = I_D/V_T$ for the transconductance of the back gate. However, eq. (2.77) cannot be used for this purpose since both I_{Do} and m depend on V_{SB}. A better approximation is (Tsividis, 1987, Ch. 8).

$$g_{mb} \approx \frac{m-1}{m} \frac{I_D}{V_T} \qquad (2.81)$$

If $m = 1.5$, then $g_{mb} \approx 0.5 g_m$.

Using eq. (2.78) to find $g_d = 1/r_d$ yields, for $V_{DS} \gg V_T$

$$g_d = \frac{dI_D}{dV_{DS}} \approx \frac{\exp(-V_{DS}/V_T)}{1 - \exp(-V_{DS}/V_T)} \frac{I_D}{V_T} \qquad (2.82)$$

predicting that g_d rapidly goes to zero with increasing V_{DS}. However, as in the case of strong inversion, the drain field directly influences the channel. Because of this effect, eq. (2.69), found for the case of strong inversion, is also a better approximation in the case of weak inversion; hence, again, $g_d = I_D/V_A$.

The subthreshold (weak inversion) region does not end abruptly. It merges smoothly with the quadratic region. The intermediate region is usually designated as the region of *moderate inversion*. Modelling of this region is difficult and is a somewhat controversial topic. However, if in the design of a circuit, compliance with the exponential behaviour is of the essence, the intermediate region should be avoided anyway. A commonly accepted practical limit for I_D ensuring operation in the exponential mode is

$$I_D < 0.1 \, \beta \, V_T^2 \tag{2.83}$$

If, for instance, $\beta = 150 \, \mu A \, V^{-2}$, eq. (2.83) yields $I_D < 10 \, nA$.

From eq. (2.83) one might conclude that the subthreshold region can be extended by making β large. However, it should be noted that, given β_{sq}, this can only be accomplished by making W/L large. Since the minimum value of L is dictated by the manufacturing process on the one hand, and by requirements concerning V_A on the other, transistors with an extended exponential range tend to consume a great deal of chip area. Moreover, enlarging W leads to increased leakage currents, thus restricting the minimum useable drain current. All in all, the possibilities to accommodate high-dynamic-range signals within the subthreshold region are restricted. In spite of the constraints associated with this mode of operation, there is an increasing interest in putting it to good use in circuit design. Where low-power operation is essential, this mode of operation has considerable advantages. The use of bipolar transistors at current levels in the nanoampere range is unattractive because of the reduction of α_0. Moreover, in principle, the maximum attainable voltage gain can be larger with a MOST in subthreshold than with a BJT. In a BJT, A_{max} is determined by V_A/V_T, in a MOST by V_A/mV_T. If one can afford to use a large-area MOST, V_A can be made larger than is possible with a BJT, since in a MOST V_A is proportional to L. Bandwidth capabilities are very much restricted in both cases, since a low value of g_m is associated with a large time constant C_L/g_m, where C_L is the unavoidable load capacitance. For low-power, low-bandwidth applications, the MOST in the subthreshold mode can be an attractive proposition. Applications where these conditions apply can be found in transducer interfaces,

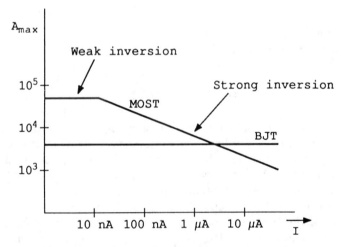

Fig. 2.29 *Asymptotic plot of A_{max} as a function of biasing current for a BJT and a MOST with large V_A.*

for instance where transducers are located in inaccessible locations making battery replacements difficult. Figure 2.29 depicts an asymptotic plot of A_{max} as a function of I_C, respectively I_D, for a BJT and a MOST with large V_A, showing that in a certain current range the MOST is the preferred device if a large value of A_{max} is desired.

2.12 Monolithic technology

Modern monolithic manufacturing technology encompasses a vast body of knowledge. In a book devoted to electronic signal-processing circuits the treatment of this subject necessarily has to be restricted to some basic notions that are essential for understanding the problems of monolithic circuit design.

In a monolithic circuit all devices are formed in a single piece of silicon, commonly designated a chip. A typical chip measures no more than some tens of mm^2. Since the minimum features that can be accurately defined are of the order of a few μm in modern monolithic processes, several thousands of devices can be accommodated within the confines of a single chip.

Devices consist of separate areas of different doping levels. In order to create local regions of defined doping, masking techniques are employed. The commonly used masking material is SiO_2. The chip is first covered with a layer of silicon oxide, after which local windows are created. These windows are formed by lithographic processes. The SiO_2 is covered with a layer of photoresist lacquer which is subsequently illuminated through a mask that defines the pattern of windows to be formed. After illumination the photoresist is fixed, so that at the window locations uncovered silicon oxide remains. Then the oxide is etched away, leaving unprotected silicon at the window locations. Local doping can be performed at these locations. By covering the wafer anew with SiO_2, this process is repeated with a second mask, defining a new pattern of local doping. This process can be repeated several times until the complete monolithic circuit is formed. The desired interconnections are also formed by a masking step, whereby an aluminum pattern is created. The number of masking steps depends on the measure of sophistication of the manufacturing process. Most processes involve ten to sixteen masking steps, but this is no more than an indication of the present state of the art.

Figure 2.30 shows, by way of example, the formation of n-type regions in a p-type substrate, according to the described lithographic process. Figure 2.30(a, b and c) shows the steps leading to the formation of the doping locations; Figure 2.30 (d, e and f) shows the steps leading to the formation of the oxide doping mask; while Figure 2.30 (g and h) shows the application of the doping and the removal of the masking oxide.

For the localized doping process two options are available: diffusion from the gas phase and ion implantation. Diffusion is done in a diffusion furnace. This is an apparatus wherein the wafer is exposed to a vapour containing the dopant material at high temperature. Due to the thermal excitation, the dopant atoms penetrate into

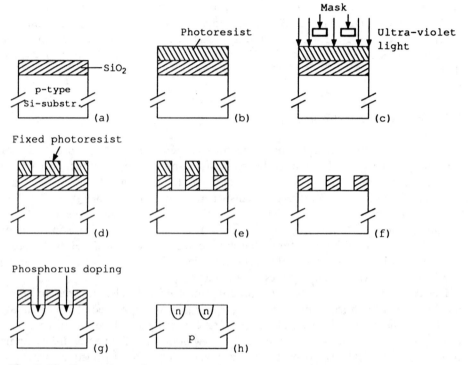

Fig. 2.30 *Formation of n-type regions in a p-type Si-substrate.*

the silicon. The doping concentration depends on the depth in the silicon crystal. Figure 2.31 sketches the doping profile for various durations of the diffusion time. The doping concentration N at a depth x in the wafer as a function of time obeys the formula

$$N(x) = N_0 \operatorname{erfc} \frac{x}{2\sqrt{Dt}} \qquad (2.84)$$

where D is the diffusion coefficient, which depends on the doping material and strongly on temperature (typically about 1200 °C). The concentration N_0 at the surface is determined by the concentration of the dopant in the vapour.

The maximum doping concentrations that can be attained differ widely for the various doping materials, due to the differences in solid solubility of the dopants in silicon. For diffusion at 1200 °C the following data apply

As: 1.5×10^{21} atoms cm^{-3}
P: 1×10^{21} atoms cm^{-3}
B: 5×10^{20} atoms cm^{-3}
Al: 2×10^{19} atoms cm^{-3}

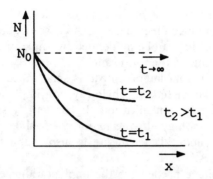

Fig. 2.31 *Doping profile N(x) for various durations of diffusion time.*

Recall that $1\,\text{cm}^3$ of silicon contains 5×10^{22} atoms. Diffusion speed, as parametrized by the diffusion coefficient D, at a given value of T, depends largely on the specific dopant. Arsenic diffuses very slowly; it is therefore suitable for doping of buried layers, which must maintain their places during subsequent diffusion cycles of shorter duration.

Making pn-junctions by diffusion is done by way of *compensation*. For example, if a pn-junction is to be formed in lightly doped p-type material, an n-type dopant is diffused into the material. At a certain depth the concentration of p- and n-dopants match. At this place the pn-junction is formed.

A disadvantage of doping by diffusion is that the doping also extends in lateral direction to a certain degree. Figure 2.32 sketches this effect which, particularly in deep diffusions, can be considerable.

The second option for accomplishing local doping is the use of *ion implantation*. An ion implanter is a complicated machine wherein a beam of accelerated ions is formed by means of a high-voltage accelerator section. The wafer is placed in a wafer chamber where it is exposed to the ion beam. Since all the ions have the same energy, they all penetrate the silicon at about the same distance. By varying the acceleration voltage the doping profile can be shaped. Even a 'reverse' doping profile can be realized, i.e. a profile whereby the doping is stronger at a greater distance from the surface. It must be noted, though, that accurate prediction of the doping profile is hampered by the phenomenon of *ion channelling*. The ions tend

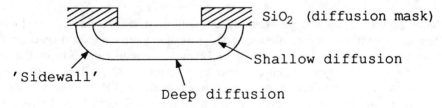

Fig. 2.32 *Lateral extension of diffused regions.*

54 ANALOGUE ELECTRONIC CIRCUIT DESIGN

to prefer paths that are determined by the crystal orientation. Channelling ions can penetrate deeper than non-channelling ions. As a consequence the doping profile exhibits a 'tail'. Channelling can be moderated by implanting through a thin oxide layer or by some skewing of the ion beam.

A further advantage is that lateral spread is small since all ions hit the surface in a nearly perpendicular direction. Obviously, ion implantation is a much more versatile doping technique than diffusion. It has largely superseded diffusion in modern monolithic technology. In the majority of MOS-processes all doping is done by implantation and this technique is also making its way into bipolar manufacturing processes.

The acceleration voltage needed for the implantation depends strongly on the mass of the ions to be implanted and, of course, on the intended implantation depth. Most industrial implanters can produce ions accelerated up to 220 kV.

A further technique that must be mentioned is *epitaxial growth*. This is a process whereby a thin layer of silicon is deposited from a vapour phase upon a silicon wafer. Ideally, the layer is a flawless continuation of the monocrystalline silicon lattice. In addition to free silicon atoms the vapour contains the dopant to be built into the layer. Epitaxial growth is a non-masked step in the processing of the wafer. Basically, doping of the layer is homogeneous. In practice, due to the finite reaction time, some outdiffusion of previously formed doped regions takes place, causing slight aberrations of the uniform doping profile. It is also possible to grow differentially doped layers on top of one another, for instance a p-type layer can be deposited on a previously deposited n-type layer. Epitaxial growth is indispensable in bipolar manufacturing processes. The collector regions of BJTs are formed in a lightly doped n-type epilayer deposited on a p-substrate.

Interconnection patterns are usually made by selective etching of an aluminum layer. The layer is deposited by vacuum evaporation in a bell jar. Because Al is an acceptor, it cannot be deposited directly on lightly doped n-type silicon, as this would give rise to the formation of unwanted pn-junctions. Fortunately, the solid solubility of Al is considerably smaller than for P, which is commonly used as a donor material. If Al is deposited on a heavily doped n-region, an ohmic contact is obtained. Doping enhancement is therefore needed for contact areas in n-type material, particularly the collector contact in bipolar transistors.

2.13 Monolithic bipolar devices

Figure 2.33(a) shows the generally used method for the formation of bipolar transistors in monolithic technology. An epitaxial, lightly doped n-type layer is grown on a lightly doped p-type substrate. By means of a deep p-diffusion, reaching into the substrate, n-type islands or pockets are formed in the epilayer. The devices constituting the monolithic circuit are formed in these islands. The islands are mutually isolated by reverse-biasing of the pn-junctions. Figure 2.33(b) shows how a transistor can be made by subsequent masked doping steps. In many bipolar

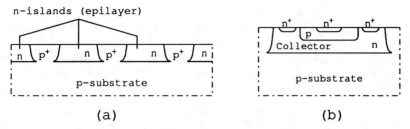

Fig. 2.33(a) *Formation of n-islands through perforation of the epilayer by the isolation diffusion (DP).* (b) *Structure of a npn-transistor in an epitaxial island.*

processes the doping steps are performed by diffusion. The subsequent steps are commonly indicated by special names. The isolation diffusion is called *deep-p diffusion* (DP), the base is formed by *shallow-p diffusion* (SP), the emitters and the collector contact regions are formed by *shallow-n diffusion* (SN).

A drawback of the transistor shown in Figure 2.33(b) is its relatively high ohmic collector resistance. In order to lower this resistance, prior to the formation of the epilayer, a low-resistance (i.e. highly doped) n-diffusion is applied at the locations where the islands are to be formed. This is called a 'buried layer'; the relevant diffusion is called the *buried-n diffusion* (BN). In order to lower the collector resistance further an additional deep n-diffusion (DN) is applied at the location of the collector contact. Figure 2.34(a) sketches the transistor obtained in this way.

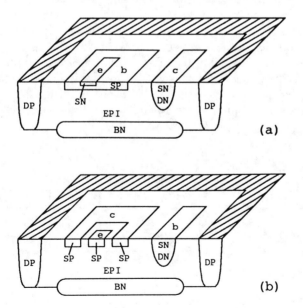

Fig. 2.34 *Sketch of monolithic transistors:* (a) *vertical npn;* (b) *lateral pnp.*

In circuit design it is often very desirable to dispose of npn- as well as pnp-transistors. One possibility is to use the pnp-structure formed by an SP-region, the epilayer and the substrate. This is, of course, a low-quality transistor, since the epilayer is much too wide to provide a good base. Moreover, the collector is formed by the substrate and hence is directly connected with the negative supply voltage. This implies that it can only be used in the common-collector configuration (see Chapter 5).

An alternative is the use of a so-called *lateral pnp-transistor*, as sketched in Figure 2.34(b). Here, the emitter and the collector are formed in the SP-step, while the epilayer functions as the base region. Obviously, current flow is lateral to the surface; hence the name. The price to be paid for the constructional simplicity of these transistors is their mediocrity in performance. The base must be made comparatively wide, in view of tolerances imposed by sideways spread of the emitter and collector diffusions and because of the proneness of the base area to punch-through. This is due to the high doping level of the collector. As a consequence, β_0 is small (often below 70–80), V_A is small, and the diffusion capacitance is large. In addition the useful current range is limited. Optimum performance is usually obtained with currents at the 0.1 mA level. An additional unpleasant effect is the drain of injected current by the vertical parasitic pnp-transistor operating with the substrate as collector. This effect is lowered by including a buried layer. This creates a potential barrier acting as a reflector for the vertically injected holes in the epilayer.

At present, bipolar manufacturing processes exist which feature the availability of high-performance, vertical pnp-transistors. These are made by applying additional masking steps. Needless to say, such complementary bipolar (COBI) processes are considerably more costly than basic bipolar processes.

Monolithic *resistors* can be made in several ways. The most common implementation uses a diffused or implanted region, isolated from its surroundings by a reverse-biased pn-junction. If the bases of the transistors are realized by an SP-diffusion, this step can also be used for the formation of resistors (Figure 2.35a). For very low resistance values the low-ohmic emitter diffusion (SN-diffusion) can

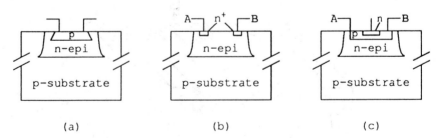

Fig. 2.35 *Various resistor implementations: (a) use of base doping (SP); (b) use of epilayer; (c) pinch resistor.*

be used. The specific conductivity σ and the specific resistivity ρ are given by

$$\sigma = \frac{1}{\rho} = q(p\mu_p + n\mu_n) \qquad (2.85)$$

In p-type materials $p \gg n$ holds, so that $\sigma \approx qp\mu_p \approx q\mu_p N_A$. The resistance of a doped area of length L and width b is

$$R = \frac{1}{qd\bar{\mu}_p \bar{N}_A} \cdot \frac{L}{b} \qquad (2.86)$$

where d is the depth of the doped area and $\bar{\mu}_p$ and $\bar{N}_A$ denote the averages of μ_p and N_A over the inhomogeneously doped zone. For the determination of $\bar{\mu}$ and $\bar{N}$ obtained with a given diffusion or implantation process, tables and graphic plots are available. The factor $1/qd\bar{\mu}_p\bar{N}_A$ is called the *sheet resistance* which is the resistance of a square section of the doped sheet ($L = b$). This quantity is uniquely determined by the doping process. The factor L/b, the aspect ratio of the doped area, is fully determined by the mask dimensions. A typical value of the sheet resistance for the base doping of a transistor is 200 Ω; for the emitter doping it is about 2–3 Ω. When the doping is done by diffusion, the lateral spread of the doped area is a cause of inaccuracy of the resistance value. The absolute value of such resistors can exhibit an uncertainty up to about 15%. The relative accuracy, i.e. the accuracy of the ratio of resistor values within a chip, can be much better, even in the order of 1%. The temperature coefficient of diffused resistors is high, in the order of 1500 ppm $°C^{-1}$.

EXAMPLE
Determine the area occupied by an SP-resistor of 4 kΩ value, when the SP-sheet resistance is 200 Ω and the minimum width that can be realized with reasonable accuracy is 6 μm. Lateral spread of the SP-region is assumed to be 1 μm.

SOLUTION
The L/b ratio must be $4000/200 = 20$. The effective width $b = 7$ μm, hence $L = 140$ μm and $Lb = 980$ μm^2. If a length of 140 μm is inappropriate in view of practical chip layout, a more compact structure can be obtained by meandering (Figure 2.36) or by series connection of a number of smaller resistors.

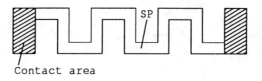

Fig. 2.36 *A meandering resistor structure.*

Resistors made by ion implantation can be more accurately realized than diffused resistors. Absolute tolerance can be in the order of 3%; matching tolerance for relatively wide resistance tracks, about 1% or even better. In the design of integrated circuits the tolerance data pertaining to the intended manufacturing process must be taken into account carefully. In general, circuits should be conceived in such a way that their properties are determined by resistor ratios, rather than by the absolute values of the resistors.

Because of the low sheet resistivity of the base doping, large resistor values require much chip area. This is unattractive. If large resistors are unavoidable, two alternative implementation methods are available. One possibility is the use of the high-resistivity epilayer. Such *bulk resistors* are, in fact, just elongated epi-islands (Figure 2.35b). The absolute tolerance of such resistors is poor, often about 30%. Resistors up to several hundreds of kΩ can be formed in this way. The second alternative is the so-called pinch-resistor (Figure 2.35c). This is basically the channel resistance of a JFET. It can be formed by the p-type base doping step. When the emitter is used as the gate, the p-channel can be pinched by reverse-biasing the pn-junction. Resistors up to about 100 kΩ can be made in this way, albeit that the resistance value is very inaccurate.

Capacitors can be formed in two ways. The simplest method is to use a reverse-biased pn-junction (Figure 2.37a). A disadvantage is that this type of capacitor is voltage-dependent. And care must be taken to avoid forward-biasing of the junction. An advantage is the relative compactness of this type of capacitor. A 10 pF capacitor takes about the same chip area as a small transistor. The second method uses a low-resistivity doped area (mostly provided by the SN-emitter doping) as one electrode, a thin SiO_2 layer as the dielectric, and an aluminum layer as the second electrode (Figure 2.37b). This type of capacitor does not exhibit the disadvantages of the junction capacitor, but it requires more chip area and the inclusion of a precision oxidation step in the manufacturing process.

The availability of JFETs is often of great advantage in the design of analogue circuits. Bipolar manufacturing processes can be enhanced relatively simply to

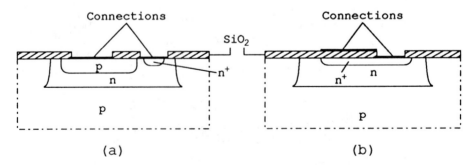

Fig. 2.37 *Two types of monolithic capacitors:* (a) *pn-junction used as a capacitor;* (b) *capacitor with SiO_2 dielectric.*

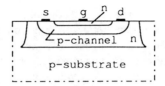

Fig. 2.38 *Monolithic p-channel JFET.*

include the possibility of the formation of these devices. Figure 2.38 shows the basic structure of a monolithic p-type JFET. The channel is formed by the SP-doping, together with the bases of the BJTs. The gate is formed by a heavily doped n-region. The structure obtained in this way closely resembles the structure of a common BJT. However, using the standard emitter diffusion does not provide the channel width needed for satisfactory operation as a JFET. The pinch-off voltage exceeds the breakdown voltage of the pn-junction and, consequently, pinch-off cannot be accomplished. Therefore, an extra processing step is needed, providing a deeper extension of the n-region. The additional step is usually done by implantation.

Figure 2.39 shows the structure of a monolithic n-channel JFET. Here the epilayer is used as the channel region. An SP-region provides the gate. Again, using the standard SP-doping that is part of the bipolar manufacturing process is not adequate. Making the channel narrow enough to be pinched-off by a voltage below the junction breakdown voltage requires an additional p-type doping extending sufficiently deep into the epilayer. Note that in general the availability of p-channel JFETs is of more value than that of n-channel devices. A p-type JFET can often be employed instead of a low-quality, lateral pnp-transistor for achieving a level shift.

Diodes can be formed by using just one pn-junction, either the emitter junction or the collector junction. The latter is rarely used in monolithic circuits because of its larger leakage current, due to the low doping levels involved. Nearly always, diodes are formed by interconnecting the base and the collector of a standard BJT. The properties of a diode formed in this way closely resemble those of a diode formed by the emitter–base junction.

Nowadays, there exists an abundance of bipolar manufacturing processes. Every major IC-manufacturer has at its disposal several standardized processes with well-

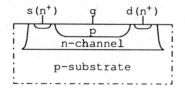

Fig. 2.39 *Monolithic n-channel JFET.*

defined design rules. The processes are optimized towards use in particular applications, for instance high-frequency or high-voltage circuits. It goes without saying that the more sophisticated a process is, the more costly the produced chips are. The description of special processes is beyond the scope of this work. IC-manufacturers and silicon foundries provide detailed data sheets concerning the applicable design rules and the device parameters. Suffice it, by way of example, to mention here some data concerning a basic non-sophisticated bipolar process using only seven masks and using vapour diffusion as the doping technique. The masks serve to define the five diffusions needed: BN, DP, DN, SP and SN, together with the mask for the formation of the contact holes and the mask for defining the interconnection pattern. Denoting the sheet resistances (also called 'square resistances') by R_{sq} and the junction depths by d_j, the data are

BN: $R_{sq} = 22\ \Omega,\ d_j = 0.6\ \mu m$
N-epi: $R_{sq} = 700\ \Omega$, thickness of epilayer 10 μm
DP: $R_{sq} = 5\ \Omega,\ d_j = 12\ \mu m$
DN: $R_{sq} = 4\ \Omega,\ d_j = 4.5\ \mu m$
SP: $R_{sq} = 220\ \Omega,\ d_j = 2.6\ \mu m$
SN: $R_{sq} = 6\ \Omega,\ d_j = 2.0\ \mu m$

Interconnection pattern: $R_{sq} = 0.03\ \Omega$, thickness of Al-layer = 1.2 μm. Substrate (p-type): $\rho \approx 5\ \Omega cm$, thickness = 380 μm. The doping concentrations associated with the given resistivities are:

Emitter (SN): $C_{max} = 5 \times 10^{20}\ cm^{-3}$
Base (SP): $C_{max} = 5 \times 10^{18}\ cm^{-3}$
Collector (epi): $C = 1 \times 10^{16}\ cm^{-3}$
Collector (BN): $C = 4 \times 10^{19}\ cm^{-3}$
Substrate: $C = 3 \times 10^{15}\ cm^{-3}$

A final topic that must not go unmentioned concerns the occurrence of parasitic devices in a monolithic IC. Since all devices are formed in the same substrate, undesired mutual coupling cannot be fully avoided. As an example, Figure 2.40 sketches two islands, each containing an npn-transistor. The base and collector of each transistor form, together with the substrate, a parasitic pnp-transistor. Since

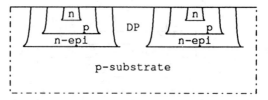

Fig. 2.40 *Two transistors in a monolithic IC.*

in analogue circuits the transistors normally operate in the active region, the collector junctions are reverse-biased, so that the parasitic pnp is in the off-state, normally making it harmless. The same applies to the parasitic npn-transistor formed by the two collectors and the DP region between them. Further, each transistor forms, together with the substrate, a npnp-structure that in principle can act as a thyristor. Normally, this parasitic element will also be non-conducting. However, the designer of monolithic circuits should always be aware of the presence of the parasitic elements, since they may cause problems in case a transistor operates in saturation. Of more concern are the abundant parasitic capacitive couplings. By far the most important of these is the parasitic capacitance between collector and substrate. Since the collector is entirely immersed in the p-substrate, this capacitance can be considerable, in spite of the low doping levels involved. Even a small transistor can be afflicted with a substrate capacitance of a few pF, which acts as an additional capacitive collector load.

If a monolithic capacitor is realized by a reverse-biased pn-junction, as sketched in Figure 2.37(a), a parasitic pnp-transistor is formed. Should the signal voltage on the capacitor drive the junction in forward bias, this transistor becomes conductive.

Resistors, formed by tracks of SP-material (Figure 2.35a) can normally be housed in one island, provided the island is connected to the positive supply voltage. In that case parasitic transistor action cannot occur. However, all resistors are capacitively coupled. If this cannot be tolerated, the resistors must be located in separate islands.

2.14 MOS-ICs

Basically, a MOST is a very simple device. An enhancement type (normally-off type) MOST can be formed by providing drain and source regions and by depositing a conductive layer acting as the gate. The channel region comes into being automatically as an inversion layer at the surface of the substrate. And as the formation of the inversion region depends on the electric field induced by the gate biasing, no conductive channels are formed where this field is absent. Hence, no measures whatsoever are needed for mutual isolation of the devices. Conversely, depletion (normally-on) devices are conductive without an external field being applied. The inversion layer is present over the entire surface area. For mutual isolation of the devices, additional measures are needed. Figure 2.41 sketches two n-type MOSTs in a p-substrate. If nothing is done, a parasitic channel exists between the drain of the first transistor and the source of the second. This must, of course, be prevented. One method is to apply a so-called channel-stop p+ doping, as indicated in the diagram. Most modern MOS-processes perform the isolation with the aid of a sinked oxide area. This is called the LOCOS-technique (local oxidation of silicon). Figure 2.42 shows how the formation of the localized oxide walls is accomplished. The mask used for localizing the oxidation consists of silicon nitride (Si_3N_4). First the substrate is covered with a layer of this material

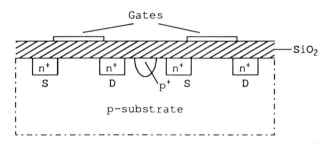

Fig. 2.41 *Channel-stop for the prevention of parasitic channels.*

(Figure 2.42a). Where the local oxide is to be formed the Si_3N_4 is etched away (Figure 2.42b). Then a shallow boron implant is applied (Figure 2.42c). This is done in order to ensure that no parasitic channel can develop under the local oxide. Subsequently, the oxidation step is performed (Figure 2.42d). Typically the depth of the local oxidation is about 750 nm. The transistors are formed in the space between the locally oxidized areas.

Early MOS-processes were of the 'p-only' variety. This is not surprising since, without the application of additional doping, p-MOSTs in an n-substrate are always enhancement types. Hence, no separate isolation step is necessary. The disadvantage is the poor high-frequency behaviour, due to low hole mobility and the need for gates overlapping the source and drain regions. For some time, p-MOS has been superseded by n-MOS and CMOS (complementary MOS). Presently, the

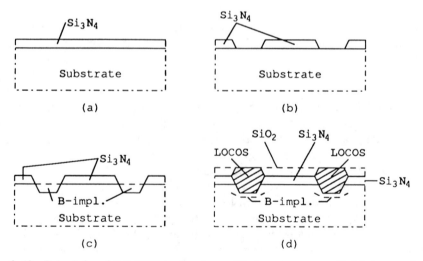

Fig. 2.42 *Principle of LOCOS-technology:* (a) *homogeneous Si_3N_4-layer;* (b) *local etching of openings;* (c) *B-implantation;* (d) *local oxidation.*

dominant technology is CMOS. N-MOS is still in use for certain digital applications. In most N-MOS processes provisions are made for the formation of enhancement as well as depletion types. The possibilities for the implementation of analogue circuits in an N-MOS process are quite limited. CMOS provides much better conditions. Three types of CMOS-processes can be distinguished. In a p-well process, the substrate is n-Si; the p-type transistors are formed in the substrate, whereas the n-type transistors are formed in p-wells (Figure 2.43a). In an n-well process, the n-type transistors are formed in the substrate, whereas the p-type transistors are formed in n-wells (Figure 2.43b). Since the wells must be formed by overdoping the substrate, the high-frequency quality of the transistors in the wells is somewhat poorer than that of those formed in the substrate. Since, due to lower hole mobility, the speed of p-MOSTs lags behind that of n-MOSTs, in a p-well process a reasonably good speed balance is achieved. Most modern processes, though, use n-wells. They provide optimal n-MOSTs, which are often preferred in circuit design. A fundamentally better approach is to use a very lightly doped (nearly intrinsic) substrate provided with p-wells as well as n-wells. Such processes are called *twin-well* or *twin-tub* processes (Figure 2.43c). An additional advantage of such processes is that they give more freedom in coping with the body-effect, since transistors can be accommodated in separate wells. The disadvantage of the twin-well option is, of course, higher cost.

Modern MOS-processes are 'self-aligned', which means that the source and drain regions are formed in such a way that their edges just fit with the associated gate regions. Often, two or more interconnection layers are provided, either poly-Si or Al, or both. If two layers are available, these can also be used for the formation of capacitors.

A general problem of MOS-circuits is their vulnerability to breakdown due to unwanted static charges. This applies in particular to the locations where the circuits are connected to the outer world, i.e. at the bonding pads. Since the input capacitance of a MOST is very small, even a very small amount of static charge can give rise to a voltage exceeding the oxide breakdown voltage. Such a static charge can easily be acquired in the process of handling and mounting MOS-chips. Usually, the external connections are provided with protection devices in the form of reach-through diodes, or zener-diodes, in combination with small resistors and capacitors.

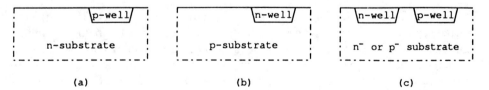

Fig. 2.43 *MOS-processes:* (a) *p-well;* (b) *n-well;* (c) *twin-tub or twin-well.*

2.15 BICMOS-technology

Both bipolar technology and CMOS-technology have their weak and strong points. Generally speaking, bipolar technology is the technology normally chosen for the implementation of analogue circuits. But for certain signal-processing functions, particularly filter functions, MOS-technology has definite advantages. Also, in the design of digital circuits, where MOS-technology is preferred for the implementation of the majority of the functions needed, the availability of BJTs for certain purposes is an advantage. Therefore, there is a growing interest in the development of processes that combine the best of the two worlds. Several such processes are in existence nowadays. The only disadvantage is their complexity. With the growing maturing of monolithic technology, the problems of coping with complex manufacturing methods are rapidly diminishing. BICMOS-processes require more masking steps than bipolar-only or MOS-only processes. A count of nineteen or twenty masking steps is not uncommon. It is anticipated that the proliferation of combination processes will continue. For the design of analogue circuits in particular, these processes hold great promise.

2.16 Design styles

In comparison with the use of discrete devices, employing monolithic technology has great advantages. In summary they encompass miniaturization of the physical volume, high reliability, high reproducibility and low mounting costs. The general trend is towards maximizing the degree of integration within a chip. The feasibility of integrating complex circuit and system functions has steadily increased from the conception of chip technology in the early 1960s. Several developments have contributed to the progress in this area:

- improvement in mastering silicon technology;
- the steady pursuit of handling smaller device dimensions;
- the development of processes offering more design freedom by increasing the assortment of available devices;
- yield improvement so that larger chips can be produced with acceptable yield;
- the development of new circuit concepts directed towards integrable circuit solutions.

The last-mentioned item has changed the art of electronic design drastically, particularly in the area of analogue circuit design. Of old, analogue circuits contained many passive components, such as inductances, large capacitances and large resistances. In the days of discrete circuitry the overwhelming design adagium was to use a minimum count of (expensive) active devices. Early analogue ICs reflected the proven state of the art of discrete design principles. The chips contained the integrable devices, whereas the non-integrable devices were externally connected to the chips. Later, new circuit concepts were gradually developed that broke with the familiar approaches. Not least because of the difficult exercise of turning over established design habits, progress in analogue chip design has been relatively slow, but at present analogue IC-design is a flowering topic.

An essential problem in IC-electronics is that the initial costs for the manufacturing of a particular chip are high. Mask making is a particularly costly affair. This is no problem when the designed chip is to be used in a mass product. In that case the initial costs hardly affect the ultimate price of a single chip. Chips that are designed for one particular application are called 'dedicated chips' or ASICs (application-specific ICs). In the development of such a full-custom chip, the following steps must be made. First, an integrable circuit is designed, taking into account the restrictions imposed by the manufacturing process to be used. This circuit is translated into a chip-layout, which in turn is translated into a set of masks. In all stages the computer is an indispensable tool. A large body of appropriate software is available. Finally, a silicon foundry produces the chips as defined by the set of masks.

The question now arises as to whether it is possible to save on the costs of these procedures if a limited amount of specimens is needed. Several approaches have been developed, leading to various design styles fitted to different design situations.

Chips that are not manufactured in full-custom fashion are generally called 'semicustom chips'. Characteristic of all semicustom design styles is that they reduce design freedom to a certain degree. The least interventional approach is the use of prefabricated semicustom device arrays. Several semiconductor manufacturers offer such arrays. An array contains a deliberately chosen amount of various transistors, resistors and capacitors. The devices and their location on the chip are chosen with a view at circuits belonging to certain broad classes. Once an array has been chosen by the circuit designer, he or she has to live with the kit of devices it contains and with their positions on the chip. Usually, not all available devices will be used. Of course, it is good practice to try to economize on silicon area by using as many devices as is possible. The chip is finalized by making the relevant interconnection pattern. This requires the definition of one or two masks, depending on the type of process involved. It will be clear that the initial costs for such a design will be much lower than for a full-custom chip. Some semicustom arrays are subdivided in 'tiles'. These tiles are arranged for fitting to frequently used subcircuits. This is particularly convenient if a chip design contains an amount of similar subcircuits, as is often the case.

The semicustom array approach can also be practised at a higher level in that complex subcircuits are prefabricated. The finalization is performed by one simple interconnection pattern that interconnects the predefined blocks. This is common practice in digital semicustom design, but in analogue design it is not often practicable. Exceptions are filter structures where many similar configurations are usually involved. Examples of standardized subcircuits for such applications are operational amplifiers, voltage and current references and buffer stages.

A design style that does not economize on mask production costs, but is capable of speeding up circuit design and layout effort, uses a predefined library of standard cells. Complete information about structure and layout of these cells is stored in a data base. A complex signal-processing function is first broken down in subfunctions at the complexity level of the standard cells. Subsequently, with the aid of CAD-tools the interconnection pattern between the cells is defined. Layout effort is restricted to the layout of this pattern, since the layout of the cells is already available in the data base. A further refinement is obtained by parametrizing the cell library. This implies that certain parameters describing cell behaviour can be chosen within given bounds. The cells specified in this way are automatically designed through the CAD-software.

It must be noted that analogue circuit design is often concerned with highly specific signal-processing functions. Moreover, high-performance requirements often have to be met. It is obvious that any approach using predefined standard circuits is bound to have to compromise as to performance aspects. To a much lesser degree this is true for the use of semicustom device arrays. By selecting an appropriate array from the catalogue, a nearly optimum design can often be accommodated. These arrays can also be put to good use as 'silicon breadboards'. First a design is implemented on the basis of an array, which greatly speeds up the 'turnaround time'. This chip is subjected to measurements and prototype trials. Often, improvements will be made leading to a final design, which is then manufactured in full-custom fashion.

Semicustom design has evolved into a viable and often attractive middle way between full-custom design and discrete-component implementations. Though full-custom design tends to become cheaper by the continuing development of advanced CAD-tools, it is probable that semicustom design approaches will persist to fill a niche in the world of microelectronics.

References

Getreu, I. E. (1978) *Modeling the Bipolar Transistor*, Elsevier, Amsterdam.
Tsividis, Y. P. (1987) *Operation and Modeling of the MOS Transistor*, McGraw-Hill, New York.
Vittoz, E. and Fellrath, J. (1977) 'CMOS analog integrated circuits based on weak inversion operation', *IEEE Journal of Solid-State Circuits*, **12** (3), 224–31.

Further reading

Allen, P. E. and Holberg, D. R. (1987) *CMOS Analog Circuit Design*, Holt, Rinehart and Winston, New York.
Gray, P. R. and Meyer, R. G. (1984) *Analog Integrated Circuits*, Wiley, New York.

Glaser, A. B. and Subak-Sharpe, G. E. (1977) *Integrated Circuit Engineering*, Addison-Wesley, Reading.

Grebene, A. B. (1984) *Bipolar and MOS Integrated Circuit Design*, Wiley-Interscience, New York.

For references to original papers on the topics dealt with in this chapter, see the extended lists in the works mentioned above.

3 Signal transfer in linear signal-processing circuits

3.1 Introduction

The idea of modelling active devices in the form of electrical networks was introduced in Chapter 2. By using these device models, network-theoretical methods can be used in calculations concerning electronic circuits. Though it must be emphasized again that the validity of any model is restricted to those aspects of the modelled structure that are incorporated in the model, modelling is a powerful tool for coping with complex matters by translating them into entities that can be handled by familiar and well-established methods.

The most important signal-processing function in analogue electronics is linear amplification. Amplifier circuits can, as far as their linear transfer characteristics are concerned, be modelled as linear electrical networks. In this way, the well-developed methods pertaining to the analysis and design of linear networks can be used to their best advantage for the analysis and design of amplifier circuits. The reader is assumed to be acquainted with basic network theory and the theory of electrical signals. However, for the convenience of users of this text whose knowledge on these topics has to a certain extent been eroded, some basic concepts are summarized in this chapter, particularly the use of poles and zeros of transfer functions in describing signal transfer properties.

3.2 Poles and zeros of linear transfer functions

Signal transfer by a linear system can be described by linear differential equations. If S_i and S_o denote the time dependent input and output signals, the general form of such an equation is

$$a_0 S_i + a_1 \frac{dS_i}{dt} + a_2 \frac{d^2 S_i}{dt^2} + \ldots = b_0 S_o + b_1 \frac{dS_o}{dt} + b_2 \frac{d^2 S_o}{dt^2} + \ldots . \tag{3.1}$$

By virtue of the properties of exponential functions, if S_i is an exponential function of time, the same applies to S_o. Putting $S_i = \exp(st)$, where s is a complex variable, eq. (3.1) takes the form of an algebraic equation:

$$\frac{S_o}{S_i} = A(s) = \frac{a_n s^n + a_{n-1} s^{n-1} + \ldots a_0}{b_m s^m + b_{m-1} s^{m-1} + \ldots b_0} \qquad (3.2)$$

The polynomials constituting the numerator and denominator can be factorized, yielding

$$A(s) = \frac{a_n}{b_m} \frac{(s-z_1)(s-z_2)\ldots(s-z_n)}{(s-p_1)(s-p_2)\ldots(s-p_m)} \qquad (3.3)$$

$A(s)$ is called the 'complex transfer function' of the system.

The roots $z_1, z_2, \ldots$ of the numerator are called the *zeros of $A(s)$*, whereas the roots $p_1, p_2, \ldots$ of the denominator are called the *poles of $A(s)$*. Together with the factor a_n/b_m, the zeros and poles contain complete information concerning the transfer properties of the network and, being complex variables, they can be represented by points in the complex plane. It will be immediately clear that representing the transfer properties of a circuit or system by a few points in a plane, rather than by differential or algebraic equations, has great advantages. Moreover, as will become apparent, the locations of poles and zeros are intimately related to certain specific circuit parameters. It is this latter feature that provides the designer with a powerful tool for relating design aspects to transfer properties.

There are several ways to give the quantity s a meaning. One of them is to interpret s as a complex frequency. Putting $s = \alpha + j\omega$, a physical interpretation of $\exp s(t)$ is possible, in that $\exp(\alpha + j\omega)t$ denotes the complex notation of a harmonic signal with increasing $(\alpha > 0)$ or decreasing $(\alpha < 0)$ amplitude $\exp(\alpha t)\cos \omega t$. The complex notation and the associated real time function are concatenated in that the latter is found by adding the conjugate complex quantity to the former, since

$$\exp[(\alpha + j\omega)t] + \exp[(\alpha - j\omega)t] = 2 \exp(\alpha t) \cdot \cos \omega t$$

Of particular interest is the case $\alpha = 0$, yielding $\exp(j\omega t)$, which is the complex notation of a harmonic signal $\cos \omega t$. In order to find the poles and zeros of a given network it is not at all necessary to write down the appropriate differential equations. A much easier way is to write down the node and mesh equations, denoting the impedance of a capacitor by $1/sC$ and that of an inductance by sL. The simple circuits of Figure 3.1 may serve as examples; in Figure 3.1(a)

$$A = \frac{V_o}{V_i} = \frac{sRC}{1+sRC} = \frac{s}{s+1/\tau}, \text{ with } \tau = RC$$

Figure 3.1(b) depicts the pole–zero plot in the complex plane. As is customary, the zero is indicated by the symbol $\bigcirc$, and the pole by the symbol $\times$.

Figure 3.1(c) depicts a simple bandpass transistor amplifier stage. If the transistor is modelled as a voltage-controlled current source, voltage transfer is given by

$$\frac{V_o}{V_i} = -g_m Z \text{ with } Z = \frac{sL}{1+s^2LC+sL/R} = \frac{1}{C} \frac{s}{(s+1/2\tau)^2 + (\omega_0^2 - 1/4\tau^2)}$$

70 ANALOGUE ELECTRONIC CIRCUIT DESIGN

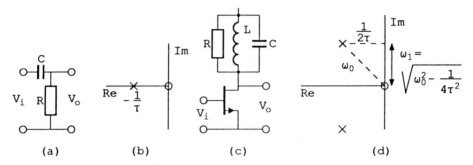

Fig. 3.1 *Two simple circuits with their pole–zero plots.*

with $\tau = RC$ and $\omega_0 = 1/\sqrt{LC}$. Figure 3.1(c) shows the pole–zero plot, containing one zero and two conjugated complex poles. These simple examples clearly show how elegantly the poles and zeros are related to essential circuit parameters in the form of time constants and characteristic frequencies that play key roles in the transfer properties of the circuits. For instance, in Figure 3.1(d), ω_0 represents the resonant frequency of the circuit, and $\omega_1 = \sqrt{\omega_0^2 - 1/4\tau^2}$ the frequency of free oscillations.

In many cases it is not even necessary to write down the node and mesh equations of the network for finding the poles and zeros. A zero is a value of s for which transfer is zero; a pole is a value of s for which transfer is infinite. Writing sL and $1/sC$ for the impedance of reactive elements, conditions for zero and infinite transfer can often be found by inspection. As an example, in Figure 3.1(a), for $s = 0$, the capacitor is an open circuit, hence $V_o = 0$. A pole occurs when transfer is infinite. Obviously, this occurs in the circuit of Figure 3.1(a) when C and R form what might be called a 'series resonant circuit' for complex frequencies. If $R + 1/sC = 0$, a finite value of V_i gives rise to an infinitely large current, hence $V_o \to \infty$. Consequently, the pole location follows from $1/sC + R = 0$, hence $s = -1/\tau$. In circuits with complex pole locations, this method is not always easily applicable, because in such cases pole locations are usually determined by a combination of easily recognizable individual time constants.

Apart from associating the complex variable s with the concept of a complex frequency, it can be interpreted in two further ways. First it can be interpreted as a differential operator: in this interpretation s stands for d/dt. And, finally, it can be interpreted as a Laplace operator. The last-mentioned interpretation is particularly useful in determining the transient response of a network. This topic will be taken up in a subsequent section.

Pole–zero plots of stable systems have to comply with a number of general rules. The most important of these are that complex poles and zeros are bound to occur in conjugate complex pairs and that poles have to be located in the left half of the complex plane. If one or more poles occur in the right half-plane, the system is unstable and acts as an oscillator. Zeros may be located in the right half-plane. A system with one or more zeros in the right half-plane is called a non-minimum phase system, as opposed to a system with all its zeros in the left half-plane, which is called a minimum phase system.

Further, in any practical system, the number of finite poles always exceeds the number of finite zeros. This is easily seen from eq. (3.2) since for $s = j\omega$ and $\omega \to \infty$, the magnitude of the transfer is of the order $(\infty)^{n-m}$. If $m < n$, the transfer would be infinite for $\omega \to \infty$, which is not possible in practical networks because of (parasitic) capacitances that act as high-frequency shorts. If such parasitic capacitive shorts are excluded in the modelling of the circuit, this restriction does not, of course, hold. An example is provided by the simple high-pass circuit of Figure 3.1(a), where the unavoidable capacitive bridging of the resistor R is omitted. This is frequently done if the associated effects are of no interest in the case under consideration.

The number of poles in the transfer function is called the *order* of the system. It is determined by the number of independent capacitor voltages and inductor currents.

A transfer function of a twoport network can be dimensionless (for instance voltage-to-voltage transfer V_o/V_i), an impedance function ($Z = V_o/I_i$) or an admittance function $Y = I_o/V_i$. The transfer of a oneport network can be expressed by either an admittance or an impedance function. In this case an additional rule applies: in such a function the number of finite poles exceeds the number of finite zeros by one. Theoretically, the number of poles and zeros can be equal, but again practical circuits will exhibit zero response at $\omega \to \infty$, due to unavoidable (parasitic) capacitances. Of course, for impedance functions as well as for admittance functions, the rule that poles have to be located in the left half-plane applies. Since, with $Z = 1/Y$, the poles of Z are the zeros of Y, impedance and admittance functions cannot possess zeros in the right half-plane.

If a system consists of a cascade of non-interacting networks, the transfer of the cascade is found by multiplication of the transfer of the constituents. Hence, the pole–zero plot of the cascade is found by the addition of the pole–zero plots of the constituting cascaded networks.

Frequently, one is interested in the response of a network to harmonic signals. With $s = j\omega$, the amplitude response of the network follows from

$$|A| = \frac{a_n}{b_m} \frac{|s - z_1| |s - z_2| \ldots |s - z_n|}{|s - p_1| |s - p_2| \ldots |s - p_m|} \qquad (3.4)$$

and the phase response from

$$\arg A = \psi_1 + \psi_2 + \ldots \psi_n - \Theta_1 - \Theta_2 - \ldots - \Theta_m \qquad (3.5)$$

with $\psi_1 = \arg(s - z_1)$, $\psi_2 = \arg(s - z_2)$, $\Theta_1 = \arg(s - p_1)$, ... with $s = j\omega$. With the aid of eqs. (3.4) and (3.5) the amplitude and phase responses can be easily found from the pole–zero plot. The factors and terms in eqs. (3.4) and (3.5) can be geometrically interpreted. If z_k is one of the zeros (Figure 3.2a), then $|j\omega - z_k|$ is the distance from the point $j\omega$ on the imaginary axis to the point z_k. The same

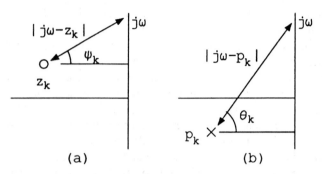

Fig. 3.2 *Geometric determination of* $|j\omega - z_k|$ *and* $|j\omega - p_k|$.

applies for $|j\omega - p_k|$ if p_k represent a pole (Figure 3.2b). The phase angles ψ_k and Θ_k can also be read from the diagram.

To obtain a quick impression of the general frequency behaviour of a circuit, it is not necessary to execute the operation described by eqs. (3.4) and (3.5) in full. Direct inspection of the pole–zero plot of Figure 3.3(a) makes it obvious that this plot represents a low-pass circuit, that of Figure 3.3(b) a bandpass circuit, and that of Figure 3.3(c) a high-pass circuit.

Using a logarithmic dB-scale, we can rewrite eq. (3.4) in the form

$$|A|(\text{dB}) = 20 \log|A| = \sum_{k=1}^{n} 20 \log|j\omega - z_k| - \sum_{k=1}^{m} 20 \log|j\omega - p_k| + C \qquad (3.6)$$

where $C = 20 \log|a_n/b_m|$.

A feature of using a log-scale is that the amplitude response is found simply by addition and subtraction of the contributions from the individual zeros and poles. The contribution of a pole p_k on the negative real axis is

$$A_k = -20 \log|j\omega - p_k| = -20 \log p_k \sqrt{1 + (\omega \tau_k)^2} \qquad (3.7)$$

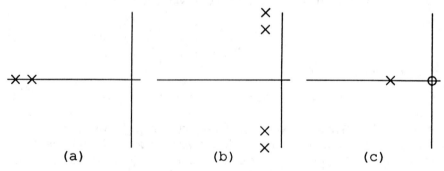

Fig. 3.3 *Examples of pole–zero plots:* (a) *low-pass circuit;* (b) *bandpass circuit;* (c) *high-pass circuit.*

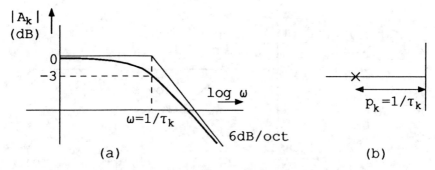

Fig. 3.4(a) *Contribution to the amplitude response of a pole on the negative real axis* (b).

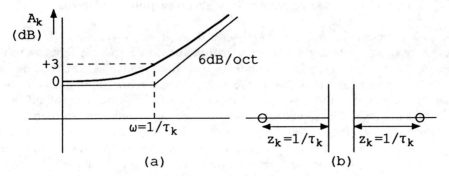

Fig. 3.5(a) *Contribution to the amplitude response of a zero on the real axis* (b).

where $\tau_k = 1/p_k$. If $\omega = 0$, then $A_k = -20 \log p_k = 20 \log \tau_k$. If $\omega = 1/\tau_k$, then $A_k = -20 \log p_k\sqrt{2}$. With respect to $\omega = 0$, at $\omega = 1/\tau_k$ the response has fallen by 3 dB. For values of $\omega \gg 1/\tau_k$, $A_k = -20 \log \omega$. Hence for $\omega \gg 1/\tau_k$ the amplitude response falls by 6 dB per octave. Figure 3.4 shows the logarithmic response on a logarithmic frequency scale. It is customary to approximate the response by the asymptotes for $\omega \to 0$ and for $\omega \to \infty$, which intersect in the 3-dB point. Such asymptotic diagrams are called Bode plots. Figure 3.5 shows the same plot for a zero located either on the positive or on the negative real axis.

EXAMPLE

Figure 3.6 shows the plot of a frequently used compensation network. The transfer is characterized by two singularities, one pole and one zero.

74 ANALOGUE ELECTRONIC CIRCUIT DESIGN

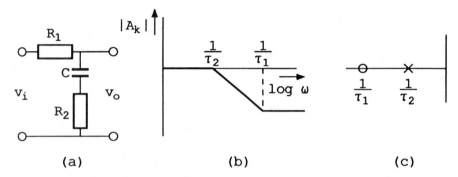

Fig. 3.6 *Bode plot of compensation network* (a) *with a pole and a zero on the real axis* $[\tau_1 = R_2C;\ \tau_2 = (R_1 + R_2)C]$.

3.3 Transient response of networks with given poles and zeros

A time-dependent signal can be described as it is: a function of time or, alternatively, by applying a Fourier transformation, as a spectral function in the frequency domain. Frequency responses and Bode plots, as discussed in the preceding section, are related to the frequency domain representation. For a rough description of such a response, quantities such as 3 dB frequency and roll-off (in db per octave) are widely used. Frequency domain representations are particularly appropriate for signals that have a more or less periodic character or that encompass a relatively small frequency band. Examples are audio signals and modulated carriers. However, in many cases, representing a signal in the frequency domain is a less obvious choice. If a signal consists of a non-repetitive sequence of transitory variations, a time domain representation is often more appropriate. Examples are video signals and pulse sequences. In order to describe the transfer of such signals by a given circuit, its *transient response* or *step response* is a useful vehicle. By definition, the transient or step response of a circuit is its time-dependent reaction on a unit step, i.e. a signal that at $t = 0$ instantaneously changes its value from zero to one (Figure 3.7).

Since the response of a linear system is fully specified by its singularities (poles

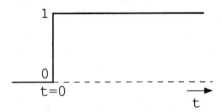

Fig. 3.7 *The unit step function.*

and zeros), it must be possible to derive its transient response from its poles and zeros. The relation between the singularities and the transient response can be elegantly expressed by using the Laplace transform. As is assumed to be known by the reader, by definition the (one-sided) Laplace transform of a function $f(t)$ is given by

$$F(s) = \int_0^\infty \exp(-st)\,dt \tag{3.8}$$

If the differential equation representing the response of a linear network or system is Laplace transformed, this equation is converted into an algebraic expression that is equivalent to the expression found by using the complex frequency concept. The Laplace transform of the unit step function is $1/s$. Hence, if a transfer function is given by

$$A(s) = \frac{a_n(s-z_1)(s-z_2)\ldots(s-z_n)}{b_m(s-p_1)(s-p_2)\ldots(s-p_m)}$$

then the transient response follows from the inverse Laplace transform of

$$F(s) = \frac{1}{s} \frac{a_n(s-z_1)(s-z_2)\ldots(s-z_n)}{b_m(s-p_1)(s-p_2)\ldots(s-p_m)} \tag{3.9}$$

The inverse transformation, i.e. the transformation from $F(s)$ to $f(t)$, can be executed by splitting up $F(s)$ in partial fractions:

$$F(s) = \frac{A_0}{s} + \frac{A_1}{s-p_1} + \frac{A_2}{s-p_2} \ldots \tag{3.10}$$

Finding the associated time domain function, i.e. the transient response sought for, is a simple matter. The inverse transformation of a term $A_n/(s-p_n)$ yields the time-dependent function $A_n(\exp p_n t)$. The method commonly used for finding the coefficients A_n is best illustrated by a simple example.

EXAMPLE
Find the transient response of the network of Figure 3.6(a). The formulation in the s-domain is

$$F(s) = c\,\frac{s-z_1}{s(s-p_1)} \tag{3.11a}$$

with $z_1 = -1/\tau_1$, $p_1 = -1/\tau_2$ and $c = \tau_1/\tau_2$. Writing $F(s)$ in the form of eq. (3.10), we obtain

$$F(s) = c\left(\frac{A_0}{s} + \frac{A_1}{s-p_1}\right)$$

This can be rewritten in the form

$$F(s) = c\,\frac{A_0(s-p_1) + A_1 s}{s(s-p_1)} \tag{3.11b}$$

For the equivalence of eqs. (3.11a) and (3.11b)

$$A_0(s - p_1) + A_1 s = s - z_1$$

most hold. Equating the coefficients of equal powers of s in the left-hand and the right-hand member, we find

$A_0 + A_1 = 1$ and $z_1 = A_0 p_1$, hence $A_0 = z_1/p_1 = \tau_2/\tau_1$ and $A_1 = 1 - A_0 = 1 - \tau_2/\tau_1$

Hence, the transient response of the network is given by

$$f(t) = \frac{\tau_1}{\tau_2}\left[\frac{\tau_2}{\tau_1} + \left(1 - \frac{\tau_2}{\tau_1}\right)\exp\left(-\frac{t}{\tau_2}\right)\right] = 1 - \left(1 - \frac{\tau_1}{\tau_2}\right)\exp\left(-\frac{t}{\tau_2}\right)$$

Figure 3.8 shows this transient response. At the onset of the step signal the capacitor does not contain any charge, hence at $t = 0$ the circuit behaves as a resistive voltage divider, so that $v_o = v_i R_2/(R_1 + R_2)$. For $t \gg \tau_2$, the capacitor behaves as an open circuit, so that $v_o/v_i = 1$.

If the set of poles includes a multiple pole, for instance a k-fold pole in $s = p_1$, then the partial fraction expansion takes the form

$$F(s) = \left[\frac{A_0}{s} + \frac{A_1'}{(s-p_1)^k} + \frac{A_1''}{(s-p_1)^{k-1}} + \ldots + \frac{A_1^{(k)}}{s-p_1} + \frac{A_2}{s-p_2}\right.$$

$$\left. + \frac{A_3}{s-p_3} + \ldots + \frac{A_m}{s-p_m}\right] \qquad (3.12)$$

The transformation of a term of the type $A_n/(s-p_n)^k$ yields

$$\frac{A_n}{(k-1)!} t^{k-1} \exp(p_n t)$$

Finding the coefficients A_n proceeds in the same way as shown in the example.

For readers who appreciate a formal mathematical approach, it is noted that the coefficient A_n is the residue of $F(s)$ in the pole p_n, hence A_0 is the residue in $s = 0$,

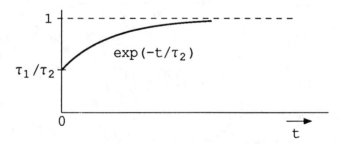

Fig. 3.8 *Transient response of the circuit of Figure 3.6(a).*

A_1 is the residue in $s = p_1$, etc. The residue of a function $F(s)$ in a pole p_n is given by

$$A_n = [(s - p_n) \cdot F(s)_{s=p_n}] \tag{3.13a}$$

Likewise, if a k-fold pole in $s = p_n$ is involved:

$$A_1' = [(s - p_n)^k f(s)]_{s=p_n}$$

$$A_1'' = \frac{1}{1!} \left[\frac{d}{ds} \{(s - p_n)^k F(s)\} \right]_{s=p_n}$$

$$A_1''' = \frac{1}{2!} \left[\frac{d^2}{ds^2} \{(s - p_n)^k F(s)\} \right]_{s=p_n} \tag{3.13b}$$

and so on.

The conclusion reached from the foregoing is that the transient response of a linear circuit is composed of a number of exponential functions that decrease in time (the real part of p_n being always negative), possibly multiplied by a factor t, t^2, etc. If p_n is real, these time functions are aperiodic. But if p_n and p_n^* form a conjugate complex pair, then simultaneously the functions $A_n \exp(p_n t)$ and $A_n^* \exp(p_n^* t)$ make part of the response. If $p_n = \alpha_n + j\omega_n$, and if for convenience we put $A_n = B + jD$, then

$$A_n \exp(p_n t) + A_n^* \exp(p_n^* t) = 2 \exp(\alpha_n t) \cdot [B \cos \omega_n t - D \sin \omega_n t] \tag{3.14}$$

Since in a stable system $\alpha_n < 0$, this corresponds with a harmonic function with decreasing amplitude. Hence, conjugate complex poles yield periodic terms in the transient response. They express mathematically the physical phenomenon that a network containing elements that store electric and/or magnetic energy respond to the disturbance of a state of equilibrium with periodic or aperiodic transient phenomena that extinguish in the course of time. For the occurrence of periodic components in the transient phenomenon, the availability of conjugate complex pole pairs is a necessary and sufficient condition. The frequency of ringing, and the time constant that characterizes the decay time, are fully determined by, respectively, the imaginary and the real part of the pole pair. Likewise, the characteristic time constant of non-periodic transient phenomena is also entirely determined by the location of the poles on the (negative) real axis. However, this does not mean that the zeros of the transfer function are of minor importance. The magnitude and the sign of the coefficients A_n depends on them, and therefore they play a decisive role in establishing the amplitude of the various components that make up the transient response. When discussing in subsequent sections the properties of basic electronic circuits, we will encounter situations in which the position of zeros has surprising, and even decisive effects, on their behaviour. An obvious illustration of the effect a zero can have is the influence of zeros situated in $s = 0$ and $s \to \infty$. With a zero $z = 0$, the term contributed by the transformation of A_0/s, vanishes. This means that the transient response lacks a d.c.-term. Likewise, with a zero in $s \to \infty$, the transient response assumes zero value for $t = 0$, expressing that a certain delay is involved in the onset of the transient response.

EXAMPLE
Find the transient response of the resonant circuit of Figure 3.1(c).

SOLUTION
$$\frac{V_o}{V_i} = -\frac{g_m}{C} \frac{s}{(s-p_1)(s-p_2)}$$

The poles are given by

$$p_{1,2} = -\frac{1}{2\tau} \pm j\omega_1, \text{ with } \tau = RC \text{ and } \omega_1 = \sqrt{\frac{1}{LC} - \frac{1}{4\tau^2}}$$

Putting

$$\frac{1}{(s-p_1)(s-p_2)} = \frac{A_1}{s-p_1} + \frac{A_2}{s-p_2}$$

we find

$$A_1 = \frac{1}{2j\omega_1} \text{ and } A_2 = -\frac{1}{2j\omega_1}$$

The transient response is

$$v_o(t) = -\frac{g_m}{C} \cdot \frac{1}{2j\omega_1} [\exp(p_1 t) - \exp(p_2 t)]$$

$$= -\frac{g_m}{\omega_1 C} \cdot \sin \omega_1 t \, \exp\left(-\frac{t}{2\tau}\right)$$

Figure 3.9 depicts the response.

When discussing the frequency response of a circuit we found it helpful to characterize its general character by a few key figures, such as bandwidth and steepness of the roll-off. In like manner it is convenient to characterize the transient response of a circuit by a few general parameters. Figure 3.10 depicts the general shape of a transient response.

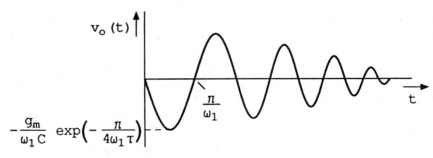

Fig. 3.9 *Transient response of the circuit of Figure 3.1(c).*

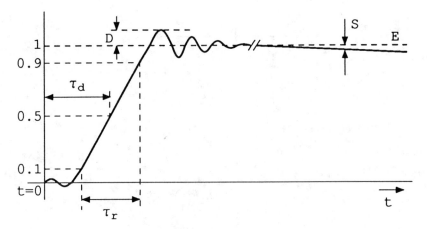

Fig. 3.10 *Shape of a typical transient response.*

The following quantities have found general acceptance as overall qualifiers:

1. The *rise time* τ_r is arbitrarily defined as the time interval between the moments where the response assumes values of 0.1 and 0.9 times the value E that occurs after the extinguishing of the initial phenomena (whether of a periodic nature or not).
2. The *overshoot* D is the percentage by which the response surpasses the value E.
3. In most cases there is one dominant *ringing frequency* that can be specified.
4. The *delay time* τ_d is the time between $t = 0$ and the moment the transient reaches half the value of E.
5. In certain cases, the value occurring a prolonged time after the extinguishing of the initial phenomena is less than that value. This phenomenon is characterized by the *sag* or *droop* (S in Figure 3.10) of the response. It is commonly expressed as a percentage of the value E with a specification of the time when the observation was made. For television signals, the usual choice is the time corresponding with one frame period ($\frac{1}{50}$ or $\frac{1}{60}$ s in the television standards in common use).

Bandwidth in the frequency domain description and rise time in the time domain description are strongly correlated quantities. Since the quantities derive from arbitrarily chosen definitions, no fixed relation can be given. But for 'well-behaved' circuits, the relation $\tau_r B = 0.35$ is a reasonable rule of thumb. For instance, with a bandwidth of about 5 MHz the corresponding rise time is about 70 ns.

EXAMPLE
Find the rise time of the transfer v_o/i_g and the sag at $t = \frac{1}{50}$ s for the interstage coupling circuit of Figure 3.11.

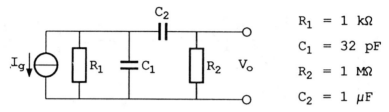

Fig. 3.11 *Interstage coupling circuit.*

SOLUTION

$$\frac{V_o}{I_g} = \frac{sR_1R_2C_2}{s(R_1+R_2)C_2 + 1 + sR_1C_1 + s^2R_2C_2R_1C_1} \quad (3.15)$$

For convenience we write $(R_1+R_2)C_2 = \tau_l$ and $R_1C_1 = \tau_h$. Since $R_2 \gg R_1$, we can simplify eq. (3.15) by putting $R_2C_2 \approx (R_1+R_2)C_2 = \tau_l$. Eq. (3.15) can then be rewritten

$$\frac{V_o}{I_g} = \frac{sR_1R_2C_2}{(s\tau_l + 1)(s\tau_h + 1)}$$

In the *s*-domain the transient response is given by

$$\frac{V_o}{I_g} = \frac{R_1R_2C_2}{\tau_h\tau_l(s + 1/\tau_l)(s + 1/\tau_h)} \quad (3.16)$$

Writing

$$\frac{1}{(s + 1/\tau_l)(s + 1/\tau_h)} = \frac{A_1}{(s + 1/\tau_l)} + \frac{A_2}{(1 + 1/\tau_h)}$$

we find

$$A_1 = \frac{\tau_h\tau_l}{\tau_l - \tau_h} \approx \tau_h \text{ and } A_2 = -A_1 \approx -\tau_h$$

Transformation of eq. (3.16) to the time domain yields

$$\frac{v_o}{i_g} = R_1[\exp(-t/\tau_l) - \exp(-t/\tau_h)]$$

$\tau_h = 32$ ns
$1 - \exp(-t/\tau_h) = 0.1$ for $t = 3.37$ ns
$1 - \exp(-t/\tau_h) = 0.9$ for $t = 73.68$ ns, hence, $\tau_r = 70.3$ ns

Further, $\tau_l = 1$ s. For $t = 0.02$ s, $\exp(-t/\tau_l) = 0.98$, hence the sag is 2%. Figure 3.12 depicts this response

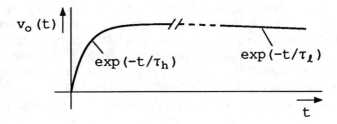

Fig. 3.12 *Transient response of the interstage coupling circuit.*

3.4 Poles and zeros of feedback systems: root loci

In the design of electronic amplifiers, a key problem is the achievement of accurate signal transfer. Most parameters of active components cannot be accurately known because of statistical spread, temperature dependence, dependence on biasing, ageing and so on. Therefore, measures must be taken to provide accuracy in spite of the use of inaccurate components. One method that is widely used is the employment of negative feedback. Figure 3.13 depicts the basic structure of an amplifier using this method. The block A denotes a high-gain inaccurate amplifier, whereas block B denotes an accurate attenuator composed of passive components. In many implementations B takes the form of a voltage or current divider consisting of passive devices. By subtracting the output signal of B from the source signal S_s, a closed loop is formed in such a way that the transfer of the complete arrangement is mainly determined by the B-transfer.

As can be easily derived from Figure 3.13, the transfer S_o/S_s is given by

$$A_f = \frac{S_o}{S_s} = \frac{A}{1 + BA} \tag{3.17}$$

If $BA \gg 1$, then $S_o/S_s \to 1/B$, so that the effects of the inaccuracy of transfer A are strongly reduced. In Chapter 4 the details of the design of feedback amplifiers will be discussed at length. At this stage, we are interested in the properties of the poles and zeros of feedback structures. For this purpose, the general model of Figure 3.13 is a sufficient basis.

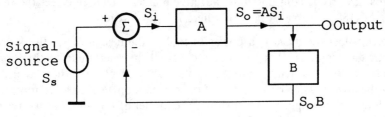

Fig. 3.13 *Principle of negative feedback.*

In the design of feedback amplifiers, the poles and zeros of the closed-loop transfer A_f have to be moved to desired locations. Obviously, these poles and zeros are somehow related to the poles and zeros of A and B, i.e the open-loop transfer, and to the magnitude of the gain available in the loop. With regard to the zeros, the situation is very simple. By virtue of eq. (3.17), the zeros of A_f correspond to the zeros of A and the poles of B. But with regard to the poles the relation is not so obvious. The poles of A_f correspond to the roots of the equation $AB = -1$. With

$$AB = A_o \frac{(s-z_1)(s-z_2)\ldots(s-z_n)}{(s-p_1)(s-p_2)\ldots(s-p_m)}$$

where $z_1, \ldots z_n$ and $p_1, \ldots p_m$ denote the poles and zeros of AB, that is, of the open-loop transfer, the equation $AB = 1$ can be written in the form

$$G(s) = \frac{(s-z_1)(s-z_2)\ldots(s-z_n)}{(s-p_1)(s-p_2)\ldots(s-p_m)} = \frac{-1}{A_o} \qquad (3.18)$$

where A_o is related to the magnitude of the open-loop gain. The poles of A_f are located in points in the complex plane where eq. (3.18) holds. If A_o is varied, the poles move along paths in the complex plane. These loci are called *root loci*. Of course, it is possible to find points complying with eq. (3.18) by trial and error, but fortunately some general rules can be given that are very helpful in finding the approximate or even the precise location of the root loci quickly. Since the present section does not pretend to be more than a compact survey of the use of poles and zeros, no attempt will be made here to be complete or to prove these rules. The most important rules are as follows:

1. When A_o increases from zero to infinite, the root loci start in the poles of AB and they end in the zeros of AB.
2. The root loci are symmetric with respect to the real axis.
3. Those parts of the real axis that are to the left of an odd number of poles and zeros form part of the root loci. By the way, note that conjugate complex singularities do not count in this respect. By occurring in pairs they do not affect the parity (even or odd) of the count.
4. If there are m finite poles and n finite zeros, there are $m - n$ zeros in infinity. The asymptotes of the root loci directed to these zeros in infinity intersect the real axis in the 'centre of gravity' of the finite poles and zeros. In this context zeros have to be considered as 'negative masses'. They intersect the real axis at angles $(180° + k360°)/(m - n)$, where k is an integer.

The root locus method is a powerful tool in the design of feedback amplifiers. A particular feature is that the root loci reveal the magnitude of the loop gain at which the poles of A_f tend to move into the right half-plane, indicating instability – that is, behaviour as an oscillator – of the closed loop. A simple example may serve as an illustration. Figure 3.14(a) shows the model of an amplifier with a feedback network consisting of three RC sections. The amplifier proper is modelled by a

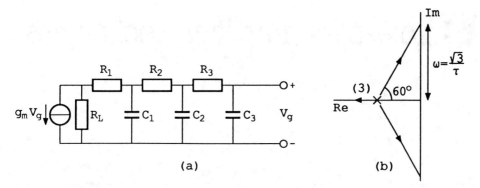

Fig. 3.14(a) *Example of an amplifier with third-order feedback network:* (b) *root locus plot.*

voltage-driven current source $g_m V_g$, loaded with a resistor R_L. If $R_3 \gg R_2 \gg R_1 \gg R_L$, the transfer of the chain can be approximated by the product of the transfers of the separate *RC*-sections. With $R_1 C_1 = R_2 C_2 = R_3 C_3 = \tau$, $A = -g_m R_L$ and

$$B = \frac{1}{\tau^3} \frac{1}{(s + 1/\tau)^3}$$

holds, so that

$$AB = -g_m - R_L \frac{1}{\tau^3} \frac{1}{(s + 1/\tau)^3}$$

Figure 3.14(b) shows the root loci. Here, the asymptotes coincide with the root loci proper. They intersect the real axis at angles of 60°. Since there are no finite zeros, all loci move to infinity. One branch coincides with the part of the real axis to the left of the poles. The root loci intersect the imaginary (frequency) axis at $\omega = \sqrt{3}/\tau$. From the root locus equation $AB = 1$, it follows that at this value of ω, $A = -8$. Hence, if $g_m R_L \geqslant 8$, the circuit is unstable and it behaves as an oscillator. If $A < 8$, but not far from this value, the transient response of the closed-loop amplifier will exhibit strong ringing at a frequency close to the oscillation frequency.

Further reading

There are several books that deal with the theory of signals, circuits and systems. Well-known examples are:

Dorf, R. C. (1980) *Modern Control Systems*, Addison-Wesley, Reading.
Papoulis, A. (1980) *Circuits and Systems: A modern approach*, Holt, Rinehart and Winston, New York.

4 Low-pass amplifier techniques

4.1 Amplification is the most important electronic signal-processing function

In virtually any electronic system some kind of signal amplification can be identified. Amplification is obviously an indispensable function. Its significance is a direct consequence of the second law of thermodynamics, which states that nature exhibits a fundamental tendency to increasing disorder. This tendency manifests itself in two ways. First, any information-carrying structure is affected by *dissipation*, i.e conversion of ordered energy into heat, which is a disorderly form

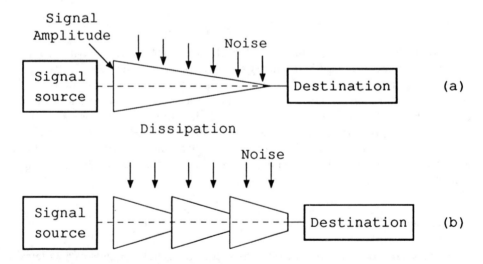

Fig. 4.1(a) *The signal is attenuated on its way through the transmission path; noise is active all along the transmission path. Without amplification the signal is drowned in the noise at arrival at the destination.* (b) *Amplification regenerates the signal power before it approaches the noise level too closely.*

of energy. In any passive system a signal will therefore undergo attenuation. Second, any system is afflicted by entirely uncontrolled fluctuations, generally designated by the term *noise*. Since noise is ubiquitous and a signal in a passive system is prone to attenuation, the ratio of signal power to noise power decreases during signal transport and processing, unless the signal is amplified before it is 'drowned in the noise', making regeneration impossible. Figure 4.1 illustrates the effects of noise and dissipation on signal transfer. An important conclusion drawn from these considerations is that minimization of noise effects is an important design criterion in any signal-processing system. This topic deserves ample attention and therefore a full chapter (Chapter 8) will be devoted to it.

4.2 The problem of accurate signal transfer: definition of quality factors

At any time within the interval in which an analogue signal is defined, its value is relevant and consequently has to be preserved. Accuracy of signal transfer is therefore a main subject of concern. The threats to be coped with are diverse, the most important ones being noise, distortion and interfering signals. In any signal-processing system three essential parts can be distinguished, namely a signal-acquisition part, a signal-processing part and a signal-delivering part. In each of these parts accuracy is of the essence, but frequently the pursuit of accurate information transfer is most difficult in the acquisition part, i.e. the interface between the signal source and the main signal-processing part.

If the primary information is of a non-electrical nature, it has first to be converted into an electrical signal. The devices used for this purpose are commonly called transducers or sensors. Examples are a microphone, which converts acoustic information into an electrical signal, and a television camera tube, which converts optical information into an electrical signal. A primary question with regard to the signal-acquisition part is that of which quantity is the best representation of the primary information. The output of any signal source can be represented either by a voltage source with a certain internal impedance or by a current source with a certain internal admittance (Figure 4.2).

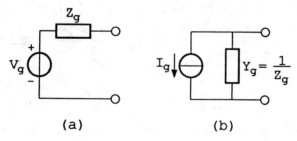

Fig. 4.2 *Thévenin-* (a) *and Norton-* (b) *representation of a signal source.*

These representations are usually indicated as the Thévenin and the Norton representations respectively. From a network theoretical viewpoint both are equivalent, but from the designer's viewpoint the choice of the representation is not irrelevant. A good example is provided by a piezoelectric transducer, used for instance as a microphone or a phonograph cartridge. The quantity that conveys the primary information is the pressure p_g acting on the transducer. The working mechanism of the transducer is such that the pressure is converted into an electrical charge q_g. The relation between p_g and q_g is, with great accuracy, linear. The internal impedance of the transducer is a capacitance C_g. Figure 4.3 shows the Thévenin and Norton representations of the transducer. The capacitance C_g is not accurately defined. It depends on constructional variations, it is non-linear and temperature-dependent. Consequently, the voltage $v_g = q_g/C_g$ is not an accurate representation of the primary information. In the Norton scheme $i_g = dq_g/dt$ holds. Hence, the short-circuit current does not depend on the unreliable value of C_g. Therefore, i_g is an accurate analogue of the primary information. Hence, for accurate information transfer, a current amplifier is required, i.e the input impedance of the amplifier has to be as low as possible, ideally zero.

The 'moving coil cartridge' may serve as a second example. Such a transducer consists of a coil that can move in a magnetic field with flux Φ. The primary information is conveyed by the velocity v of a membrane or needle. The voltage v_g generated in the coil is proportional to $d\Phi/dt$. In this case the amplifier has to sense and to transfer the voltage of the source. The latter should therefore be represented by its Thévenin equivalent. Hence, a voltage amplifier is required, i.e. the input impedance of the amplifier has to be high, ideally infinite.

Let us now look at the delivery part of the electronic signal-processing chain. The output transducer or the display system has to be driven in such a way that it is capable of accurately representing the information conveyed by the driving signal. If the quantity delivered is accurately related to the voltage of the driving source, i.e. the output of the signal-processing system, then this output has to behave as a voltage source. Hence, the internal impedance should be low, ideally zero. Similarly, if the driving quantity has to be a current, the output of the signal-

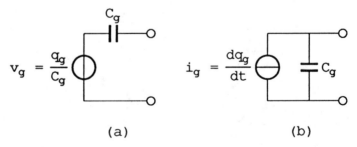

Fig. 4.3 *Thévenin-* (a) *and Norton-* (b) *representations of a piezoelectric transducer.*

processing chain has to behave as a current source: its internal impedance has to be high, ideally infinite.

It must be noted, first, that judging the relevant output quantity is not always an obvious matter. Frequently, it requires careful analysis of the underlying physical working mechanism of a particular type of transducer. A second remark that must be made is that practical constraints or certain properties of a transducer sometimes impede the optimum implementation. Consider as an example a television display tube. In such a tube an 'electron gun' produces an electron beam that can be intensity-modulated by a voltage between its cathode, and a control electrode that is commonly called the control grid. The amount of light produced is, within a wide dynamic range, proportional to the beam current. Hence, the logical conclusion is that the display tube should preferably be current driven. However, the internal capacitance C_{cg} between cathode and grid causes the impedance between the input terminals to be low for high-frequency signals. To this capacitance the output capacitance C_o of the driver stage has to be added. Let us assume, by way of example, that the bandwidth of the video signal is 4.5 MHz and that $C_{cg} + C_o = 20$ pF. For the 3 dB-bandwidth B of the driving circuit to be 4.5 MHz, its time constant $\tau = RC$ then has to be $1/2\pi B = 35$ ns, hence $R_L = 1.75$ kΩ. Figure 4.4 depicts the situation. Obviously, practical constraints force the driving signal to be a voltage. Since a display tube is a field-effect device, its current–voltage characteristic obeys a power law: $I = cV_d^\gamma$, where V_d is the driving voltage and c and γ are constants. In most display tubes γ assumes a value of approximately 2.5, dictated by the geometry of the electron-gun structure. Consequently, there is no linear relationship between the driving quantity and the light output. In order to linearize the transfer, the non-linearity has to be compensated for prior to the application of the signal to the display tube. This compensation is called 'gamma correction'. By the way, it may be mentioned that in broadcast television this predistortion is implemented in the camera chain in order to preclude the necessity of providing it in every receiver.

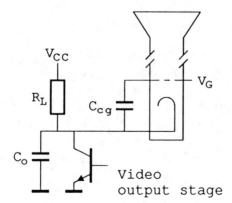

Fig. 4.4 *Driving a television display tube with a video signal.*

In the examples discussed in the foregoing paragraphs, accuracy requirements with regard to the signal acquisition and signal delivery were found to have consequences for the input and output impedances of, respectively, the input amplifier and the output amplifier of the signal-processing chain. In the examples these impedances should ideally be either zero or infinite. The examples are representative of many, but not all, practical design situations. In communication networks, signals are transmitted over cable circuits that have to be loaded with an accurate impedance that matches the characteristic impedance of the cable section. Similarly, the driving source delivering a signal to a cable section has to possess an internal impedance matching this characteristic impedance. Hence, in these situations the input and/or output impedance of the amplifiers have to possess finite and accurately determined impedances.

Up to now we have only considered accuracy requirements imposed by the properties of the signal sources and loads. It goes without saying that signal transfer through the signal-processing chain proper also has to be accurate. As has already been stated, the main waylayers are noise and distortion. At any location in the signal chain a certain amount of noise is present within the signal bandwidth. We will call this the 'noise threshold'. Distortion is due to the non-linearity of signal transfer in active devices, and sometimes to a lesser extent in passive devices such as inductances. In addition, there are parametric uncertainties, for instance invoked by temperature variations, power supply variations and variations of device parameters during their lifetimes. And finally, spurious signals can be picked up somewhere in the signal chain. Noise is essentially unavoidable, though it can be minimized by careful design (Chapter 8). The noise threshold determines the minimum signal that can be handled. The maximum signal level that can be handled is essentially limited by the finite magnitude of power-supply voltage and current. The ratio of the maximum signal and the minimum signal that can be handled is called the *dynamic range* (DR) of the system. It is a key design parameter in the design of any system. Commonly it is expressed in dB, hence

$$DR = 10 \log P_{max}/P_{min}$$

where P_{max} and P_{min} denote the maximum and the possible minimum values of the signal power.

If there is distortion in the system, and in any practical system some amount of distortion will be present, a second important design parameter has to be specified. Distortion leads to the occurrence of higher harmonics and of intermodulation components in the output signal. Intermodulation is the process of generation of signal components at frequencies which are the sum or difference of multiples of harmonics of primary signal components. If, for instance, signal transfer is described by a third-degree polynomial and the input signal consists of two harmonic components at frequencies f_1 and f_2, then the output signal contains components at frequencies f_1, f_2, $2f_1$, $2f_2$, $3f_1$, $3f_2$, $f_1 \pm f_2$, $2f_1 \pm f_2$ and $2f_2 \pm f_1$. Components at $f_1 \pm f_2$ are called second-order intermodulation products, those at $2f_1 \pm f_2$ and $2f_2 \pm f_1$ are called third-order intermodulation products. In order to

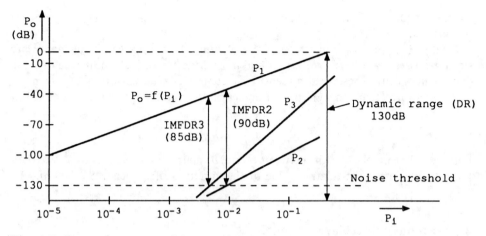

Fig. 4.5 *Dynamic range and intermodulaton-free dynamic ranges in a practical situation. The dynamic range amounts to 130 dB, IMFDR2 is 90 dB and IMFDR3 is 85 dB.*

qualify second-order intermodulation distortion (IM2), the power P_2 of one of the second-order intermodulation products is determined when two harmonic signals of different frequencies and equal amplitudes are fed to the circuit. Let P_1 be the output power at each of the input frequencies. As long as P_2 is below the noise power threshold, IM2 will be negligible. When it reaches the noise threshold its value P_{2n} can no longer be neglected. Let the magnitude of P_1 provoking this amount of distortion be denoted by P_{1m}.

The *intermodulation-free dynamic range* with regard to IM2 is defined as

$$IMFDR2 = 10 \log P_{1m}/P_{2n} \text{ [dB]}$$

Similarly for the third-order intermodulation IM3:

$$IMFDR3 = 10 \log P_{1m}/P_{3n} \text{ [dB]}$$

where P_{3n} equals P_3 when it reaches the noise threshold. Figure 4.5 illustrates the definitions of DR, IMFDR2 and IMFDR3 for a practical circuit. The circuit has been designed such that second-order distortion is suppressed by appropriate circuit design, for instance by a balanced circuit arrangement. Consequently, in this case IM3 is the dominating distortion effect.

Apart from intermodulation distortion, harmonic distortion occurs, caused by the generation of harmonics of the frequency of the input signal. A commonly used measure of this type of distortion is the total harmonic distortion (THD), defined as

$$D = \sqrt{D_1^2 + D_2^2 + \ldots D_n^2}$$

where $D_n = A_n/A_1 \times 100\%$, where A_1 is the amplitude of the input signal

$A_1 \cos \omega_1 t$ and A_n is the amplitude of the distortion component at the nth harmonic $A_n \cos \omega_n t$.

Finally some remarks must be made concerning the further causes of inaccurate signal transfer: parametric variations and interfering signals. The influence of these disturbances can be described by sensitivity quantities. The sensitivity of a transfer quantity q with regard to a circuit parameter or an interfering signal p, is defined as

$$S_p^q = \frac{\partial q/q}{\partial p/p} = \frac{p}{q}\frac{\partial q}{\partial p}$$

If there is linear relationship between q and p, hence $q = q_0 + \alpha p = q_0 + \Delta q$, then S_p^q is equivalent to the relative value of the variation of q, hence $S_p^q = \Delta q/q$.

4.3 Methods for achieving accuracy

At this stage the question arises as to how accurate signal transfer in an amplifier can be achieved. To summarize the findings of the preceding section: the requirements are correct adaptation of the amplifier to source and load, and minimization of noise and distortion. Figure 4.6 shows schematically the transfer of the signal from a source S to a load L.

The signal source – represented by either a current source, a voltage source or a characteristic impedance source – delivers its signal to the input terminals of the amplifier. As was explained before, the input impedance Z_i of the amplifier should (ideally) be zero, infinite or characteristic, depending on the nature of the driving source. Similarly, the load L dictates whether the output should behave like a current source, a voltage source or a characteristic impedance source. The amplifier transfer A can therefore behave in nine different ways. For instance, when the input has to sense voltage ($Z_i \to \infty$) and the output has to behave as a current source ($Z_o \to \infty$), A should behave as a voltage-to-current amplifier with transadmittance $A = I_o/V_i$. When the input senses current ($Z_i \to 0$) and the output behaves as a current source, A should behave as a current amplifier ($A = I_o/I_i$), and so on. If

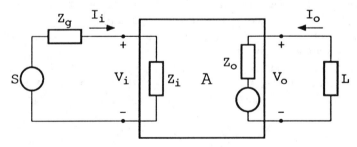

Fig. 4.6 *Basic schematic of signal transfer between source S and load L.*

Table 4.1 *The nine different types of amplifier-transfer functions*

1.	$Z_i \to \infty$	$Z_o \to \infty$	$V_o/V_i = A_v$ (voltage amplifier)
2.	$Z_i \to 0$	$Z_o \to \infty$	$I_o/I_i = A_i$ (current amplifier)
3.	$Z_i \to 0$	$Z_o \to 0$	$V_o/I_i = Z_t$ (transimpedance amplifier)
4.	$Z_i \to \infty$	$Z_o \to \infty$	$I_o/V_i = Y_t$ (transadmittance amplifier)
5.	$Z_i \to \infty$	$Z_o = Z_{char}$	P_o/V_i
6.	$Z_i \to 0$	$Z_o = Z_{char}$	P_o/I_i
7.	$Z_i = Z_{char}$	$Z_o = 0$	V_o/P_i
8.	$Z_i = Z_{char}$	$Z_o \to \infty$	I_o/P_i
9.	$Z_i = Z_{char}$	$Z_o = Z_{char}$	$P_o/P_i = A_p$ (power amplifier)

a finite (characteristic) impedance is required, either a voltage or a current can be chosen as the relevant quantity. The most logical quantity to be used in this case is the power P. Table 4.1 lists the nine possible situations.

The requirements concerning the impedance levels, discussed in the foregoing, are dictated by the inherent properties of source and load. In addition, the transfer A must be accurate, i.e. insensitive to parameter variations, non-linearities and interferences. Fortunately, there is a method that is capable of coping with all causes of inaccuracy, whatever their origin. This is the application of negative feedback, the essence of which is that the quantity that is required to be accurate is constantly or intermittently measured and compared with a reference quantity. Any deviation of the required value results in an error signal that is used for correction of the transfer.

Figure 4.7 shows a widely applied arrangement for implementing this method. The required value of the transfer S_o/S_s is A_f. The transfer of block A is inaccurate. The output signal S_o is fed to the block B, with $B = 1/A_f$. The resulting error signal BS_o is subtracted from S_s. Hence $S_o = AS_i = A(S_s - BS_o)$, so that

$$\frac{S_o}{S_s} = \frac{A}{1 + AB} \tag{4.1}$$

If $AB \gg 1$, then $S_o/S_s \to 1/B = A_f$. The transfer quantity B acts as the reference

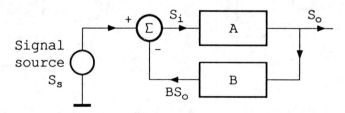

Fig. 4.7 *Principle of negative feedback.*

quantity. It is accurately implemented, for instance by a ratio of accurate passive components. The transfer becomes independent of A if $A \to \infty$. Accuracy is then entirely determined by the accuracy of the implementation of B.

EXAMPLE
Let $A = 10^6$ and $B = 10^{-3}$, then $AB = 10^3$ and $A_f = S_o/S_i = 10^3$. If A deviates as much as 100% from its typical value, i.e. if $A = 5.10^5$, then A_f changes by only 1‰.

Let us now consider specifically the application of feedback in amplifier design. In order to systematize our design considerations we first assume that the block A, as indicated in Figure 4.7, behaves ideally, that is, $A \to \infty$. This means that for any finite output signal, the input signal S_i is zero. In terms of network theory, the block behaves as a *nullor*. As long as this approximation is valid, the transfer A_f is entirely determined by the feedback network (the block B in Figure 4.7). If there is one feedback loop, it can be connected in four different ways to the nullor. The input signal to the feedback network can be either a voltage or a current, and the same applies to its output signal. Figure 4.8 lists the four possibilities schematically. In these models A is the high-gain amplifier acting as a nullor, whereas B represents the feedback path with accurate transfer. Voltage and current at the input terminals of the feedback configuration are denoted by V_g and I_g. The symbol Σ_v denotes addition of voltages, while Σ_i denotes addition of currents. In the idealized models

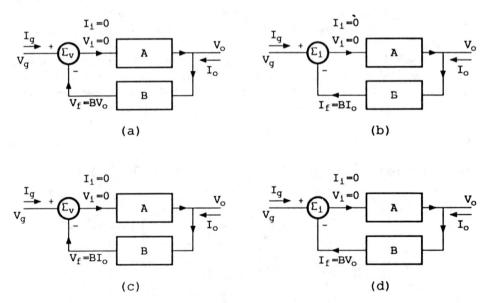

Fig. 4.8(a) *Voltage amplifier.* (b) *Current amplifier.* (c) *Transadmittance amplifier.* (d) *Transimpedance amplifier.*

of Figure 4.8, the feedback network is assumed to influence neither the output, nor the input quantities of the feedback configuration. If, in addition, A has the properties of a nullor, so that $I_i = 0$ and $V_i = 0$, very simple relations apply for the transfer properties of the configurations.

In Figure 4.8(a), $V_i = 0$ implies $V_g = BV_o$, or $A_{vf} = V_o/V_g = 1/B$. And $I_i = 0$ implies $I_g = 0$, hence $Z_{if} = V_g/I_g \to \infty$. Likewise: if $V_g = 0$, then $V_i = 0$ implies $V_f = BV_o = 0$, hence $Z_{of} = V_o/I_o = 0$. This configuration behaves as a voltage amplifier with $Z_i \to \infty$, $Z_o = 0$ and $A_{vf} = 1/B$.

In Figure 4.8(b), a similar reasoning yields $A_{if} = I_o/I_g = 1/B$ and $Z_i = 0$, whereas $Z_o \to \infty$. This configuration behaves as a current amplifier.

For the configuration of Figure 4.8(c) we find $Y_f = I_o/V_g = 1/B$ and $Z_i \to \infty$, $Z_o \to \infty$. The configuration behaves as a transadmittance amplifier.

Finally the configuration of Figure 4.8(d) yields $Z_f = V_o/I_g = 1/B$ and $Z_i \to 0$, $Z_o \to 0$. The configuration behaves as a transimpedance amplifier.

As long as the amplifying block behaves as a nullor, Z_i and Z_o are either zero of infinite. Obviously, if a finite (characteristic) input and/or output impedance is required, the four possible configurations with one feedback loop cannot meet the requirements. However, taking into account that current and voltage feedback have opposite effects on the impedance levels, it stands to reason that by employing two dissimilar feedback loops, finite and accurate impedance levels can be obtained. The actual value of the impedance depends on the numerical parameters of the feedback networks. A formal proof will be omitted here, but the point will be resumed in the discussion of practical feedback amplifier implementations.

Note that the use of negative feedback is a much better method for realizing accurate input and output impedances than placing resistors in parallel or in series with the amplifier terminals. Such resistors act as noise sources and they cause signal attenuation. Hence, they degrade power efficiency and dynamic range.

In the foregoing, the feedback path has been schematically represented by the quantity B that can take the form of either a dimensionless voltage or current ratio or an immittance. We must now face the question of how to implement the feedback network. A basic consideration is that the feedback network is responsible for the ultimate accuracy of the amplifier. Since active components are much more afflicted by inaccuracies, particularly parameter variations, non-linearity and noise, than are good passive components, these networks should be realized by using passive components. These can be divided into three classes. The transformer and the gyrator are 'non-energic': in their ideal form they transfer power, but they neither dissipate nor store energy. The remaining network elements are 'energic'. Capacitances and inductances do not dissipate but they do store energy. Resistors, on the other hand, do not store energy but they dissipate. Moreover, they are sources of additional thermal noise.

As we have concluded in the foregoing, the use of series or shunt impedances at the input and/or the output terminals of the nullor network has adverse consequences for the dynamic range that can be accommodated. If the feedback network is implemented with the aid of 'energic' elements (resistors, capacitors and inductors), series and/or shunt impedances at the nullor terminals show up. And, in addition, the use of resistors implies extra noise sources. Ideally, the use of 'energic' elements should be avoided. This leaves the transformer and the gyrator as the preferable elements.

Although from a network-theoretical point of view, the gyrator is a non-energic device, it can in practice only be synthesized by using active components. Hence, in a practical realization the gyrator dissipates power, it is not free of non-linearity and it contains noise sources. For these reasons the gyrator has to be dropped from the list of useful feedback elements.

In many cases the use of transformers is not attractive either. They are afflicted with parasitic resistances and capacitances, and if they must transfer low-frequency signals, they are expensive and bulky components. Only in cases where relatively narrow-band, high-frequency signals are involved can they play a useful role in a feedback path. And, of course, they can only be used where voltage-to-voltage or current-to-current transfer is desired.

It must therefore be concluded that in the majority of practical situations, the use of capacitors, inductors and resistors must be accepted, notwithstanding the formal objections that were raised against them. However, in the next section it will be shown that the adverse effects introduced by their employment can be largely minimized by judicious design of the feedback configuration. Consequently, in the overwhelming majority of practical feedback amplifiers, capacitors and resistors will be the components of choice in their feedback paths. In principle, inductors could also be taken into consideration, but in modern IC-electronics the use of inductors is unattractive because of their non-integratability. Moreover, inductors, even if they can be implemented, suffer from non-ideal characteristics much more than do resistors and capacitors. Fortunately, abandoning the use of inductors rarely hampers the design freedom of the designer of feedback amplifiers.

4.4 Practical implementation of feedback amplifiers

In the preceding section, resistors and capacitors were shown to be the preferred building blocks of feedback networks. We will now consider how these networks can be configured. For the time being we maintain the assumption that the active block of the feedback configuration behaves as a nullor. In a subsequent section the practical implementation of this block will be taken up.

For reasons of interference rejection, it is generally required that source, load and amplifier have one terminal in common. In order to provide for current sensing at the output and for voltage comparison at the input, as must be possible for the implementation of the various negative-feedback configurations, the active block has to be equipped with 'floating' ports. By a floating port is meant a port that is not directly connected to the common terminal. Figure 4.9 shows the symbol indicating the active block. The sign convention is as usually adopted in network theory. The terminals are labelled + and −, whereby transfer between terminals

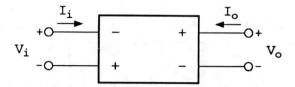

Fig. 4.9 *General symbol for twoport network.*

with opposite labels is inverting and between equally labelled terminals is non-inverting.

In the preceding sections we have already formally classified the four single-loop feedback configurations. Figure 4.10(a)–(d) shows practical implementations, using immittances in the feedback loop.

The configuration shown in Figure 4.10(a) provides accurate voltage transfer. Since $V_i = 0$, $V_{if} = Z_1 V_{of}/(Z_1 + Z_2)$, hence

$$A_{vf} = \frac{V_{of}}{V_{if}} = \frac{Z_1 + Z_2}{Z_1} = 1 + \frac{Z_2}{Z_1}$$

If $Z_2/Z_1 \to 0$, then $A_{vf} = 1$. In this special case the configuration is called a *voltage follower*. The feedback is then non-energic, but no voltage amplification is available. Nevertheless, this configuration is widely used because of the transformation of the impedance level it provides: $Z_{if} \to \infty$, whereas $Z_{of} \to 0$.

The configuration, according to Figure 4.10(b), acts as a transadmittance amplifier: $Z_{if} \to \infty$; $Z_{of} \to \infty$ and $Y_f = I_{of}/V_{if} = Y$. This is easily seen by noting that $V_i = 0$, and $I_i = 0$, so that $V_{if} = I_{of}/Y$, or $I_{of}/V_{if} = Y$. Similarly it is seen that in Figure 4.10(c) the transimpedance is accurately realized: $Z_{if} \to 0$, $Z_{of} \to 0$ and $Z_t = V_{of}/I_{if} = -Z$.

Figure 4.10(d) shows the configuration with accurate current transfer:

$$Z_{if} \to 0, \; Z_{of} \to \infty \; \text{and} \; A_{if} = I_{of}/I_{if} = -\left(1 + \frac{Z_2}{Z_1}\right)$$

If $Z_2/Z_1 \to 0$ the configuration is called a 'current follower'.

Figure 4.10(e) shows a dual-loop configuration. This configuration is capable of providing characteristic impedance matching. If the load and source impedances are denoted by Z_L and Z_g and $Z_L = Z_g = Z_{char}$, then $Z_{if} = Z_{of} = Z_{char} = \sqrt{Z/Y}$. This relation can be proved as follows. For the determination of Z_{if} consider Figure 4.11(a). Since $V_i = 0$:

$$V_{if} = I_{if} Z + V_{of} \tag{4.2}$$

$$V_{of} = -I_{of} Z_{char} \tag{4.3}$$

$$V_{if} = (I_{if} + I_{of})/Y \tag{4.4}$$

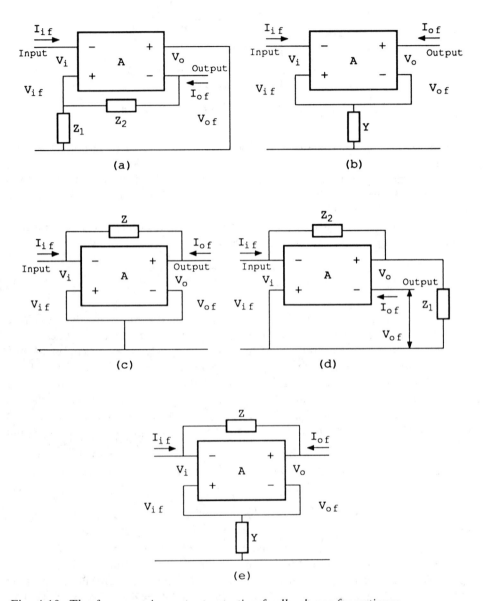

Fig. 4.10 *The five most important negative-feedback configurations:* (a) *through* (d) *single loop,* (e), *dual loop;* (a) *voltage amplifier,* (b) *transadmittance amplifier,* (c) *transimpedance amplifier,* (d) *current amplifier,* (e) *characteristic impedance amplifier.*

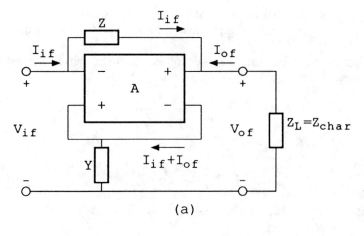

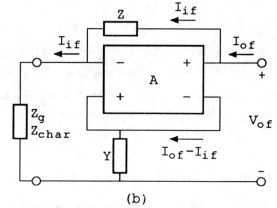

Fig. 4.11 *Schematic for the determination of Z_{if} (a) and Z_{of} (b) of the dual-feedback loop configuration.*

Elimination of V_{of} and I_{of} from these equations yields

$$Z_{if} = V_{if}/I_{if} = (Z + Z_{char})/(1 + YZ_{char}) \qquad (4.5)$$

For the determination of Z_{of} consider Figure 4.11(b). Since $V_i = 0$ and $I_i = 0$:

$$V_{of} = I_{if}(Z + Z_{char}) \qquad (4.6)$$

$$I_{if} Z_{char} = (I_{of} - I_{if})/Y \qquad (4.7)$$

Elimination of I_{if} yields

$$Z_{of} = V_{of}/I_{of} = (Z + Z_{char})/(1 + YZ_{char}) \qquad (4.8)$$

The requirement for $Z_{if} = Z_{of} = Z_{char}$ is $Z = YZ_{char}^2$ or

$$Z_{char}^2 = Z/Y \qquad (4.9)$$

If Z^2_{char} is resistive, which is usually the case, Z and Y can be resistors. Alternatively, Z and Y can be implemented by an inductor and a capacitor or vice versa.

It is interesting to investigate the voltage and the current transfer of this configuration. From eqs. (4.2)–(4.4) we find

$$A_{vf} = \frac{V_{of}}{V_{if}} = \frac{1 - YZ}{1 + \dfrac{Z}{Z_{char}}} \tag{4.10}$$

and

$$A_{if} = \frac{I_{of}}{I_{if}} = -\frac{1 - YZ}{1 + YZ_{char}} \tag{4.11}$$

If Y and Z are implemented by resistors with $Y = 1/R_2$ and $Z = R_1$ and putting $Z_{char} = R = \sqrt{R_1 R_2}$,

$$A_{vf} = -\left(\frac{R_1}{R} - 1\right) \text{ and } A_{if} = \left(\frac{R}{R_2} - 1\right)$$

Introducing the dimensionless quantity α, defined by $\alpha = R_1/R = R/R_2$, we obtain

$$A_{vf} = -(\alpha - 1) \text{ and } A_{if} = (\alpha - 1)$$

Hence, the power gain $A_{pf} = A_{vf} A_{if} = (\alpha - 1)^2$. Of course, if voltage gain is determined with respect to the source voltage V_g, by virtue of $R_g = R = R_{if}$ we have $V_{if} = \frac{1}{2} V_g$ and $A_{vf} = -\frac{1}{2}(\alpha - 1)$.

EXAMPLE
Let $R_1 = 10$ kΩ and $R_2 = 1$ Ω, then $R_{of} = R_{if} = 100$ Ω and $|A_{vf}| = \frac{1}{2}(\alpha - 1) = 49.5$.

As has already been stated in Section 4.3, maximum dynamic range is obtained if the feedback is 'non-energic', which can be accomplished by using a transformer in the feedback path. Hence, if practical considerations do not exclude the use of transformers, it is worth while contemplating this option. Since a transformer can only transfer voltage into voltage and current into current, two single-loop configurations are possible, implementing, respectively, a voltage amplifier and a current amplifier. Figure 4.12(a) and (b) depicts the appropriate arrangements. In the configuration shown in Figure 4.12(a) $V_{of}/V_{if} = -n$ and in that depicted in Figure 4.12(b) $I_{of}/I_{if} = n$.

The results obtained thus far can be summarized by stating that each feedback loop around a nullor accurately fixes one particular relation between input and output quantities. The four single-loop types of feedback fix voltage transfer, current transfer, transimpedance and transadmittance respectively. For purposes of formal analysis a network-theoretical formulation can be helpful in supporting the acquired insight. From the point of view of network theory an amplifier is a linear twoport (Figure 4.13). The transfer can be fully described by a set of two linear relations between the input and output quantities.

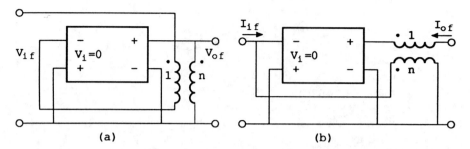

Fig. 4.12 *Feedback configurations with a transformer in the feedback loop: (a) voltage amplifier; (b) current amplifier.*

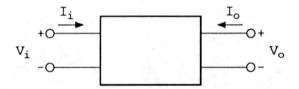

Fig. 4.13 *A linear twoport.*

Six different sets of such relations can be written down. A well-known and widely used set is the one expressing I_i and I_o in V_i and V_o:

$$I_i = y_i V_i + y_r V_o \quad (4.12a)$$

$$I_o = y_f V_i + y_o V_o \quad (4.12b)$$

The quantities y_i, y_r, y_f and y_o are admittances. The set is commonly designated as the y-matrix. The definitions of these quantities follow straightforwardly from eqs. (4.12a) and (14.12b):

$$y_i = \left(\frac{I_i}{V_i}\right)_{V_o=0}, \quad y_r = \left(\frac{I_i}{V_o}\right)_{V_i=0}, \quad y_f = \left(\frac{I_o}{V_i}\right)_{V_o=0} \quad \text{and} \quad y_o = \left(\frac{I_o}{V_o}\right)_{V_i=0}$$

For the purpose of describing the properties of feedback amplifiers the set commonly called the transmission matrix or chain matrix is more suitable. The defining equations are

$$V_i = a_i V_o + a_r I_o \quad (4.13a)$$

$$I_i = a_f V_o + a_o I_o \quad (4.13b)$$

At a first glance it seems illogical to express input quantities as functions of output quantities. This way of describing signal transfer can be called 'anticausal'. The quantities a_i through a_o are defined by the relations

$$a_i = \left(\frac{V_i}{V_o}\right)_{I_o=0}, \quad a_r = \left(\frac{V_i}{I_o}\right)_{V_o=0}, \quad a_f = \left(\frac{I_i}{V_o}\right)_{I_o=0}, \quad a_o = \left(\frac{I_i}{I_o}\right)_{V_o=0}$$

For a true nullor, all these quantities take zero value.

It may be that the transmission matrix, due to its 'anticausal' nature, lacks appeal to the electronic designer, who feels quite at home, though, when using the reciprocal quantities. We will indicate these by the name 'transfer parameters'. They are defined as follows:

Voltage-gain factor $\quad \mu = \dfrac{1}{a_i} = \left(\dfrac{V_o}{V_i}\right)_{I_o=0}$

Transadmittance $\quad \gamma = \dfrac{1}{a_r} = \left(\dfrac{I_o}{V_i}\right)_{V_o=0}$

Transimpedance $\quad \zeta = \dfrac{1}{a_f} = \left(\dfrac{V_o}{I_i}\right)_{I_o=0}$

Current-gain factor $\quad \alpha = \dfrac{1}{a_o} = \left(\dfrac{I_o}{I_i}\right)_{V_o=0}$

For a true nullor, all these quantities are infinite.

Any feedback loop causes one of these parameters to obtain a finite and accurately defined value, whereby the input and output impedances of the feedback amplifier are either zero or infinite, depending on the type of feedback. As a consequence, the accurately defined transfer quantity is independent of the source and load impedances. The validity of this statement can easily be verified. Figure 4.14 shows an amplifier between source and load, driven from a voltage source, while in Figure 4.15 the amplifier is driven from a current source. Together with the equations (4.13), the equations

$$V_g = I_i Z_g + V_i \tag{4.14a}$$

$$V_o = -I_o Z_o \tag{4.14b}$$

and

$$I_g = I_i + V_i/Z_g$$

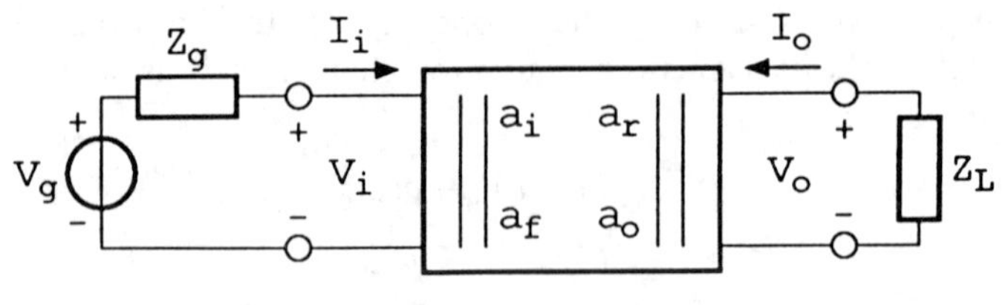

Fig. 4.14 *Amplifier, represented by a-matrix between source and load, driven from a voltage source.*

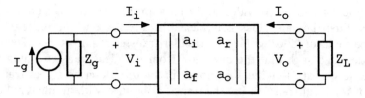

Fig. 4.15 *As Figure 4.14, but driven from a current source.*

are valid. From this set we obtain,

$$A_v = \frac{V_o}{V_g} = \frac{Z_L}{D}, \quad Y_t = \frac{I_o}{V_g} = -\frac{1}{D}, \quad Z_t = \frac{V_o}{I_g} = \frac{Z_L Z_g}{D}$$

$$A_i = \frac{I_o}{I_g} = -\frac{Z_g}{D}, \quad \text{where } D = a_i Z_L - a_r + a_f Z_g Z_L - a_o Z_g$$

These equations reveal that, if all *a*-parameters except one are zero, one transfer is accurately defined by the non-zero parameter. This transfer is independent of Z_g and Z_L. The non-zero parameter determines which transfer quantity will be fixed. Obviously, if $a_i \ne 0$, then $A_v = 1/a_i$, and so on.

As we have seen, for a nullor all *a*-parameters are zero. In order to obtain an accurate voltage transfer A_v, a feedback network connected in series with the input and in parallel with the output of the nullor is required. We conclude that this type of connection, depicted in Figure 4.10(a), fixes the parameter a_i (or, alternatively, the transfer parameter μ), and so on. If two loops are present, two parameters are fixed. And if all four parameters must obtain a well-defined value, four feedback loops are necessary.

4.5 Consequences of the imperfections of the active block; asymptotic-gain model of feedback amplifiers

Up to now we have treated the active block in the feedback amplifier as an ideal network element: the nullor. It has been shown that under this assumption signal transfer is entirely determined by the feedback loop(s). By judiciously designing the feedback loop(s), all desired types of transfer can be realized. The accuracy of the information transfer is entirely determined by the accuracy of the feedback elements. Since the feedback circuitry can be implemented with accurate passive elements, the accuracy can meet high standards. In practice the active blocks cannot be made to comply with the ideal nullor characteristics. Since active components are afflicted with non-ideal characteristics, the imperfect nullor implementation has consequences for the accuracy of the feedback amplifier. We shall now investigate how the properties of the active block can be accounted for. Our goal is to find a model that makes the deviations between the ideal and the practical circuitry explicit. Such a model provides a direct approach to the design of feedback amplifiers. First the gain block is modelled as a nullor and the feedback network is designed in accordance with the particular transfer requirements. Then the gain block is designed such that the deviation from the ideal transfer is minimized.

The model we are looking for should preferably describe signal transfer as a product of the ideal transfer, i.e. the transfer obtained with an ideal nullor, and a correction factor. The basic model of Figure 4.7 cannot meet this objective. It is perfectly suited for investigating the fundamental properties of feedback configurations, but not for design purposes. The problem is that in this model A and B are assumed to be precisely distinguishable blocks. In practical feedback amplifiers the gain block and the feedback blocks are interwoven. Hence, a model is needed that handles the complete feedback amplifier, including source and load impedances, as one entity. Fortunately, such a model can be formulated.

Figure 4.16 depicts a generalized model of an amplifier, including its feedback circuitry (either single or multiple). The amplifier is driven by a signal source that can be represented either by a current source (Figure 4.16a) or a voltage source (Figure 4.16b) and loaded with an impedance Z_L.

The amplifier must contain at least one active component, any model of which contains at least one controlled source, either a current or a voltage source. It is therefore always possible to indicate one or several controlled sources in the amplifier. Let us now take one of these sources and denote it Q_c, where Q is used

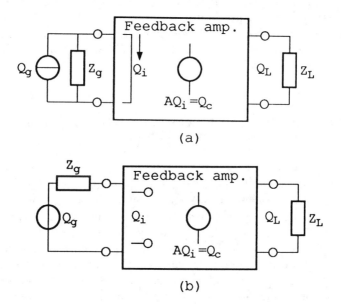

Fig. 4.16 *Generalized feedback amplifier driven by current source (a) or by voltage source (b). The subscripts g, i, c and L denote source (generator), input, controlled and load.*

as a general symbol that can either be related to a current or a voltage. The subscript c reminds us of the fact that we are referring to a controlled source. The source Q_c is related to a quantity Q_i pertaining to a voltage or current elsewhere in the circuit. This can be the input voltage or current, and often it is convenient to choose for Q_c a source that is related to an input quantity (voltage or current), but this is not a necessity. Further, let Q_L denote the output quantity. By using the superposition principle, which is valid for any linear network, we can write Q_L and Q_i as functions of Q_g (the driving source) and Q_c:

$$Q_L = \rho Q_g + \nu Q_c \qquad (4.15a)$$

$$Q_i = \xi Q_g + \beta Q_c \qquad (4.15b)$$

Let the relation between Q_c and Q_i be given by $Q_c = AQ_i$. Using this relation and eqs. (4.15), we find for the transfer from source to load, i.e. the transfer of the amplifier (including its feedback loops),

$$A_f = \frac{Q_L}{Q_g} = \rho + \nu \xi \frac{A}{1 - A\beta} \qquad (4.16)$$

The parameter A is called the *reference variable*. A non-zero value of β indicates the presence of feedback in the amplifier. A non-zero value of the parameter ρ indicates a direct path between input and output, since even with $Q_c = 0$ a finite transfer between input and output results. A non-zero value of ρ is usually caused by a non-unilateral feedback network. The term 'non-unilateral' refers to a network wherein signal transfer can occur as well from input to output as from output to input. If signal transfer is only possible from the input terminals to the output terminals, the network is said to be unilateral. Practical feedback networks are nearly always non-unilateral, since they commonly consist of passive components.

The product $A\beta$ is called the *loop gain with respect to the reference variable A*. If $A\beta \to \infty$, eq. (4.16) yields

$$A_{f\infty} = \lim_{A\beta \to \infty} A_f = \rho - \frac{\nu \xi}{\beta} \qquad (4.17)$$

Substitution of eq. (4.17) into eq. (4.16) yields

$$A_f = \frac{Q_L}{Q_g} = \frac{\rho}{1 - A\beta} - A_{f\infty} \frac{A\beta}{1 - A\beta} \qquad (4.18)$$

In nearly all practical situations the contribution of $\rho/(1 - A\beta)$ can be neglected in comparison to the second term in the right-hand part of eq (4.18). In this case eq. (4.18) can be approximated by

$$A_f \approx A_{f\infty} \frac{-A\beta}{1 - A\beta} \qquad (4.19)$$

$A_{f\infty}$ is the idealized transfer, i.e the transfer if the amplifying block has nullor properties. The conclusion is therefore that eq. (4.19) obeys the requirements stated previously. Obviously, the correction factor describing the influence of the non-ideal nature of the amplifier is given by $-A\beta/(1 - A\beta)$. Ideally, this factor should assume unity value as closely as possible.

It should be noted that $A_f = Q_L/Q_g$ can denote any of the relevant transfer parameters of the amplifier (voltage gain, current gain, transadmittance, transimpedance), since Q_L and Q_g can represent voltages as well as currents. Hence, the model we have conceived is capable of making explicit the deviations of all transfer parameters pertaining to the feedback amplifier, in accordance with the requirements formulated previously.

In certain design situations not only signal transfer is a relevant quantity, but also the input and output impedances of the amplifier. The input impedance Z_i is the load the amplifier presents to the preceding stage, and the output impedance is the source impedance presented to the next stage. If needed, these quantities can also be simply calculated with the aid of our model. The derivation of the relevant formulas is straightforward and will be omitted here. The result is

$$Z_i \approx Z_{io}(1 - A\beta_{sc}) \tag{4.20}$$

for the case of series connection of the feedback network at the input terminals and

$$Z_i \approx \frac{Z_{io}}{1 - A\beta_{op}} \tag{4.21}$$

for the case of parallel connection. In these equations, Z_{io} is the input impedance when $A = 0$, while β_{sc} and β_{op} are the values of β when the *input* is, respectively, short circuited or open. Likewise,

$$Z_o \approx Z_{oo}(1 - A\beta_{sc}) \tag{4.22}$$

for the case of series connection at the output and

$$Z_o \approx \frac{Z_{oo}}{1 - A\beta_{op}} \tag{4.23}$$

for the case of parallel connection at the output. Z_{oo} is the output impedance when $A = 0$. For the calculation of Z_o, the parameters β_{sc} and β_{op} are the values of β when the *output* is short circuited, respectively open. Since $A\beta$ must be negative for negative feedback, with series connection the impedance level increases, whereas with parallel connection the impedance level decreases. If $-A\beta \gg 1$, the impedance levels take values ∞ or 0, in accordance with our findings using the nullor approximation.

For the sake of completeness it is noted that in the case of two feedback loops of different types the relevant impedance is given by

$$Z = Z_o \frac{1 - A\beta_{sc}}{1 - A\beta_{op}}$$

The four types of feedback with series or parallel connection at the input, respectively at the output, are sometimes shortly designated as series–series, series–shunt, shunt–series and shunt–shunt.

The model introduced here is called the *asymptotic-gain model*. The power of this model can be appreciated by the following observations:

- It has general validity; there is no requirement to distinguish between the amplifier block proper (the non-ideal nullor) and the feedback circuitry. This is a very desirable feature, since in many cases the active block and the feedback network are so much interwoven that discrimination between the two is difficult.
- The model clearly separates the idealized design objective and the deviation from this objective.

A disadvantage of the model is that the calculation of $A\beta$, which is needed for the determination of the error factor $A\beta/(1 - A\beta)$, can be rather cumbersome in practical cases. However, the design goal is clear: $|A\beta|$ should be maximized in order to let the transfer accuracy be determined primarily by $A_{f\infty}$, the 'asymptotic gain'.

4.6 A single-stage feedback amplifier

Let us now examine some practical configurations. Figure 4.17(a) shows a single-stage feedback amplifier. The feedback is provided by Z_f. The feedback connection is of the type parallel at the input and parallel at the output (shunt–shunt). The configuration acts as a transimpedance amplifier, hence the quantity that is accurately determined is $Z_t = V_o/I_g$.

With regard to the application of the asymptotic-gain model, it is irrelevant what type of transistor model is used in the execution of calculations. Since our present objective is to gain insight by resorting to 'manual' calculation, a simple model, such as the hybrid-π model, is appropriate.

In order to simplify the calculation we neglect the finite value of the base resistance r_b of the transistor. The simplified hybrid-π equivalent circuit can then be drawn in the general form given in Figure 4.17(b). In this generalized form Z_π stands for the combination of C_π and r_π, Z_{cb} stands for the impedance of $C_{b'c}$ and Z_c for the collector impedance. Replacing the parallel connection of Z_g and Z_π by Z_i', the parallel connection of Z_{cb} and Z_f by Z_f', and the parallel connection of Z_L and Z_c by Z_L', the equivalent circuit of Figure 4.17(c) results. An obvious choice for the controlled source is the current cource $g_m V_{be}$, hence g_m is the reference variable of the asymptotic gain model. For the calculation of $A_{f\infty}$ we put $g_m \to \infty$. This implies that for V_o finite, V_{be} must be zero, and hence also $I_i = 0$. From Figure 4.17(c) now follows $I_g Z_f' + V_o = 0$, hence

$$A_{f\infty} = \lim_{g_m \to \infty} \frac{V_o}{I_g} = -Z_f' \qquad (4.24)$$

Further, $\beta = (V_{be}/I_c)_{I_g = 0}$, where I_c is the controlled current source. With $I_g = 0$, from Figure 4.17(c) we find

$$\beta = \frac{V_{be}}{I_c} = \frac{-Z_L' Z_i'}{Z_L' + Z_i' + Z_f'}$$

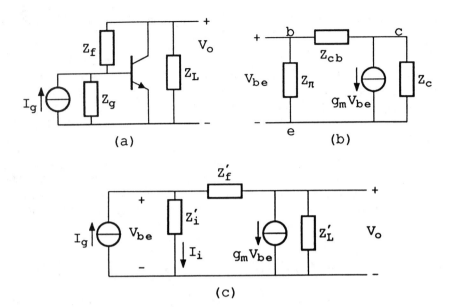

Fig. 4.17(a) *Single-stage transimpedance amplifier.* (b) *Simplified equivalent circuit of the transistor.* (c) *Equivalent circuit of the amplifier.*

and

$$A\beta = g_m\beta = -g_m \frac{Z_i' Z_L'}{Z_L' + Z_i' + Z_f'} \quad (4.25)$$

In order to find a condition for maximizing $A\beta$, we have to take into account that Z_i' cannot increase beyond the input impedance Z_π of the transistor. For any regular bipolar transistor $Z_c \gg Z_\pi$ is valid, so by taking Z_L large it is in principle possible to make $Z_L' \gg Z_i' + Z_f'$, leading to $A\beta \approx -g_m Z_i'$ and

$$\frac{A\beta}{1 - A\beta} \approx \frac{-g_m Z_i'}{1 + g_m Z_i'} \quad (4.26)$$

If $Z_g \gg Z_\pi$, the right-hand side of eq. (4.26) obtains the value $-g_m Z_\pi/(1 + g_m Z_\pi)$. Since $g_m r_\pi = \beta_0$, where β_0 is the low-frequency current gain of the transistor, in the frequency range where $1/\omega C_\pi \gg r_\pi$, the correction factor is $\beta_0/(1 + \beta_0)$. As could be expected, the finite current gain of the transistor determines the magnitude of the correction factor $-A\beta/(1 - A\beta)$.

In the specific case under discussion the loop gain is modest ($\approx \beta_0$) because only one active component is involved. Hence, it might be anticipated that in this case possibly the direct feedthrough, which is accounted for by the parameter ρ in eq. (4.16), has a non-negligible effect on signal transfer. Using eq. (4.15a) we find

$$\rho = \left(\frac{V_o}{I_g}\right)_{g_m = 0}$$

By inspection of Figure 4.17(c) we find

$$\rho = \frac{Z_i' Z_L'}{Z_i' + Z_f' + Z_L'}, \text{ so that } \frac{\rho}{1 - A\beta} \approx \frac{-\rho}{A\beta} \approx \frac{1}{g_m}$$

Hence if

$$A\beta \gg 1, \quad A_f \approx A_{f\infty} + \rho/(1 - A\beta) \approx -Z_f' + 1/g_m \qquad (4.27)$$

The conclusion is, therefore, that the correction term accounting for direct feedthrough can be neglected if $g_m \gg 1/Z_f$.

For determining the input and output impedance of the amplifier, the equations (4.20) through (4.23) can be used. Since in the circuit under discussion the connections of the feedback network are parallel both at the input and output, eqs. (4.21) and (4.23) apply. By inspection of Figure 4.17(c) we find $Z_{io} = Z_\pi \| (Z_f' + Z_L')$ and $Z_{oo} = Z_o \| (Z_f' + Z_i')$, where $\|$ denotes parallel connection. Further,

$$\beta_{0p} = \left(\frac{V_{be}}{I_c} \right)_{I_g = 0; \ Z_g \to \infty} = \frac{-Z_L' Z_\pi}{Z_\pi + Z_f' + Z_L'}$$

Substituting these values into (4.21) we find

$$Z_i = \frac{Z_\pi (Z_f' + Z_L')}{Z_\pi + Z_f' + Z_L' + g_m Z_\pi Z_L'}$$

Usually $g_m Z_L' Z_\pi \gg Z_\pi + Z_f' + Z_L'$ and $Z_L' \gg Z_f'$, so that $Z_i \to 1/g_m$. Similarly $Z_o = Z_{oo}/(1 - g_m \beta_{0p})$ with $\beta_{0p} = (V_{be}/I_c)_{I_g = 0; \ Z_i \to \infty}$. If $|g_m \beta_{0p}| \gg 1$, we obtain $Z_o = (Z_f' + Z_i')/g_m Z_i'$. Apparently, Z_i and Z_o are small, but finite as a consequence of the finite loop gain.

The reader may be inclined to the opinion that calculations using the asymptotic-gain model as presented here, though straightforward, tend to be rather cumbersome. An alternative approach is to find A_f, Z_i and Z_o by solving the mesh-and-node equations pertaining to the circuit. In a one-transistor circuit, such as the one we have just discussed, this is manageable, though hardly less clumsy. As an example, reference is made to Chapter 5, where the same configuration is analyzed in this way. If more than one stage is involved in the feedback loop, the mesh-and-node approach is hardly manageable. And above all, using the asymptotic-gain model offers the unique benefit that the ideal behaviour and the deviations from that behaviour are clearly separated and brought into relation with relevant design parameters.

4.7 A voltage amplifier with feedback over two stages

Figure 4.18 shows an amplifier configuration that is widely used. The active block consists of two transistor stages in CE-configuration. As in the preceding case, biasing circuitry is left out of the consideration; only the components responsible for signal transfer are shown.

108 ANALOGUE ELECTRONIC CIRCUIT DESIGN

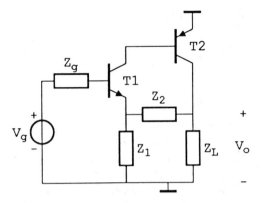

Fig. 4.18 *A two-stage voltage amplifier with overall feedback.*

Negative feedback is implemented by the passive network consisting of Z_2 and Z_1. The connection of the feedback network with the active block is of the type: series at input, parallel at output (series –shunt). Hence, the output impedance is low: the output acts as a voltage source. The input impedance is high, so the input senses voltage. The transfer that is accurately determined by the feedback arrangement is the voltage-to-voltage transfer $A_v = V_o/V_g$.

First of all, we calculate the transfer $A_{f\infty}$ if the loop gain is infinite.

$$A_{f\infty} = \lim_{A\beta \to \infty} \frac{V_o}{V_g}$$

Since there are two stages, there are also two controlled sources. Both can be used for defining the reference variable A. It is convenient to choose the controlled source associated with the transistor with grounded emitter, hence $T2$. In accordance with this choice, Figure 4.19 shows a schematic representation that can be used for our calculations. As in the preceding example, the base resistances r_b are omitted. The reference variable is g_{m2}. If $g_{m2} \to \infty$, the amplifier acts as a nullor. From $V_{be1} = 0$ and $I_{b1} = 0$ then follows

$$V_g = V_{Z_1} = \frac{Z_1}{Z_1 + Z_2} V_o \text{ and } A_{f\infty} = 1 + \frac{Z_2}{Z_1} \tag{4.28}$$

In order to find the loop gain $A\beta$, with $A = g_{m2}$, we replace the controlled source $g_m V_{be2}$ by an independent source I_c. Subsequently, we calculate $\beta = V_{be\,2}/I_c$. In order to simplify the calculation, we assume that the impedance, seen in the emitter of $T1$, is $1/g_{m1}$. This is permitted if the effect of Z_g on this impedance is negligible, which is the case if $Z_g/\beta_o \ll 1/g_{m1}$. The current I_c splits up into I_{Z2} and I_L, whereas I_{Z2} splits up into I_{Z1} and I_1. Due to this twofold current division we have

$$I_1 = \frac{-Z_L}{Z_L + Z_2 + Z_1 \| g_{m1}^{-1}} \cdot \frac{Z_1}{Z_1 + g_{m1}^{-1}} I_c$$

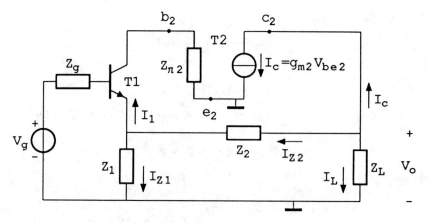

Fig. 4.19 *Schematic representation of the amplifier of Figure 4.18 for the calculation of the loop gain; g_{m2} is chosen as the reference variable.*

and

$$V_{be2} = I_1 Z_{\pi 2} = -I_c Z_{\pi 2} \cdot \frac{Z_1}{Z_1 + g_{m1}^{-1}} \cdot \frac{Z_L}{Z_L + Z_2 + Z_1 \| g_{m1}^{-1}}$$

hence

$$A\beta = g_{m2} \frac{V_{be2}}{I_c} = \frac{-g_{m2} Z_{\pi 2} Z_L}{Z_L + Z_2 + Z_1 \| g_{m1}^{-1}} \cdot \frac{Z_1}{Z_1 + g_{m1}^{-1}} \tag{4.29}$$

Obviously, $A\beta$ can be made large by choosing Z_L, Z_1 and Z_2 such that

$$Z_1 \gg 1/g_{m1} \text{ and } Z_L \gg Z_2 + 1/g_{m1}$$

In that case we obtain

$$A\beta \to -g_{m2} Z_{\pi 2} = -\beta_{t2} \tag{4.30}$$

where β_{t2} denotes the complex current gain of $T2$. Note that transistor β is here denoted by β_t in order to avoid confusion with the feedback quantity β.

Total transfer can then be written in the form

$$\frac{V_o}{V_g} = A_{f\infty} \left(\frac{\beta_{t2}}{1 + \beta_{t2}} \right) \tag{4.31}$$

Again, it may be noted that writing the voltage transfer in the form given in eq. (4.31), makes the design objective transparent. By judicious design of the active block the factor $A\beta/(1 - A\beta)$ can be optimized with respect to certain properties that are considered to be of particular interest in a certain application, such as bandwidth, distortion, effect of circuit parameter tolerances, and so on.

In order to find the input and output impedances of the circuit we can use eqs.

(4.20) and (4.23). Hence,

$$Z_i \approx Z_{io}(1 - A\beta_{sc})$$

and

$$Z_o \approx Z_{oo}/(1 - A\beta_{op})$$

Z_{io} is the input impedance if $g_{m2} = 0$. By inspection of Figure 4.19 and assuming $Z_1 \gg 1/g_{m1}$:

$$Z_{io} = \beta_{t1}[Z_1 \| (Z_2 + Z_L)]$$

and

$$\beta_{sc} = \beta(V_g = 0;\ Z_g = 0) = \frac{-Z_L}{Z_L + Z_2 + Z_1} \frac{Z_1 Z_{\pi 2}}{Z_1 + 1/g_{m1}} \quad (4.32)$$

Introducing $\beta_{t2} = g_{m2} Z_{\pi 2}$, and assuming again $Z_1 \gg 1/g_{m1}$, we find

$$A\beta_{sc} = \frac{-Z_L \beta_{t2}}{Z_1 + Z_2 + Z_L}$$

If in addition $Z_L \gg Z_1 + Z_2$ holds, we obtain $A\beta_{sc} \approx -\beta_{t2}$. Hence

$$Z_i \approx \beta_{t1}\beta_{t2}[Z_1 \| (Z_2 + Z_L)] \quad (4.33)$$

Likewise, putting $g_{m2} = 0$,

$$Z_{oo} = Z_2 + (Z_1 \| g_{m1}^{-1}) \approx Z_2 + 1/g_{m1}$$

Assuming again $Z_g \ll \beta_{t1}/g_{m1}$, we obtain $\beta_{op} = \beta(V_g = 0;\ Z_L \to \infty) \approx -Z_{\pi 2}$, so that $A\beta_{op} = g_{m2}\beta_{op} \approx -\beta_{t2}$. Hence

$$Z_o \approx \frac{Z_2 + 1/g_{m1}}{\beta_{t2}} \quad (4.34)$$

As was expected from the nature of the configuration (series connection at the input side, parallel connection at the output side), Z_i has increased, whereas Z_o has decreased with respect to the impedance levels of the active block without the feedback network connected.

4.8 Some general remarks concerning the application of negative feedback

At the end of our discussion of the application of negative feedback as a means of obtaining accurate signal transfer, some general observations concerning this implement are in order.

If it is assumed that the feedback network exhibits accurate transfer, accuracy of the feedback amplifier is determined by the 'correction factor' $A\beta/(1 - A\beta)$. Obviously, the loop gain $A\beta$ should be made as large as possible for optimum transfer. Of course, the loop gain can be larger if more stages are involved in the feedback loop. If the feedback is confined to one single stage, it is called *local feedback*, if two or more stages are involved, it is denoted *overall feedback*. From the point of view of optimization of accuracy, overall feedback is to be preferred over local feedback. A large value of $A\beta$ implies that the non-ideal properties of active elements do not significantly affect signal transfer accuracy. A further advantage is the reduction of spurious signals occurring in the amplifier. The dominant source of such signals is distortion, which can be represented by a spurious signal that is evoked by the desired signal due to the non-linearity of the active components that constitute the amplifying network. Since distortion is greater the larger the signal that is handled by an active component, it is mainly generated in the final stages of the amplifier.

As an example consider two amplifiers, as schematically depicted in Figure 4.20(a) and (b). Let S_d denote the desired input signal, and S_{ui}, S_{uo} undesired signals acting respectively at the input and output terminals of the amplifier. The amplifier of Figure 4.20(a) realizes an amplification $S_o/S_d = A_a$ without using feedback. The amplifier of Figure 4.20(b) realizes the same amplification by using a high-gain amplifier with gain G that is equipped with a feedback network B. If $B = 1/A_a$, both amplifiers provide equal gain for the desired signal S_d.

For the configuration of Figure 4.20(a)

$$S_o = A_a(S_d + S_{ui}) + S_{uo} \tag{4.35}$$

holds. For the configuration of Figure 4.20(b)

$$S_o = \frac{G}{1 + GB}(S_d + S_{ui}) + \frac{S_{uo}}{1 + GB} = A_a(S_d + S_{ui}) + \frac{S_{uo}}{1 + GB} \tag{4.36}$$

holds. Comparison of eqs. (4.35) and (4.36) reveals that in the feedback amplifier the spurious signal S_{uo} is reduced by the large factor $1 + GB$. Hence, spurious signals occurring in the final stages or at the output terminals of the amplifier are suppressed to an extent that is determined by the loop gain.

The obvious question arising at this stage is: how far can the loop gain be increased without penalty? Up to now we have assumed that the general model of

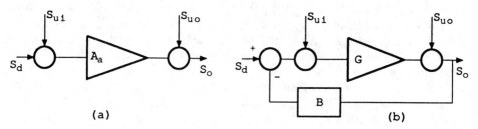

Fig. 4.20 *Spurious signals S_{ui} and S_{uo} in an amplifier without feedback (a) and with (b) feedback.*

Figure 4.7 is valid without restrictions. An essential element of this arrangement is that the feedback signal that is added to the input node is inverted with respect to this input signal. In the model of Figure 4.7 this is expressed by the negative sign at the summation point. Alternatively, it can be implemented by using an inverting amplifying block, and this is the usual implementation. It is this aspect of the arrangement that is responsible for the designation *negative feedback*. Let us now consider what happens when a non-inverted signal is fed back. Then, the transfer equation for the feedback systems reads

$$\frac{S_o}{S_s} = \frac{A}{1 - AB} \qquad (4.37)$$

In this case, which is designated by the name *positive feedback*, if $AB = 1$, then $S_o/S_s \to \infty$. If this were true, the output signal would increase beyond any limit, even if the input signal were infinitesimally small. This is, of course, impossible because the finite bias currents and voltages will set a natural limit to the output signal. Under this condition, the circuit is unstable and the output signal is no longer predictably related to the input signal. In a practical amplifier, parasitic capacitances introduce time constants, implying phase shifts for harmonic signals. If the total phase shift amounts to π radians (180°), a configuration that was intended to provide negative feedback exhibits positive feedback with the inherent consequence of instability. The conclusion is that the feedback loop must be prevented from exhibiting phase shift to a degree that will provoke instability. Since every amplifier stage adds phase shift, this requirement sets a limit on the number of amplifier stages involved in the feedback loop, and thereby to the extent in which the beneficial effects of negative feedback can be pocketed.

4.9 Error-feedforward techniques

As we have seen in the preceding sections, negative feedback is a powerful means of achieving accuracy, in spite of the inaccurate transfer properties of active devices. However, the method is not without its drawbacks. For one thing, the gain of the closed-loop configuration is much lower than that of the open-loop amplifier. In the past, when active devices were expensive and possessed modest gain capabilities, this disadvantage had considerable weight, but this changed with the advent of semiconductor devices and, particularly, monolithic technology. A more fundamental drawback was indicated at the end of the preceding section, viz. the inherent proneness to instability due to unavoidable phase shifts within the feedback loop. As we have concluded, this puts a limit on the amount of feedback that can be applied. A complementary approach to achieving accuracy, lacking this particular drawback, is the application of a method that is designated by the name of *error feedforward*. Figure 4.21 depicts the basic idea underlying this method. It is assumed that the desired value of the gain is G. The amplifier A_1 is designed such that its gain is approximately G.

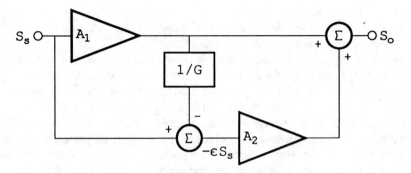

Fig. 4.21 *Principle of the method of error feedforward.*

Let us denote the gain of A_1 by $G(1+\varepsilon)$, where $\varepsilon \ll 1$. The output signal is divided by G with the aid of an accurate passive divider. The resulting signal $(1+\varepsilon)S_s$ is subtracted from the input signal S_s, yielding the error signal $-\varepsilon S_s$. This small error signal is amplified by A_2, whose gain is assumed to be $G(1+\varepsilon')$. The output signal $-\varepsilon(1+\varepsilon')S_sG$ is added to the output signal of the main amplifier, yielding $G(1-\varepsilon\varepsilon')S_s$. Thus the relative error in the total amplification is $\varepsilon\varepsilon'$. If ε and ε' both are small, this method accomplishes a considerable improvement in accuracy. If the output signal of A_1 contains a spurious signal component S_u, this signal is reduced by a factor ε'. Since distortion in A_1 can be interpreted as the generation of signal-dependent spurious intermodulation and higher harmonics, distortion is also reduced. Spurious signals and distortion in A_2 are not reduced, but since the error signal that has to be handled by A_2 is small, distortion in A_2 can in general be kept low. The transfer of A_2 can be made accurate for instance by the application of a moderate amount of feedback, hence ε' can be made quite small. Since the method does not rely on the formation of a closed loop, the problem of instability is essentially absent. But, of course, this method also has its weaknesses. First, it is essential that the inaccuracy of A_1 is kept low. Applying a moderate amount of feedback can be helpful in fulfilling this requirement. A more serious problem is that frequency-dependent phase shifts in the main amplifier can cause serious errors in signal transfer. And a further cause of concern is the correct realization of the subtraction and adding operations.

4.10 Compensation techniques

Situations where accurate signal transfer is required but where neither feedback nor feedforward methods can be applied, are not uncommon. In certain cases a non-electrical quantity is involved in the signal-processing chain, hampering the application of these methods. In other cases the power dissipation involved in the use of several additional active devices cannot be tolerated. Another reason may be that the preferred implementation technology excludes the use of components that are not compatible with this technological substrate. In such cases the application of compensation techniques may be considered.

Compensation of an erroneous transfer comes down to adding a compensating signal with the inverse effect. The problem of compensation methods is that accurate knowledge is required about the error that is to be compensated. If the prediction of the error fails, the remedy can be worse than the ailment. This is particularly irksome if the error to be compensated for is fluctuating. A good example of a useful compensation technique is the so-called gamma correction that is widely used in television systems. The problem has already been indicated in Section 4.2 in the discussion of the formation of the driving signal for a television display tube. As has been explained there, a display tube exhibits a non-linear transfer characteristic. The light output Φ is related to the driving voltage V_d by the relation $\Phi = cV_d^\gamma$, with $\gamma \approx 2.5$. For correct display of the picture information the light output must be proportional to the video signal V_v. Since the output of the display tube is a non-electrical quantity, neither feedback nor feedforward techniques are usefully applicable. The required proportionality can be obtained by inserting a circuit with the inverse non-linear transfer: $V_d = c' V_v^{1/\gamma}$ (Figure 4.22a). Such a transfer can be implemented in several ways. One method is to search for a device or a combination of devices approximating the desired non-linearity, for instance a combination of properly biased diodes. Another method is indicated in Figure 4.22(b). This method has the advantage of flexibility and of compatibility with monolithic implementation techniques. First, V_v is logarithmically converted, for instance by using a bipolar transistor in the feedback path of an amplifier. For an example see Section 6.16. The signal log V_v is attenuated by an adjustable linear attenuator, yielding $(1/\gamma) \log V_v$, which is subsequently fed to an exponential (antilog) amplifier. The latter can consist of a voltage driven bipolar transistor, since this device exhibits exponential $V \rightarrow I$ transfer.

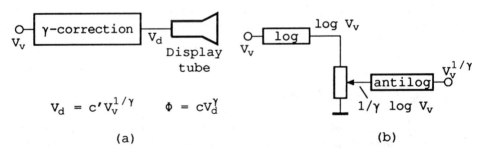

Fig. 4.22 *Example of compensation: gamma correction in television:* (a) *principle of the correction;* (b) *example of implementation.*

Further reading

Negative-feedback techniques are discussed in most books on basic electronics and on control theory. A notable variety of approaches to the subject can be found. The approach followed in this chapter, emphasizing the use of the asymptotic-gain model, is mainly based on the specialist work of Nordholt.

Nordholt, E. H. (1983) *Design of High-performance Negative-feedback Amplifiers*, Elsevier, Amsterdam.

Also of a specialist nature is
Waldhauer, F. D. (1982) *Feedback*, Wiley, New York.

General texts:
Cherry, E. M. and Hooper, D. E. (1968) *Amplifying Devices and Low-pass Amplifier Design*, Wiley, New York.
Shea, R. F. (1961) *Amplifier Handbook*, Chapter 6 (by S. K. Ghandhi), McGraw Hill, New York.
Smith, R. J. (1980) *Electronics: circuits and devices*, Wiley, New York.

Network theory and control theory approaches:
Chua, L. O., Desoer, C. A. and Kuh, E. S. (1987) *Linear and Nonlinear Circuits*, McGraw Hill, New York.
Dorf, R. C. (1980) *Modern Control Systems*, Addison Wesley, Reading.

5 Basic amplifying circuit configurations

5.1 Introduction

This chapter is concerned with the structure and the properties of amplifying building blocks. In the preceding chapter the key role of the amplification function was pointed out. In addition, methods for achieving accurate signal transfer were discussed. Among these, the use of negative feedback is by far the most important. The application of this method calls for the availability of high-gain building blocks. In order to implement such blocks, knowledge of basic amplifier configurations is indispensable and these will form the substance of the present chapter.

5.2 CE- and CS-amplifier stages

Up to now, whenever the practical implementation of amplification was involved, we have resorted to common-emitter (CE) or common-source (CS) amplifier stages. This is a logical approach as long as no more than the gain function as such is involved. The CE- and CS-configurations can be considered as the most elementary gain providing configurations. This statement is based on the view that only these configurations do not possess any inherent feedback. All other configurations with one active device can be considered as local feedback versions of CE- or CS-stages. In agreement with this statement is the fact that their low-frequency power gain, obtained with a given resistive load, is larger than in any of the other basic configurations. But on the other hand, due to the absence of negative feedback, none of their transfer parameters is accurately determined.

To start with we will derive the transfer parameters of the CE- and CS-configurations in the low-frequency range, where the capacitances showing up in the equivalent circuits of the transistors have negligible effect. Figure 5.1 shows the simple hybrid-π equivalent circuit of a BJT for low-frequency signals.

BASIC AMPLIFYING CIRCUIT CONFIGURATIONS 117

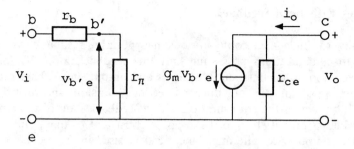

Fig. 5.1 *Low-frequency hybrid-π equivalent circuit of a BJT.*

Using the relations $\beta_0 = g_e r_\pi$ and $\mu_{e0} = g_e r_{ce}$ with

$$g_e = g_m = \frac{qI_C}{kT}$$

we find

$$\mu_{lf} = \left(\frac{v_o}{v_i}\right)_{i_o=0} = -\mu_{e0}\frac{r_\pi}{r_b + r_\pi} \approx -\mu_{e0} \tag{5.1}$$

$$\gamma_{lf} = \left(\frac{i_o}{v_i}\right)_{v_o=0} = g_e\frac{r_\pi}{r_b + r_\pi} \approx g_e \tag{5.2}$$

$$\zeta_{lf} = \left(\frac{v_o}{i_i}\right)_{i_o=0} = -\beta_0 r_{ce} \tag{5.3}$$

$$\alpha_{lf} = \left(\frac{i_o}{i_i}\right)_{v_o=0} = \beta_0 \tag{5.4}$$

Hence, all transfer parameters are finite but inaccurate, since they rely entirely on the small-signal parameters of the transistor.

These transfer parameters pertain to the transfer of signals from the input to the output terminals. Frequently the relation between voltage and current at a given terminal pair must also be known. These relations are specified by the input and output impedances. In the case of low-frequency signals, where the effects of capacitances can be neglected, these quantities are for the CE-configuration given by $R_i = r_b + r_\pi$ and $R_o = r_{ce}$. The input and output resistances are neither very low nor very high. In other words, the input behaves neither as an ideal voltage input nor as an ideal current input; the output behaves neither as an ideal voltage source, nor as an ideal current source. Both R_i and R_o are finite and inaccurate, as is to be expected from a non-feedback arrangement.

5.3 Stages with local feedback

Coming now to single-stage configurations incorporating feedback, we will restrict ourselves to configurations possessing only one feedback loop. If, for practical reasons, we impose the constraint that input and output should have one terminal in common, we obtain the four different feedback versions shown in Figure 5.2.

Figure 5.2(b) shows the configuration with a feedback loop of the type series at input, parallel at output. Due to the imposed constraint of one terminal common to input and output, no finite impedances can be used in the feedback path. The configuration is known under the name of common-collector (CC) configuration or emitter follower. The transfer parameters are easily found from the hybrid-π equivalent circuit:

$$\mu_{lf} = \frac{r_{ce}}{r_{ce} + r_e + r_b/\beta_0} \approx 1 \tag{5.5}$$

$$\zeta_{lf} = r_{ce}(1 + \beta_0) \tag{5.6}$$

$$\gamma_{lf} = \frac{1}{r_e + r_b/\beta_0} \approx -g_e \tag{5.7}$$

$$\alpha_{lf} = -(1 + \beta_0) \tag{5.8}$$

In agreement with our expectation, the parameter accurately determined by the feedback arrangement is the voltage transfer. Because of the absence of external impedances in the feedback path, the configuration has unity voltage gain. This property has given rise to the generally used name of 'emitter follower'.

As to the input and output resistance: $R_i \approx r_b + r_\pi + \beta_0 r_{ce} \approx \beta_0 r_{ce}$ and $R_o \approx r_e + r_b/\beta_0$. As expected, R_i is high, whereas R_o is low. The input behaves as a voltage input; the output behaves as a voltage source.

It should be noted that the transfer parameters as well as the input and output resistances assume different values if the circuit is driven from a source with finite

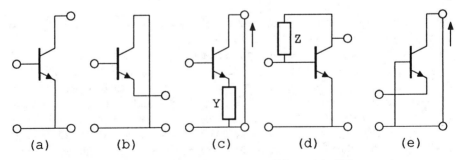

Fig. 5.2 Feedback versions of single-stage amplifiers: (a) CE-stage (non-feedback starting configuration); (b) CC-stage or emitter follower; (c) series stage; (d) shunt stage; (e) CB-stage or current follower.

impedance and is loaded with a finite impedance. For instance, it will be immediately clear that with a finite load R_L with $r_e \ll R_L \ll r_{ce}$, then $R_i \approx \beta_0 R_L$. And if the internal resistance of the driving source is R_g, then $R_o \approx r_e + (r_b + R_g)/\beta_0$.

Figure 5.2(c) shows the single-element feedback configuration with series feedback at the input as well as at the output. The configuration is known as the 'series stage'. The parameter that is accurately fixed is the transadmittance. If $Y \ll g_e$, then $\gamma_{lf} \to Y$. In agreement with the expectation, both the input resistance and the output resistance are high: $R_i \approx r_b + \beta_0(r_e + R)$ with $R = 1/Y$, and, when driven from a voltage source with low internal impedance, then $R_o \approx R' + r_{ce}(1 + g_e R')$, where R' stands for the parallel connection of R and $r_\pi + r_b$.

Figure 5.2(d) depicts the configuration with feedback of the parallel–parallel type. The configuration is known under the name of 'shunt stage'. Here the transimpedance is fixed by the feedback arrangement. If $Z/r_{ce} \ll 1$ and $\beta_0 Z \gg r_b + r_\pi$, then $\zeta_{lf} \approx -Z$. Both the input and output impedance are low, as should be expected with this type of feedback. Straightforward calculation yields

$$Z_i = \frac{(Z + R'_L)(r_b + r_\pi)}{r_b + r_\pi + Z + \beta_0 R'_L} \tag{5.9}$$

where R'_L is the parallel connection of r_{ce} and the load resistance R_L.

If $R_L \gg r_{ce}$ and $R_L \gg Z$ (approximating the open output situation):

$$Z_i \to \frac{r_b + r_\pi}{\beta_0} = r_e + \frac{r_b}{\beta_0}$$

And if $R_L \ll Z$ and $R_L \ll (r_b + r_\pi)/\beta_0$:

$$Z_i \to Z \| (r_b + r_\pi) \tag{5.10}$$

If we further assume $r_{ce} \gg Z + r_\pi$, we find

$$Z_o = \frac{(r_b + r_\pi)(Z + R_g) + Z R_g}{(1 + \beta_0)R_g + r_\pi + r_b} \tag{5.11}$$

where R_g denotes the internal resistance of the driving source. If $R_g \to 0$ (driving from ideal voltage source), then $Z_o \to Z$. And if $R_g \to \infty$ (driving from current source), then $Z_o \approx (r_\pi + r_b + Z)/\beta_0$.

EXAMPLE
Let
$$r_b = 50\ \Omega,\ r_\pi = 5\ \text{k}\Omega,\ Z = 5\ \text{k}\Omega,\ \beta_0 = 200$$

With
$$R_g = 0 : Z_o = 5\ \text{k}\Omega$$

With
$$R_g = 100\ \Omega : Z_o = 1.05\ \text{k}\Omega$$

With

$$R_g \to \infty : Z_o = 50 \ \Omega$$

Figure 5.2(e) depicts the configuration with the feedback loop of the parallel-series type. The transfer parameter that is accurately fixed is the current gain. Because of the absence of external impedances dictated by the assumed constraint of one terminal common to input and output, the configuration has (approximately) unity current gain. Straightforward calculation yields

$$\alpha_{lf} = \left(\frac{i_o}{i_i}\right)_{v_o=0} = \frac{r_{ce}}{r_\pi + r_b + (\beta_0 + 1)r_{ce}} - 1 \tag{5.12}$$

With $\beta_0 \gg 1$ and $\beta_0 r_{ce} \gg r_b + r_\pi$ this formula simplifies to $\alpha = -1$. The configuration has the properties of a current follower, but it is usually referred to as the common-base (CB) configuration. The input impedance is low, whereas the output impedance is high:

$$Z_i = \frac{r_b + r_\pi}{\beta_0}\left(1 + \frac{Z_L}{r_{ce}}\right) \tag{5.13}$$

If $Z_L \ll r_{ce}$, then $Z_i \to r_e + r_b/\beta_0$. And

$$Z_o = r_{ce} + \beta_0 r_{ce}\left(1 - \frac{r_b + r_\pi}{R_g + r_b + r_\pi}\right) + \frac{R_g(r_b + r_\pi)}{R_g + r_b + r_\pi} \tag{5.14}$$

If

$$R_g \to 0, \text{ then } Z_o \to r_{ce}$$

and if

$$R_g \to \infty, \text{ then } Z_o \to \beta_0 r_{ce}$$

EXAMPLE
If

$$\beta_0 = 200, \ R_g \to \infty \text{ and } r_{ce} = \frac{V_A}{I_C} = 150 \text{ k}\Omega, \text{ then } Z_o = 30 \text{ M}\Omega$$

5.4 High-frequency properties of single-transistor stages

In the preceding section we have studied the properties of single-transistor stages under the assumption that the nature of the driving signals is such that the effects caused by the capacitances showing up in the equivalent circuit of the transistor can be neglected. This is, of course, the case for low-frequency signals only. The advantage of starting with this simplification is that the essential properties of the configuration are revealed with clarity. We must now consider the situation wherein the effects of the capacitances are taken into account. It will be clear that searching for analytical expressions is only practical if the transistor is represented by a simple first-order equivalent circuit. The results of deriving such expressions are useful for two reasons. First, in this way analytical expressions are found that are simple enough to be useful for acquiring general insight into the signal transfer properties of the various configurations, and hence into the possibilities of appropriate combinations. Second, they provide a starting point for more elaborate studies of signal transfer with the aid of computer calculations employing complex models. The small-signal models we will use in our first-order analysis were introduced in Chapter 2. They are redrawn in Figure 5.3 (for the BJT) and 5.4 (for the FET).

In Chapter 2 some basic relations have already been derived in order to provide a basis for the illustration of the general methods for achieving accurate signal processing discussed in Chapter 4. In the present chapter we will elaborate on the properties of the elementary circuit configurations.

Since, as has been determined, the CE- and CS-configuations can be considered to be the basic amplifier configurations, they provide a logical starting point. By inspection of the equivalent circuits of Figures 5.3 and 5.4 we note that the presence of the feedback capacitances $C_{b'c}$(BJT) and C_{gd}(FET) seriously thwarts the calculation of the transfer properties. We will therefore first unilateralize the stages by omitting these capacitances and in a separate discussion consider their effect on signal transfer.

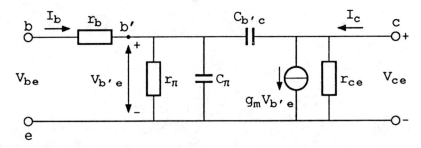

Fig. 5.3 *Hybrid-π model of BJT.*

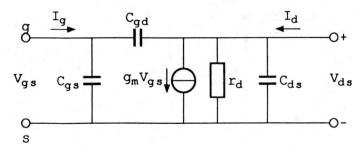

Fig. 5.4 *Elementary model of FET.*

In practice an amplifier stage is always embedded in an environment: it is driven by a signal source or by a preceding stage, and it is loaded. The driving source can be modelled either as a voltage or as a current source. Though the transfer properties of the amplifier stage as such can be adequately characterized by the transfer matrix, in practice interest is focused on the transfer from signal source to amplifier output. We will therefore concentrate our attention on this transfer. Since the input and output impedances of a CE-stage are neither very low nor very high, there is no obvious reason to focus attention on either voltage or current transfer. Therefore both transfers will be considered.

Figure 5.5 shows the equivalent circuit of the CE-stage when driven by a voltage source V_g with internal impedance R_g. In order to keep the calculations digestible we will assume that this source impedance is resistive. In practice this approximation is frequently applicable since C_π is often much larger than a possible capacitive part of Z_g.

The load is represented by Z_L which also incorporates the internal resistance r_{ce}. Denoting, as we have done previously, the parallel combination of r_π and C_π by Z_π, the voltage transfer is given by

$$\frac{V_o}{V_g} = \frac{-Z_L g_m Z_\pi}{R_g + r_b + Z_\pi} \tag{5.15}$$

Where $g_m r_\pi = \beta_0$ eq. (5.15) can be written

$$\frac{V_o}{V_g} = -Z_L \frac{\beta_0}{s(R_g + r_b)\tau_\pi + r_\pi + (R_g + r_b)} \tag{5.16}$$

where $\tau_\pi = r_\pi C_\pi$.

If Z_L is resistive ($Z_L = R_L$), the transfer is characterized by one pole:

$$p = \frac{-1}{C_\pi[(R_g + r_b) \,\|\, r_\pi]} = \frac{-1}{\tau_h} \tag{5.17}$$

If $Z_L = R_L \,\|\, C_L = R_L/(1 + s\tau_L)$ with $\tau_L = R_L C_L$, a second pole at $p = -1/\tau_L$ arises. In practice, $\tau_L \ll \tau_h$ will frequently hold, so that the frequency dependent behaviour is approximately described by one pole. We will call this pole the *dominant pole*.

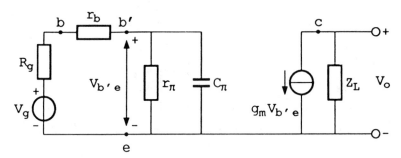

Fig. 5.5 *Unilateral CE-stage.*

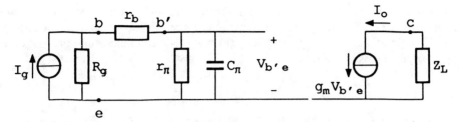

Fig. 5.6 *CE-stage driven by a current source.*

Figure 5.6 shows the stage when driven by a current source, under the assumption that $r_{ce} \gg Z_L$ within the frequency band of interest. Straightforward calculation yields

$$\frac{I_o}{I_g} = \frac{\beta_0 R_g}{R_g + r_b} \cdot \frac{1}{s\tau_\pi + (r_\pi + R_g + r_b)/(R_g + r_b)} \tag{5.18}$$

Again, the transfer is characterized by one dominant pole, in the same location as in the case of the voltage amplifier.

As will become clear in the discussion on the design of wideband amplifiers, a useful figure of merit for an active device in such an amplifier is the product of bandwidth and low-frequency gain. According to eq. (5.16), the low-frequency voltage gain is

$$|G| = \frac{\beta_0 R_L}{r_\pi + R_g + r_b} \tag{5.19}$$

Since the bandwidth, in terms of angular frequency ω, is $B_\omega = p = 1/\tau_h$ and $g_m = 1/r_e = \beta_0/r_\pi$,

$$GB_\omega = \frac{R_L}{r_e + (R_g + r_b)/\beta_0} \cdot \frac{1}{C_\pi[(R_g + r_b) \| r_\pi]} \tag{5.20}$$

If we consider a cascade of identical stages, and Z_L can be approximated by its resistive part R_L, then $R_g = R_L$ and

$$GB_\omega = \frac{R_L}{C_\pi r_e} \cdot \frac{1}{R_L + r_b} \tag{5.21}$$

Applying the same routine to the current transfer function, eq. (5.18), the same value of GB_ω is found. From eq. (5.21) we derive that for a BJT the wideband figure of merit is not independent of the driving and loading conditions. For optimum GB_ω the relation $R_L \gg r_b$ must hold. This is an important conclusion that will be taken up in the discussion on wideband amplification (Chapter 9).

Since the GB-product is not independent of the circuit environment of the transistor, it is somewhat impractical to use this quantity as a figure of merit for the high-frequency properties of a transistor. A quantity serving as a figure of merit should depend solely on the device to be qualified. In addition, it should be conveniently measurable. The quantity that has found general acceptance is the transit frequency f_T, defined as the frequency where the current gain of a CE-stage with zero load has unity value. Using the hybrid-π equivalent circuit, including $C_{b'c}$, f_T can be expressed as

$$f_T = \frac{1}{2\pi r_e (C_\pi + C_{b'c})}$$

which corresponds with GB if $R_L \to \infty$. A drawback of this figure of merit is that is does not account for the effects of r_b on high-frequency response. Transistors with the same value of f_T can therefore behave differently in high-frequency applications.

Figure 5.7 shows the equivalent circuit of a CS-stage driven by a voltage source, with the omission of C_{gd}. The internal resistance of the drain r_d is considered to be included in R_L. Contrary to the case of the BJT, the output capacitance (C_{ds}) is made explicit. This is realistic because in an FET, either of the junction or the MOST-variety, the values of the input and output capacitances are of the same order of magnitude. Since in a cascade of equal stages the output impedance of the preceding stage is the driving source impedance of the stage under consideration, a realistic comparison with the case of the CE-stage requires the representation of the driving source impedance by an R_g–C_g combination. Denoting $R_g \| C_g$ by Z_g and $R_L \| C_{gd}$ by Z_L, the voltage transfer is found to be

$$\frac{V_o}{V_g} = -g_m \frac{Z_L}{1 + Z_g s C_{gs}} \qquad (5.22)$$

With $Z_g = Z_L$, eq. (5.22) becomes

$$\frac{V_o}{V_g} = -g_m \frac{R_L}{1 + sR_L(C_{ds} + C_{gs})} \qquad (5.23)$$

The transfer is characterized by one pole $p = -1/\tau_h$ with $\tau_h = R_L(C_{ds} + C_{gd})$.

If the voltage source is replaced by a current source I_g with internal admittance

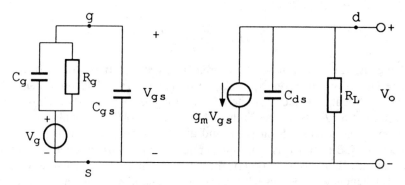

Fig. 5.7 *CS-stage driven by a voltage source.*

$Y_g = 1/R_g + sC_g \approx 1/R_L + sC_{ds}$ and it is assumed that the stage is loaded with an impedance that is small compared to its output impedance, the current transfer is given by

$$\frac{I_o}{I_g} = g_m \frac{R_L}{1 + sR_L(C_{ds} + C_{gs})} \qquad (5.24)$$

indicating that voltage and current transfer have the same characteristics. The gain–bandwidth product amounts to $GB_\omega = g_m/(C_{ds} + C_{gs})$, which is independent of R_L. The difference from the case of the BJT is due to the finite value of r_b in the BJT-case.

Applying the definition for f_T given previously, for the FET,

$$f_T = \frac{g_m}{2\pi(C_{gs} + C_{gd})}$$

holds, as can easily be verified with the aid of the equivalent circuit including C_{gd}.

We must now direct our attention to the effect of the finite value of $C_{b'c}$ and C_{gd}. If we include this capacitance in the equivalent circuit, the calculation of the voltage or current transfer leads to complicated expressions. Though formally correct, these expressions are useless for obtaining analytical insight into the relations between transistor and circuit parameters on the one hand and transfer properties on the other hand. Of course, such expressions can be used for deriving numerical results, but these can be obtained more easily by straightforward computer simulation. However, by considering a few special cases some general insight can still be obtained. In the design of transistors, minimization of feedback capacitance is one of the primary concerns. Therefore, in many practical situations the condition $sC_{b'c} \ll Y_L$ will hold true in the frequency band of interest. In this case a simple approximation can be applied. Figure 5.8 shows the relevant part of the amplifier stage.

If $Z_L \ll 1/sC_{b'c}$, then the current from the source $g_m V_{b'e}$ is largely absorbed by Z_L, hence $I_f \ll I_o$. Consequently, $V_o \approx -g_m V_{b'e} Z_L$. If Z_L is approximately resistive in the relevant frequency band, then $V_o \approx -g_m V_{b'e} R_L$. The voltage across $C_{b'c}$ is approximately $(1 + g_m R_L) V_{b'e}$. Consequently, the current through $C_{b'e}$ is increased

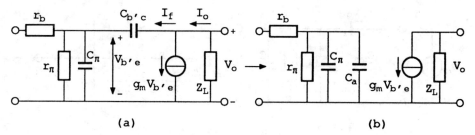

Fig. 5.8 Unilateralization of the equivalent circuit of a CE-stage: (a) non-unilateral diagram; (b) unilateral approximation.

by a factor $(1 + g_m R_L)$ in comparison with the situation that no voltage at the output side is present. The effect of this capacitive current on the loading of the input circuit can be represented by an additional capacitance $C_a = (1 + g_m R_L)C_{b'c}$ in parallel with C_π. In this way the circuit is unilateralized and in the formulae for voltage and current transfer the effect of the feedback capacitance is accounted for by replacing C_π by $C_\pi + C_a$. Likewise, in the case of the FET, C_{gs} is replaced by $C_{gs} + C_a$, with $C_a = (1 + g_m R_L)C_{gd}$. Essentially, the above reasoning comes down to the use of Miller's theorem. The capacitance $C_{b'c}$ (or C_{gd}) is sometimes called the 'Miller capacitance'.

EXAMPLE

If $g_m R_L = 100$, which is a quite common value, and $C_{b'c} = 0.5$ pF, then $C_a = 50.5$ pF.

This simple way of coping with the presence of a feedback capacitance in active components is not always applicable. In certain cases this approximation will prove to be too simple. In order to acquire a general understanding of the effects due to the presence of the feedback capacitance, let us first consider the case where the driving source is an ideal voltage source. Figure 5.9 shows the equivalent circuit, using a FET as the active component.

Calculation of the voltage transfer yields

$$\frac{V_o}{V_g} = \frac{R_L(sC_{gd} - g_m)}{1 + sR_L(C_L + C_{gd})} \quad (5.25)$$

Figure 5.9(b) shows the p–z plot. It includes a zero in the right half-plane. This zero has little effect on the characteristics of the voltage transfer. For the stage to be useful, R_L must be $>1/g_m$ and, obviously, $C_L + C_{gd} > C_{gd}$. Hence, the time constant responsible for the zero is much smaller than the one responsible for the pole. The zero thus has negligible effect in the passband of the stage. But it should not be neglected altogether, since it introduces some excess phase shift that, if occurring in the loop transfer of a feedback amplifier, can to some extent affect its response and can even give rise to instability.

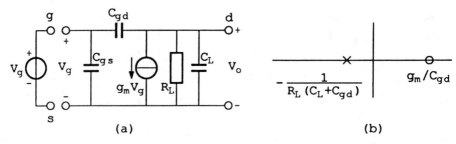

Fig. 5.9 (a) *Non-unilateral FET-stage driven from ideal voltage source.* (b) *p–z plot pertaining to the circuit.*

The pole is somewhat closer to the origin than if C_{gd} were not present. At first glance the effect of C_{gd} seems rather harmless. But it should be noted that the case we are considering is rather unrealistic. Normally, the driving source will possess a certain internal impedance and this incurs a drastic change in the situation. The input impedance of the stage affects the transfer considerably. With some calculation effort we find for the input admittance

$$Y_i = \frac{1}{Z_i} = sC_{gs} + \frac{s^2 R_L C_{gd}^2 a + sC_{gd}(1 + g_m R_L)}{1 + sR_L(1 + a)C_{gd}} \quad (5.26)$$

with $a = C_L/C_{gd}$. The right-hand member represents a parallel connection of a resistance R_{in} and a capacitance C_{in}, both highly frequency-dependent. Assuming $g_m R_L (a + 1) \gg 1$, and putting $s = j\omega$, eq. (5.26) yields

$$R_{in} = \frac{1 + \{\omega(a+1)C_{gd}R_L\}^2}{\{\omega(a+1)C_{gd}R_L\}^2} \cdot \frac{a+1}{g_m} \quad (5.27)$$

and

$$C_{in} = C_{gs} + \frac{(1 + g_m R_L) + \omega^2(a+1)aC_{gd}^2 R_L^2}{1 + \{\omega(a+1)C_{gd}R_L\}^2} \cdot C_{gd} \quad (5.28)$$

if $R_L \ll 1/\omega C_{gd}$, then eqs (5.27) and (5.28) simplify to $R_{in} \to \infty$ and $C_{in} = C_{gs} + (1 + g_m R_L)C_{gd}$, in accordance with the unilateral approximation.

Since R_{in} and C_{in} are frequency-dependent, no fixed values can be assigned to these quantities. If we assume that the highest frequency of interest is determined by the pole that was found earlier, it makes sense to calculate the values of R_{in} and C_{in} for this frequency. With $\omega = 1/R_L(C_L + C_{gd})$ we find $R_{in} = 2(a+1)/g_m$ and $C_{in} = C_{gs} + \frac{1}{2}C_{gd}(2 + g_m R_L)$. It turns out that for high frequencies R_{in} assumes a very low value. For instance, if $a = 5$ and $g_m = 5$ mA V^{-1}, then $R_{in} = 2400\,\Omega$. Hence, the approximation of the transfer by eq. (5.25) is only useable if R_g is much smaller than this value.

In order to obtain a more realistic idea about the effect of the presence of C_{gd}, let us consider the case that the stage is driven by a source with an internal impedance Z_g that is identical to the effective load consisting of R_L and C_L.

Figure 5.10 gives the equivalent circuit for this situation. The transfer V_o/V_g is found by writing down the node equations of the nodes A and B:

$$g_m V_g + Y_g V_2 + sC_{gd}V_2 - sC_{gd}V_o = 0 \quad (5.29)$$

$$g_m V_2 + Y_L V_o - sC_{gd}V_2 + sC_{gd}V_o = 0 \quad (5.30)$$

Eliminating V_2 and putting $Y_L = Y_g$, yields

$$\frac{V_o}{V_g} = \frac{g_m(g_m - sC_{gd})}{sC_{gd}(2Y_L + g_m) + Y_L^2} \quad (5.31)$$

Once again we find the right half-plane zero. From the denominator of eq. (5.31)

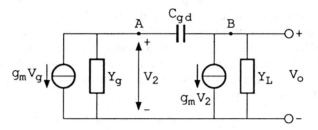

Fig. 5.10 *CS-stage with feedback capacitance, driven from a unilateral stage with the same load as the stage under consideration.*

we find the equation for the pole positions

$$s^2(\tau_p^2 + 2C_{gd}\tau_p R_L) + s(C_{gd}g_m R_L^2 + 2\tau_p + 2C_{gd}R_L) + 1 = 0 \tag{5.32}$$

with $\tau_p = R_L C_L$.

From the theory of quadratic equations we find, for the product of both poles p_1 and p_2,

$$p_1 p_2 = \frac{1}{\tau_p^2 + 2C_{gd}\tau_p R_L} \tag{5.33}$$

and for their sum

$$p_1 + p_2 = \frac{C_{gd}g_m R_L^2 + 2\tau_p + 2C_{gd}R_L}{\tau_p^2 + 2C_{gd}\tau_p R_L} \tag{5.34}$$

If $C_{gd} = 0$, these relations simplify to $p_1 p_2 = 1/\tau_p^2$ and $p_1 + p_2 = -2/\tau_p$, as should be the case.

From these relations we derive that, if $C_{gd} \ll C_L$, which is normally valid, the product of the poles decreases slightly, whereas their sum increases rather more. The poles separate. This effect is called 'pole splitting'. Since in feedback amplifiers as a rule, poles close to one another are undesirable, pole splitting is sometimes introduced intentionally in such amplifiers. But it should be noted that the gain–bandwidth product of a stage is unfavourably affected by the presence of C_{gd}. While we have considered the case of an FET-stage here, it will be clear that bipolar transistors behave similarly.

As a general conclusion it can be stated that with regard to amplifier stages that form part of a broadband amplifier, measures to enhance unilaterality are highly desirable. Obviously, such measures must reduce the effect of the feedback capacitance C_{gd} (or $C_{b'c}$). This can be achieved by adding buffer stages. It is true that such methods involve the addition of extra transistors, but, particularly in monolithic implementation, this is affordable. Two implementations of this method have found wide application, the so-called *cascode* circuit and the so-called *long-tailed* pair. Figure 5.11(a) shows the cascode configuration. It consists of a CS-stage separated from its load by a common gate (CG)-stage. The input impedance of the CG-stage is low, amounting to $1/g_m$. Since the drain current equals the source

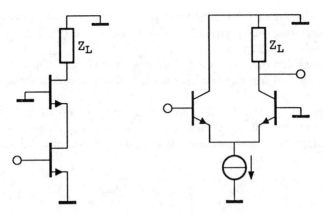

Fig. 5.11(a) *Cascode circuit (biasing elements omitted).* (b) *Long-tailed pair with BJTs.*

current, the voltage transfer of the cascode equals that of the CS-stage. But C_{gd} of the CS-stage is largely inactivated thanks to the low load impedance presented by the CG-stage. Obviously, the CS-stage possesses unity voltage gain, so that the simple unilateralization method is applicable. Consequently, the effect of C_{gd} can be modelled by adding a capacitance $2C_{gd}$ to the input capacitance C_{gs}. And C_{gd} of the CG-stage is simply in parallel with the load capacitance of the cascode. An additional feature is that the output impedance of the cascode is large. If both transistors have equal r_d, the output impedance of the cascode equals $r_d(1 + g_m r_d)$, due to the series feedback arrangement, since r_d of the CS-stage is in series with the source of the CG-stage.

If the cascode configuration is implemented with BJTs, much the same applies. With regard to the output impedance (the internal impedance of the CB-collector) straightforward calculation yields $R_o = r_\pi + \beta_0 r_{ce} \approx \beta_0 r_{ce}$. Hence, for low-frequency signals, the cascode behaves as a nearly ideal current source.

Figure 5.11(b) shows a long-tailed pair (LTP) configuration, equipped with BJTs. This circuit consists of an emitter follower (CC) and a CB-stage. The feedback capacitance of the CC is simply in parallel with its input capacitance, whereas $C_{b'c}$ of the CB is in parallel with Z_L. Except for being non-inverting, the configuration roughly behaves as a CE-stage. Since the input impedance of the CB-stage acts as the load for the CC-stage, the voltage transfer of the latter is $\frac{1}{2}$ and the voltage transfer of the LTP is $\frac{1}{2}g_m Z_L$. Both the low-frequency input impedance and the output impedance are about twice those of a CE-stage.

5.5 The CB- and CG-configurations

Next we investigate the behaviour of the configurations with local feedback, taking into account the effects of the internal capacitances on the transfer properties. Since local feedback is applied for improving the accuracy of one transfer parameter, which one depending on the type of feedback, we can restrict transfer calculations to that particular quantity.

First, we consider the CB- and CG-configurations. The transfer parameter of interest is the current transfer (I_o/I_g). Figure 5.12 depicts the equivalent circuit for a CB-stage.

The capacitance $C_{b'c}$ is, to a first approximation, in parallel with Z_L and can be considered to be part of it. Assuming $\mu_{e0} \gg 1$, $r_{ce} \gg Z_L$, $r_{ce} \gg r_b$, $\mu_{e0} Z_\pi \gg r_b$ and $g_m Z_g \gg 1$, where Z_π stands as usual for r_π and C_π in parallel, calculation of I_o/I_g yields

$$-\frac{I_o}{I_g} = Z_g \frac{g_m Z_\pi}{g_m Z_g Z_\pi + r_b + Z_g} \tag{5.35}$$

If the stage is driven from an ideal current source ($Z_g \to \infty$) this equation simplifies to

$$-\frac{I_o}{I_g} = \frac{g_m Z_\pi}{1 + g_m Z_\pi} = \alpha_0 \frac{1}{1 + sr_e C_\pi} \tag{5.36}$$

with $\alpha_0 = \beta_0/(1 + \beta_0)$.

For low frequencies the current transfer is α_0; the stage behaves as a current follower. The frequency dependent behaviour is characterized by one dominant pole $p = -1/C_\pi r_e \approx -\omega_T$. If the stage is driven from a source with finite Z_g, but yet $|Z_g| \gg r_b$ holds, eq. (5.36) still applies. If the condition $R_g \gg r_b$ does not hold, but yet $R_g \gg 1/g_m$, then the pole shifts to approximately

$$p = -\frac{1}{C_\pi r_e} \cdot \frac{1 + r_b/\beta_0 R_g}{1 + r_b/R_g} \tag{5.37}$$

The GB-product amounts to

$$GB_\omega = \frac{1}{C_\pi r_e} \cdot \frac{1}{1 + r_b/R_g} \tag{5.38}$$

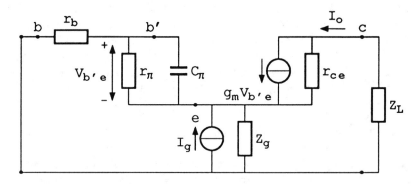

Fig. 5.12 *Equivalent circuit of a CB-stage.*

Hence, the presence of r_b causes the GB-product to depend on R_g. In this respect the situation is similar to that of the transfer of the CE-stage.

Since, to a first approximation, $C_{b'c}$ is in parallel with Z_L, the collector current splits up in a current through $C_{b'c}$ and the current I_o through Z_L. Assuming Z_L to consist of the parallel connection of R_L and C_L, this effect introduces an additional pole at $s = -1/R_L(C_{b'c} + C_L)$, together with a zero at $-1/R_L C_L$. Since the pole at $-1/C_\pi r_e \approx -1/\omega_T$ is far from the origin, this second pole can turn out to be dominant. However, in good circuit design a CB-stage will usually be loaded with low R_L.

The low-frequency resistance is low: $R_i = r_e + r_b/\beta_0$. When the presence of C_π is taken into account, the input impedance tends to assume an inductive character. This implies that a CB-stage can present an inductive load to its driving stage. If the output impedance of the driving stage has a capacitive character, a resonant circuit may be formed. Though this effect may be used to good advantage in certain cases, it is more often undesirable. Anyway, it is advisable to be watchful for the occurrence of this effect.

With the common-gate configuration similar results are obtained. In comparison with the BJT, differences occur through the absence of r_b and because the output capacitance C_{ds} is relatively large so that it cannot always be neglected. Figure 5.13 depicts the relevant equivalent circuit. Current transfer is given by

$$-\frac{I_o}{I_g} = \frac{Z_g'(1 + g_m Z_d)}{Z_L + Z_g' + (1 + g_m Z_g')Z_d} \tag{5.39}$$

where Z_g' is the parallel connection of Z_g and C_{gs}, while Z_d is the parallel connection of r_d and C_{ds}.

Taking into account that $g_m r_d \gg 1$ always and assuming that Z_g' is dominated by C_{gs}, hence $Z_g' \approx 1/sC_{gs}$, and further assuming that Z_L can be represented by a parallel connection of R_L and C_L, eq. (5.39) yields

$$-\frac{I_o}{I_g} = \frac{(g_m + sC_{ds})(1 + sR_L C_L)}{s^2 R_L(C_{ds}C_L + C_{ds}C_{gs} + C_{gs}C_L) + s[C_{ds} + (C_L + C_{gs})(R_L/r_d) + C_{gs} + g_m R_L C_L] + g_m}$$

(5.40)

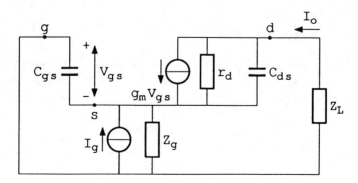

Fig. 5.13 *Equivalent circuit of a CG-stage.*

The transfer is characterized by two zeros and two poles. The zeros are

$$z_1 = -g_m/C_{ds} \text{ and } z_2 = -1/R_L C_L$$

The poles depend in a rather complicated way on the circuit parameters. Usually, in good circuit design, a CG-stage is loaded with a small load impedance (ideally zero). If $R_L \to 0$, eq. (5.40) turns into the very simple equation

$$-\frac{I_o}{I_g} = \frac{g_m + sC_{ds}}{s(C_{ds} + C_{gs}) + g_m} \tag{5.41}$$

In this case the transfer is characterized by a zero $z = -g_m/C_{ds}$ and one pole $p = -g_m/(C_{gs} + C_{ds})$. Since at $\omega = 0$, $|I_o/I_g| = 1$, the GB-product is also $g_m/(C_{gs} + C_{ds})$.

If $R_L/r_d \ll 1$ holds, the poles are given by

$$p_{1,2} = \frac{-(C_t + g_m R_L C_L) \pm [(C_t + g_m R_L C_L)^2 - 4g_m R_L (C_L C_t + C_{gs} C_{ds})]^{1/2}}{2R_L (C_L C_t + C_{gs} C_{ds})} \tag{5.42}$$

where $C_t = C_{gs} + C_{ds}$.

To get an idea about the influence of R_L on the location of the dominant pole, a numerical example is instructive. Let $C_{gs} = 2C$, $C_{ds} = C$, $C_L = 3C$, $g_m R_L = 4$. Using these data we find

$$p_1 = -0.75 \, g_m/(C_{gs} + C_{ds}) \text{ and } p_2 = -0.27 \, g_m/(C_{gs} + C_{ds})$$

which reveals that even a moderate value of R_L has considerable effect on the location of the dominant pole. This should be taken into account when a CG-stage is used as a wideband buffer stage, which is a common application.

Finally, note that, if desired, the effect of the finite value of C_{gd} can easily be taken into account since C_{gd} is in parallel with Z_L.

Note that in MOS-circuits the various substrate capacitances will have considerable effect on the high-frequency transfer properties. Since calculations with a more elaborate equivalent circuit tend to be quite cumbersome, in relevant cases resort should be made to computer simulation.

5.6 Emitter follower and source follower

The relevant transfer of the emitter follower and the source follower is the voltage transfer. First we consider the emitter follower. Figure 5.14 shows the equivalent circuit. The source impedance and the load are assumed to be resistive; the case of a capacitive load will be investigated later.

For the sake of simplicity $C_{b'c}$ has been omitted from the impedance circuit. Its effects on signal transfer is usually slight since $C_{b'c}$ is effectively in parallel with the input impedance. Further, r_{ce} is in parallel with R_L and can be considered to be included therein.

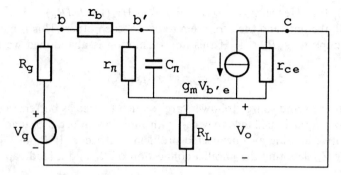

Fig. 5.14 *Equivalent circuit of the emitter follower.*

$$\frac{V_o}{V_g} = \frac{R_L(1 + sr_eC_\pi)}{r_e + (R_g + r_b)/\beta_0 + R_L + sr_eC_\pi(R_g + r_b + R_L)} \quad (5.43)$$

The transfer is characterized by a zero at $z = -1/C_\pi r_e$ and a pole

$$p = -\frac{r_e + (R_g + r_b)/\beta_0 + R_L}{r_e C_\pi (R_g + r_b + R_L)}$$

Since the pole is closer to the origin than the zero, the frequency dependency of the response is dominated by the pole.

If the stage is used for the appropriate purpose, that is, the implementation of a voltage-to-voltage transactor, R_g is low and R_L is high. In that case the pole comes close to the zero.

Figure 5.15 shows the equivalent circuit of the FET-counterpart, commonly designated as the source follower.

Assuming $g_m R_L \gg 1$ again, we find

$$\frac{V_o}{V_g} = \frac{1 + s\tau_g}{1 + s\tau_g(1 + R_g/R_L)} \quad (5.44)$$

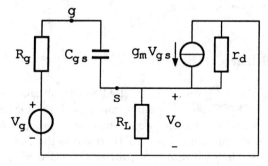

Fig. 5.15 *Equivalent circuit of the source follower.*

with $\tau_g = C_{gs}/g_m$. Hence, $z = -1/\tau_g$ and $p = -1/\tau_g(1 + R_g/R_L)$.

If $R_g \to 0$, the pole and the zero seem to cancel each other. This is a consequence of the assumption $g_m R_L \gg 1$. If this approximation is abandoned we find

$$p = -\frac{1}{\tau_g}(1 + 1/g_m R_L)$$

Emitter followers and source followers are frequently used as buffer stages. Because of their low output impedance they can withstand low-impedance loads. However, with capacitive loading an unexpected problem can occur. If R_L in Figure 5.14 is replaced by Z_L denoting the parallel connection of R_L and a load capacitance C_L, we find

$$\frac{V_o}{V_g} = \frac{Z_L(1 + g_m Z_\pi)}{R_g + r_b + Z_\pi + (1 + g_m Z_\pi)Z_L} \qquad (5.45)$$

If we now substitute $Z_L = R_L/(1 + s\tau_e)$ and $Z_\pi = r_\pi/(1 + s\tau_\pi)$, with $\tau_e = R_L C_L$, we find

$$\frac{V_o}{V_g} = \frac{R_L(1 + \beta_0 + s\tau_\pi)}{s^2 \tau_\pi \tau_e (R_g + r_b) + s[\tau_\pi (R_g + r_b + R_L) + \tau_e(R_g + r_b + r_\pi)] + r_\pi + (1 + \beta_0)R_L + R_g + r_b}$$

(5.46)

The transfer is characterized by a zero and two poles. A relevant question is whether these poles can form a complex pair. If this is so, the capacitive loading gives rise to a transient response with ringing and overshoot. Analysis of the denominator of eq. (5.46) reveals that such a situation is not at all improbable. Putting $\tau_\pi = n\tau_e$ we find that for a wide range of values of the ratio factor n the poles can be complex. The boundaries of this range depend on the values of the resistances R_g, R_L and r_π. For very small values of n, as well as for very large values, the poles will always be real. Due to the absence of r_π in its equivalent circuit, the source follower is less prone to the occurrence of complex poles. However, if it is driven from a high-impedance driving circuit, complex poles can occur. In practical circuits the occurrence of heavy overshoots due to capacitively loaded emitter followers can be surprising and quite annoying, the more since the effect is generally underestimated or even unknown.

5.7 Long-tailed pair

In Section 5.4, the long-tailed pair (LTP) was introduced as a unilateral configuration with CE(CS)-like properties. Since the LTP is essentially a cascade of an emitter follower and a CB-stage, we are now well equipped to take a look at its frequency-dependent behaviour. Figure 5.16(a) shows the configuration and Figure 5.16(b) the corresponding equivalent circuit. For the sake of simplicity the internal collector resistances r_{ce} have been omitted.

BASIC AMPLIFYING CIRCUIT CONFIGURATIONS   135

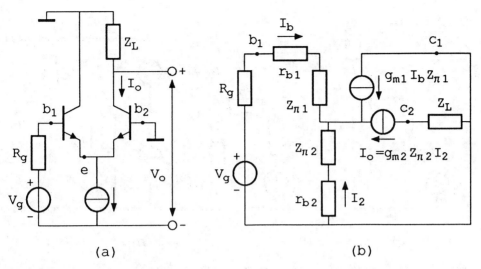

Fig. 5.16 *Long-tailed pair* (a) *and its equivalent circuit* (b).

Assuming $Z_{\pi 1} = Z_{\pi 2} = Z_\pi$, $r_{b1} = r_{b2} = r_b$ and $g_{m1} = g_{m2} = g_m$, we find the transadmittance of the circuit

$$Y_t = \frac{I_o}{V_g} = -\frac{g_m}{2 + g_m(R_g + 2r_b)(1 + s\tau_\pi)/\beta_0} \qquad (5.47)$$

The position of the pole is given by

$$p = -\frac{2r_\pi + R_g + 2r_b}{C_\pi r_\pi (R_g + 2r_b)}$$

It must be noted that with small values of R_g, the pole is so far away from the origin that possibly the pole introduced by Z_L becomes dominant.

For the corresponding circuit with FETs, the transadmittance is given by

$$Y_t = -\frac{g_m}{2 + sR_g C_{gs}} \qquad (5.48)$$

so that the position of the pole is given by $p = -2/R_g C_{gs}$.

5.8 Cascode circuit

The second combination circuit that is widely used is the so called cascode circuit. It can be looked upon as a cascade of a CE- and CB-configuration. Figure 5.17(a) shows the circuit and Figure 5.17(b) its equivalent circuit. Again, for simplicity the internal resistances r_{ce} are omitted.

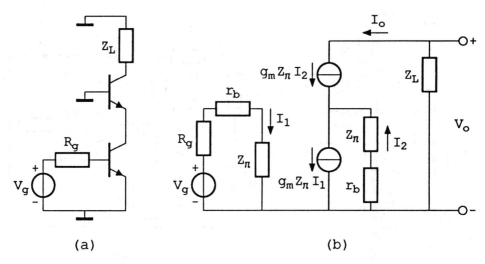

Fig. 5.17 Cascode circuit (a) and its equivalent circuit (b).

Assuming identical transistors and $\beta_0 \gg 1$, we obtain

$$Y_t = \frac{I_o}{V_g} = \frac{g_m}{1 + g_m(r_b + R_g)/\beta_0 + s^2(r_b + R_g)C_\pi^2/g_m + sC_\pi(r_b + R_g + 1/g_m)} \quad (5.49)$$

There are two poles: $p_1 \approx -g_m/C_\pi$ and $p_2 \approx -1/C_\pi(r_b + R_g)$.

If in a practical situation $r_b + R_g \gg 1/g_m$ holds, the pole p_2 will be the dominating one in Y_t. And, as already has been pointed out in the discussion of the CB-configuration, the pole introduced by Z_L can move into dominance if R_L is large.

The cascode connection is a very useful configuration. It is also as widely used in its bipolar version as in its FET-version. In the latter case, $p_1 = -g_m/C_{gs}$ and $p_2 = -1/C_{gs}R_g$.

5.9 Series stage

The next local feedback stage to be investigated is the series stage. Figure 5.18(a) shows this configuration; Figure 5.18(b) gives the equivalent circuit wherein $C_{b'c}$ and r_{ce} have been omitted, while r_b is considered to be incorporated in R_g.

The relevant transfer quantity of the series stage is its transadmittance. Assuming $\beta_0 \gg 1$, straightforward calculation yields

$$Y_t = \frac{I_o}{V_g} = \frac{1}{R_g/\beta_0 + 1/g_m + R_E + sC_\pi(R_g + R_E)/g_m} \quad (5.50)$$

The position of the pole is given by

$$p = -\frac{1/g_m + R_g/\beta_0 + R_E}{C_\pi(R_g + R_E)/g_m}$$

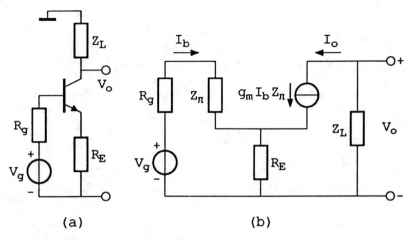

Fig. 5.18 Series stage (a) and its equivalent circuit (b).

If $R_E = 0$, the circuit turns into a CE-configuration. In comparison with the CE-configuration, the GB-product of the series stage, used as a voltage amplifier, is smaller. The reduction is stronger the larger is R_E. This can be compensated for by connecting a capacitance C_E in parallel with R_E. Replacing R_E by $Z_E = R_E/(1 + s\tau_E)$ with $\tau_E = C_E R_E$, eq. (5.50) reads

$$Y_t = \frac{1 + s\tau_E}{(R_g/\beta_0 + 1/g_m)(1 + s\tau_E) + R_E(1 + sC_\pi/g_m) + sR_g C_\pi(1 + s\tau_E)/g_m} \quad (5.51)$$

The transfer is characterized by a zero at $z = -1/\tau_E$ and two poles. If we choose $\tau_E = C_\pi/g_m$, one of the poles coincides with the zero, leaving one pole at

$$p = -\frac{1/g_m + R_g/\beta_0 + R_E}{C_\pi R_g/g_m} \quad (5.52)$$

In this case the GB-product equals that of the CE-stage. This special case is referred to as emitter compensation. It is a useful method for improving the bandwidth of a series stage without compromising low-frequency gain. However, if the emitter capacitor is larger than is needed for compensation, the zero is located to the right of the poles. As a consequence the transient response can exhibit overshoot, even in the case where the poles are located on the real axis. Moreover, depending on the ratio of the two relevant time constants τ_E and C_π/g_m, the poles can assume complex values, giving rise to ringing in the transient response. The adverse effect on the transient response of a zero to the right of the poles should not be underestimated, particularly if the signal involved does not tolerate even small overshoot, as for instance applies to video signals.

In the foregoing analysis the effect of $C_{b'c}$ was not taken into consideration. Since a series stage behaves as a current source, the load impedance will usually be of low value. In that case the effect of $C_{b'c}$ on signal transfer will be slight. If the

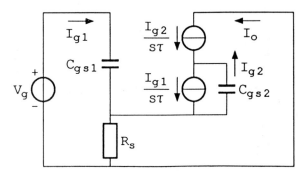

Fig. 5.19 *Simplified equivalent circuit of the cascoded FET-series stage.*

value of the load does not justify the neglect of $C_{b'c}$, the series stage can be cascoded, in the same way as is done with the CE-configuration. However, it should be noted that the cascoded series stage is prone to exhibit complex poles. Figure 5.19 shows the equivalent circuit of the cascoded series stage equipped with FETs. In order to keep the analysis simple, r_d is left out and the circuit is assumed to be driven from an ideal voltage source ($R_g \to 0$). Assuming $C_{gs1} = C_{gs2} = C$, and putting $C/g_m = \tau$, straightforward analysis yields

$$Y_t = \frac{I_o}{V_g} = \frac{g_m}{(1 + g_m R_s)(1 + s\tau) + g_m R_s s^2 \tau^2} \qquad (5.53)$$

There are two poles that form a complex pair if $3 g_m R_s > 1$. If $g_m R_s \gg 1$, the poles are given by

$$p_{1,2} = \frac{-1 \pm j\sqrt{3}}{2\tau}$$

And with $g_m R_s = 1$:

$$p_{1,2} = \frac{-1 \pm j}{\tau}$$

With $g_m R_s = \frac{1}{3}$, two coinciding real poles occur at $p = -2/\tau$.

5.10 Shunt stage

The last basic configuration to be studied is the shunt stage. Figure 5.20(a) shows the circuit and Figure 5.20(b) the equivalent circuit.

The relevant transfer function is the transimpedance Z_t. Due to the parallel feedback the input impedance of the stage is low. Neglect of the internal resistance R_g of the current source is thus justified. Then,

$$Z_t = \frac{V_o}{I_g} = -R_L \frac{\beta_0 R_f - r_\pi}{(R_L + R_f)(1 + s\tau_\pi) + r_\pi + \beta_0 R_L} \qquad (5.54)$$

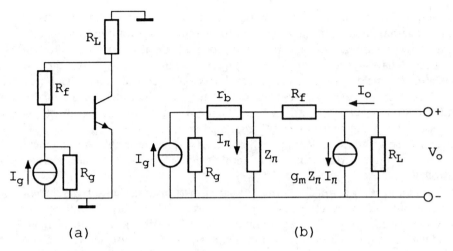

Fig. 5.20 *Shunt stage (a) and its equivalent circuit (b).*

If $R_L \gg R_f$ and $R_L \gg r_\pi$, the pole is located in the position that is by now familiar, $p = -g_m/C_\pi$.

In practice a purely resistive value of the feedback impedance is impossible, since R_f is in parallel with $C_{b'c}$. If we replace R_f by Z_f the transfer is characterized by a zero and two poles. The zero follows from

$$z = \frac{\beta_0 R_f - r_\pi}{r_\pi R_f C_{b'c}}$$

For any realistic value of R_f, the zero is located in the right half-plane, which is not surprising in the light of the results concerning the effect of $C_{b'c}$ in a CE-stage. A general expression for the pole positions turns out to be cumbersome. As could be expected, pole splitting occurs. In order to prevent excessive pole splitting, R_f must not be valued too large. Further complications arise if the load also contains an appreciable capacitive component. In cases where the capacitive character of Z_f and Z_L can be suspected of affecting the transfer of the stage appreciably, computer simulation should be resorted to for finding numerical data.

5.11 Compound active components

In the foregoing sections single-transistor amplifier stages have been studied. Some properties of such stages fall short of what they ideally should be. Among the diverse causes of these shortcomings is the modest gain that can be achieved with one transistor and the presence of capacitances in the equivalent circuit. The adverse effects produced by these imperfections can be counteracted in several ways. First it may be noted that single-transistor stages are rarely used in an isolated fashion. They nearly always form part of a larger aggregation, for instance a multistage amplifier. One possibility is to conquer the simultaneous effect of all individual imperfections in an overall approach. Several general methods that are applicable have been discussed in Chapter 4. But in some cases their use is precluded or it does not provide satisfactory results. In such cases local measures for improvement have to be considered. Some useful methods have already been indicated in preceding sections, such as the improvement of single-transistor stages by cascoding or replacing them by long-tailed pairs. These examples show that the judicious use of additional components can accomplish desirable improvements. In this section some further commonly employed configurations in this class will be introduced.

A fundamental drawback of the bipolar transistor is its finite base current and the concomitant finite input resistance. In cases where a large input resistance is required, using an FET is an obvious choice. But if the available technology precludes the use of an FET, a bipolar alternative must be looked for. A configuration that is in widespread use is the so-called Darlington pair, as shown in Figure 5.21. The combination can be looked upon as a single transistor with enhanced β. Since the base current of $T2$ equals the emitter current of $T1$, the compound current gain factor β_c equals $\beta_1\beta_2$. Further, $I_{E2} = \beta_{F2} I_{E1}$, hence $g_{m1} \approx g_{m2}/\beta_2$ (assuming $\beta_F \approx \beta$). Hence, the compound transconductance is given by $g_{mc} \approx \frac{1}{2} g_{m2}$. The input resistance of the pair, when used in the CE-configuration (Figure 5.21) is $R_{ic} \approx 2\beta_c/g_{m2}$. If a larger input resistance is desired, the Darlington pair can be used in the emitter-follower configuration (Figure 5.22).

In most present-day bipolar monolithic processes, the pnp-transistors are of the lateral variety. These transistors are of poor quality. They exhibit low β and the current they can handle is moderate. By combining such a transistor with an npn-transistor in the fashion depicted in Figure 5.23, a compound transistor is obtained that can be looked upon as an improved npn-transistor. It should be noted that the emitter of $T2$ acts as the collector of the compound pnp. For the compound transistor $\beta_c \approx \beta_p \beta_n$, where β_p and β_n denote the βs of the constituting transistors.

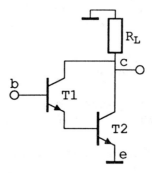

Fig. 5.21 *Darlington pair in CE-configuration.*

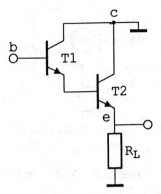

Fig. 5.22 *Darlington pair as emitter follower.*

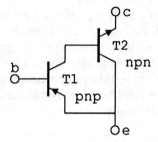

Fig. 5.23 *Compound pnp-transistor.*

The transconductance of the combination equals the transconductance of the npn. At first glance, the use of the emitter of $T2$ as the collector of the compound transistor implies a very low output resistance. But it should be noted that $T2$ is essentially current-driven, hence $R_{oc} \approx r_{ce\,1}/\beta_2$ And since $T1$ is operated on a relatively low bias current level, $r_{ce\,1}$ can take a reasonably high value.

In certain applications the use of FETs is preferred in view of their high input resistance. Moreover, when used as a source follower, the output resistance does not depend on the value of the impedance of the gate circuit (usually the driving source impedance). These are important advantages in comparison with the BJT. But a drawback in many applications is that small-geometry FETs exhibit relatively low transconductance. By combining an FET with a BJT in the way indicated in Figure 5.24, a compound component is obtained, combining to a large extent the best of the FET- and the BJT-worlds. The combination shown behaves as an n-channel FET. The transconductance of the combination is $\beta_2 g_{m1}$. The output resistance of the combination, i.e the internal resistance seen into the emitter of $T2$, is relatively low, since $R_o \approx r_{e2} + r_{d1}/\beta_2$. Particularly when used as a source follower this need not be a serious drawback. The combination is colloquially referred to as a 'superFET'.

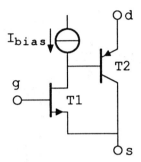

Fig. 5.24 *'Super FET'*.

In the foregoing several configurations were proposed consisting of two transistors but intended to operate as one transistor of enhanced quality. These configurations incorporate an interesting widening of the assortment of elementary building materials for amplifiers. However, it should be noted that, in essence, they remain two-transistor amplifier blocks and as a consequence their high-frequency behaviour is characterized by two dominant poles. They can therefore be the cause of excess phase shift with all the associated consequences. In critical situations, for instance if they are part of an amplifier with overall feedback, their high-frequency behaviour should be carefully analyzed.

5.12 Biasing of active devices

An active device that is employed as an amplifier of small signals must be appropriately biased. The d.c.-current and the d.c.-voltage in the absence of a driving signal constitute what is called the d.c.-operating point of the device. The relevant quantities are I_C and V_{CE} for the BJT and I_D and V_{DS} for the FET. We will first survey the considerations to be taken into account with regard to the selection of the operating point:

1. The bias current and voltage must be large enough to accommodate the a.c.-current and a.c.-voltage excursions generated by the driving signal. The momentary values of current and voltage are the superposition of the d.c.- and the a.c.-values and it is obvious that at any time current and voltage values should remain within the active region. Figure 5.25 shows the output characteristics of a bipolar transistor. With P as the operating point, the maximum voltage and current excursions are indicated. In addition, it must be noted that under any driving condition, current and voltage should never exceed their maximum allowable values as specified for a given transistor. The same applies for the power dissipation. The portion of the output characteristics (I_C versus V_{CE} for a BJT) encompassing the allowed values of current and voltage is commonly denoted as the 'safe operating area'.

BASIC AMPLIFYING CIRCUIT CONFIGURATIONS   143

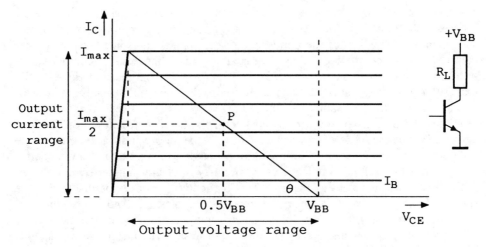

Fig. 5.25 *Biasing adapted to maximum driving conditions; cotan $\theta = R_L$.*

As for the collector dissipation P_C, note that it is maximum in the no-drive situation. This can easily be seen as follows. Let the collector be loaded by a resistor R_L. If the signal current is $i_c = \hat{i}_c \sin \omega t$, then $v_{ce} = -\hat{i}_c R_L \sin \omega t = -\hat{v}_{ce} \sin \omega t$. For the instantaneous value of i_C we can write $i_C = I_C + i_c$, and likewise $v_{CE} = V_{CE} + v_{ce}$. The collector dissipation P_C is

$$P_C = \frac{1}{T} \int_0^T (I_C + \hat{i}_c \sin \omega t)(V_{CE} - \hat{v}_{ce} \sin \omega t)\, dt = I_C V_{CE} - \tfrac{1}{2}\hat{i}_c \hat{v}_{ce}$$

This result shows that dissipation is maximum in the absence of a driving signal.
2. Current and voltage should not be taken larger than is necessary, in order to minimize the power delivered by the power supply and transistor dissipation.
3. The small-signal parameters of active devices depend on the d.c.-operating point and this can therefore be used as a means to adjust these parameters. One example of this possibility is in optimization of noise characteristics (see Chapter 8).

To establish a certain operating point, additional circuit elements are required. The biasing circuitry should not spoil the transfer characteristics of the amplifier. And it should be designed in such a way that the operating point does not shift due to variations in transistor parameters that can be expected to occur, for instance due to temperature changes or to parameter spread.

5.13 Biasing circuits

Proper biasing of a bipolar transistor implies fixation of I_c and V_{CE}. Theoretically, it would be possible to fix I_B instead of I_C, since $I_C = \beta_F I_B$. In practice the parameter β_F exhibits considerable spread and it is also temperature-dependent. This excludes indirect fixation of I_C through I_B. Since the 'output quantities' I_C and V_{CE} are not independent of the 'input quantities' I_B and V_{BE}, some kind of control mechanism must be provided. This can take the form of feedback or compensation or a combination of these methods. Figure 5.26 shows a generalized method for the fixation of I_C and V_{CE} of a CE-stage using feedback. The amplifier A is a high-gain differential amplifier: it amplifies the voltage difference between the two input terminals. Its output voltage (or current) controls the current source for the base current; in the case of an FET it controls V_{GS}. If the loop gain is large, the voltage difference between both input terminals of A tends to zero value. Consequently, $I_C = I_{ref}$ and $V_{CE} = V_{ref}$. Practical biasing circuitry is usually simpler than is shown in the formal arrangement of Figure 5.26. Frequently, it is also possible to put the signal circuitry to good use for the purpose of biasing. And particularly in monolithic circuits, instead of using feedback, compensation methods can be used. Moreover, it is often possible to combine the biasing circuits of several stages. Figure 5.27(a) shows a practical configuration of a shunt stage, where the signal feedback resistor is also used for biasing the stage. Here $I_C = I/(1 + 1/\beta_F)$ and $V_{CE} = V_{BE} + I_C R_F/\beta_F$. The strong dependence of V_{CE} on the ill-defined parameter β_F can be avoided by adding an additional current source as shown in Figure 5.27(b). Here,

$$I_C = \frac{I - I_B'}{1 + 1/\beta_F}$$

and

$$V_{CE} = V_{BE} + (I_B + I_B')R_F$$

By choosing $I_B' \gg I_B$, the dependence of V_{CE} on β_F is minimized.

With a series stage a comparable arrangement can be devised, as shown in Figure 5.28(a). Here $I_C = (V_{ref} - V_{BE})/R_E$ and $V_{CE} = V_{BB} - I_C R_E$. Figure 5.28(b) shows a

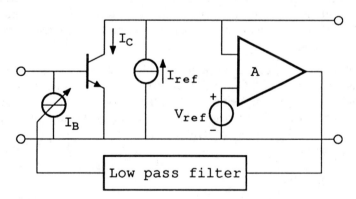

Fig. 5.26 *Generalized schematic for the realization of correct biasing through feedback.*

BASIC AMPLIFYING CIRCUIT CONFIGURATIONS 145

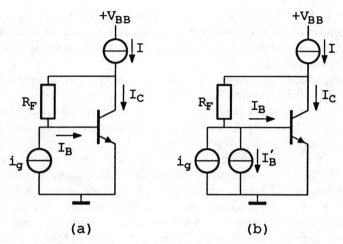

Fig. 5.27(a) *Simple biasing of a shunt stage.* (b) *Adding I'_B for better fixation of V_{CE}.*

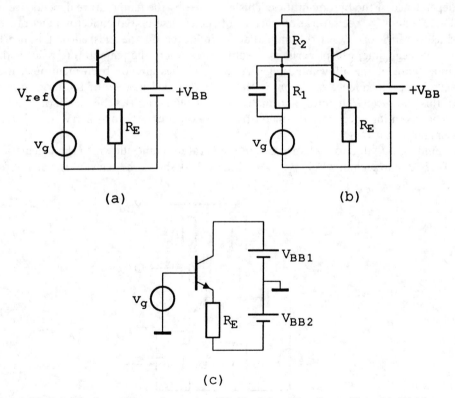

Fig. 5.28 (a) *Biasing of a series stage.* (b) *Deriving V_{ref} from V_{BB}.* (c) *Biasing of series stage when dual polarity power supply is available.*

simple method for obtaining V_{ref} from the power-supply voltage V_{BB}. In order to prevent attenuation of the input signal v_g, R_1 is bypassed by a capacitor. This capacitor is not needed in the case where a symmetrical power supply is used, as shown in Figure 5.28(c).

A CE-stage can be biased in the same way as a series stage by capacitively bypassing the feedback resistor R_E (Figure 5.29). In monolithic circuits this method cannot be used, but the availability of nearly identical transistors makes the use of compensation by balancing feasible. Biasing of symmetrical monolithic amplifier stages will be dealt with in the next chapter.

Biasing of FETs is accomplished in much the same way as with bipolar transistors. FETs of the normally-on (depletion) variety must be biased such that the gate is negative with respect to the source. In the case of a single supply voltage this implies that the source should be on a positive voltage since the gate voltage cannot be below ground potential. A simple method is the use of a source resistor R_S. If $V_G = 0$, then $V_{GS} = I_S R_S$ (Figure 5.30a). Given R_S and the I–V characteristic of the FET, V_{GS} as well as I_D are fixated (Figure 5.30b).

In any design of biasing circuitry, attention must be given to the temperature dependence of the bias quantities. This is caused by the temperature dependence of V_{BE}, β_F and I_{co}, the saturation current of the reverse-biased collector junction. In silicon transistors I_{co} is so small that its influence on the bias current is nearly always negligible, except perhaps in some cases where the temperature reaches the upper limit of the allowable range. As to β_F: it has already been mentioned that significant dependence of I_C on β_F should be avoided in view of the large spread of this parameter. Therefore, normally only the temperature dependence of V_{BE} has to be taken into account. In silicon transistors it amounts to $-2\,\text{mV}\,°\text{C}^{-1}$ for I_C remaining constant.

Finding ΔI_C due to a change ΔT in temperature comes down to the calculation of ΔI_C due to a change ΔV_{BE} in V_{BE}. Figure 5.31 shows the relevant schematic for

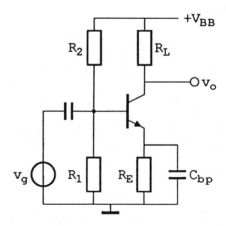

Fig. 5.29 *Simple method for biasing a CE-stage.*

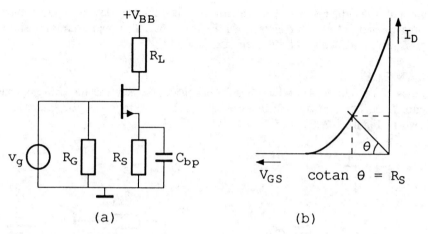

Fig. 5.30(a) *Simple method for biasing a depletion type FET.* (b) *I_D and V_{GS} are related through R_S.*

the calculation of ΔI_C. It is generally valid since, for proper operation of a BJT, d.c.-current must flow into all three transistor terminals. Though possibly superfluous, note that only d.c.-paths are relevant in constituting the schematic according to Figure 5.31. By inspection of this schematic we can write the following equation:

$$I_C = \frac{V_{BB} - V_{BE}}{R_B/\beta_F + R_E(1 + 1/\beta_F)} \approx \frac{V_{BB} - V_{BE}}{R_E + R_B/\beta_F}$$

Hence,

$$\Delta I_C = \frac{-\Delta V_{BE}}{R_E + R_B/\beta_F}$$

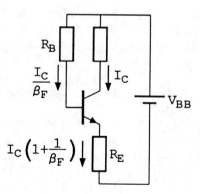

Fig. 5.31 *General schematic for the calculation of ΔI_C.*

Note that in R_B the extrinsic base resistance r_b is included. Any biasing circuitry can be rearranged into the schematic of Figure 5.31. The two examples shown in Figures 5.32 and 5.33 will elucidate this point.

EXAMPLE 1
Figure 5.32(a) shows the widely used biasing circuit that we have already met in Figure 5.29. By applying Thévenin's theorem the circuit can be rearranged, as shown in Figure 5.32(b). Hence $R_B = R_1 \| R_2$.

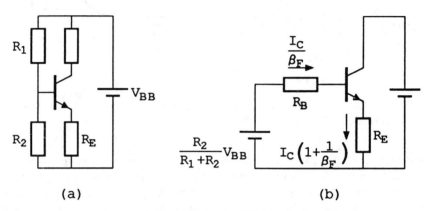

(a)   (b)

Fig. 5.32(a) *A practical biasing circuit.* (b) *Equivalent circuit.*

EXAMPLE 2
Figure 5.33(a) shows another commonly used biasing arrangement. The circuit can be redrawn as shown in Figure 5.33(b). By shifting the location of R_1, the equivalence with Figure 5.31 is made more explicit. Hence, $R_E \rightarrow R_1$ and $R_B \rightarrow R_2$. Note that the location of the 'ground point' is fully irrelevant for fixing the d.c. currents.

The mechanism behind the stabilization qualities of the circuit arrangement of Figure 5.31 is, of course, the d.c.-feedback provided by the presence of R_E and R_B. Any method of d.c.-feedback is in principle useful for stabilization of d.c.-quantities. If the feedback path must be inactive for the signals to be handled by the circuit, the resistances can be capacitively bypassed. In monolithic implementations this method is usually unattractive, but here compensation methods are readily applicable.

With regard to FET-circuits, as far as the stabilization methods are concerned, the same principles are applicable. Again d.c.-feedback and compensation methods provide the appropriate tools. In general, with FETs, temperature dependence of bias quantities is less of a problem than with BJTs. The quantity that is chiefly responsible for the temperature dependence of I_D is the electron (or hole) mobility μ. A fortunate circumstance is that the temperature coefficient of μ is negative. This

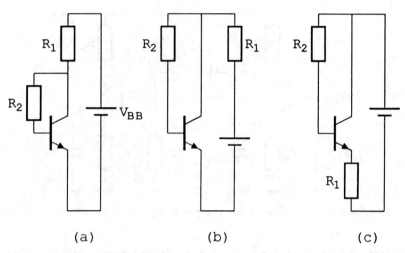

Fig. 5.33(a) *Another practical biasing arrangement.* (b) *and* (c) *Redrawn versions of Figure 5.33(a).*

causes I_D to decrease when T rises. Therefore, possible overheating effects are counteracted. This is in contrast to BJT behaviour. When T rises, so does I_C, and therefore usually the dissipation also. Increasing dissipation causes a further rise of T, so that thermal positive feedback may occur, which is potentially hazardous.

5.14 Interstage level shifts

In a multistage amplifier each stage has to be properly biased. When the output of a stage is connected to the input of the succeeding stage, the d.c.-level will be transferred along with the signal, unless measures are taken to separate the respective d.c.-levels. In an implementation using discrete components this can be simply accomplished by capacitive coupling (Figure 5.34), provided that the additional pole and the zero in $s = 0$ will not affect the transfer of relevant low-frequency spectral components of the signal. However, most present-day implementation technologies, particularly monolithic technology, preclude the use of large capacitors. Hence, as a rule d.c.-coupling is indicated. This introduces two problems. First, if the desired d.c.-levels at both sides of the coupling are different, a d.c.-*level shift* must be provided. Second, measures have to be taken to prevent drift effects. 'Drift' is the term used to describe slow changes of parameters, for instance due to temperature effects, ageing or power-supply variations.

Level shift can be accomplished in several ways. Figure 5.35 shows some obvious methods. In Figure 5.35(a) a resistive voltage divider is used. The disadvantage of this solution is the ensuing signal attenuation by a factor of $R_2/(R_1 + R_2)$. As R_2 is loaded capacitively (C_2) by the input of the next stage, an additional pole is

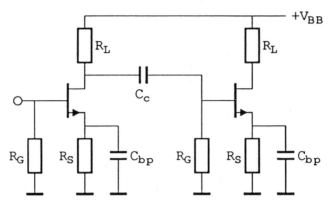

Fig. 5.34 *Interstage coupling using a capacitor for separating biasing levels.*

introduced affecting high-frequency response. Compensation of this effect can be achieved by bridging R_1 with a (usually small) capacitance C_1 so that $R_1C_1 = R_2C_2$. In Figure 5.35(b), signal attenuation is avoided by replacing R_1 by a Zener diode. In monolithic circuits the use of a Zener diode, though possible, is rarely practical. The only pn-junction that is available for this purpose is the emitter–base junction. In most bipolar processes the avalanche breakdown voltage of this junction is of the order of 5–6 V, which is usually too high to be practical. As an alternative, a string of forward-biased diodes can be used (Figure 5.35c). This solution has the disadvantage that all diodes have to be formed in separate islands; moreover the substrate capacitances associated with these islands can affect high-frequency signal response. If more than two or three diodes are involved, the circuit shown in Figure 5.35(d) is a better solution. The internal resistance of $T2$ is easily found to be $R_t = r_e(1 + R_1/R_2)$. The signal attenuation is therefore $R_3/(R_t + R_3)$. As an alternative to lowering the incremental resistance of R_1 in Figure 5.35(a), as shown in the circuits of Figure 5.35(b,c,d) the resistor R_2 in Figure 5.35(a) can be replaced by an element with a high incremental resistance operating at low d.c.-voltage, i.e. a current source. Figure 5.35(e) shows this alternative. Of course this is only feasible if the next stage possesses high input impedance.

If pnp-transistors of acceptable quality are available, these provide an obvious implement for achieving level shifts, thanks to their opposite biasing polarity. In most monolithic processes the pnp-transistors are of the lateral variety. These transistors have poor signal transfer qualities. In particular, their values of β and f_T are poor. These poor properties show up least when the transistors are used in the CB-configuration. Figure 5.36(a) shows a level-shift arrangement operating in this way. Frequently, the low input resistance of a CB-stage will be unwelcome. If so, an npn-emitter follower can be used as a buffer, as shown in Figure 5.36(b). In the given form, the circuit is not very suitable for monolithic implementation. In the next chapter a biasing method that is suitable for monolithic implementation will be dealt with.

BASIC AMPLIFYING CIRCUIT CONFIGURATIONS 151

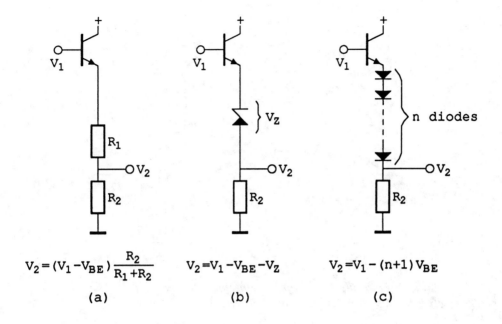
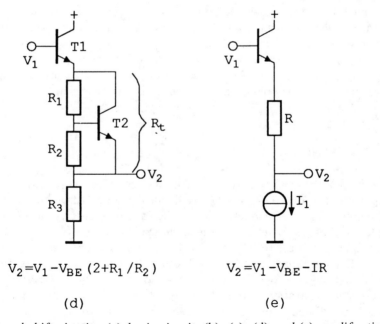

Fig. 5.35 *Level-shift circuits:* (a) *basic circuit;* (b), (c), (d) *and* (e) *modifications suitable for IC-implementation.*

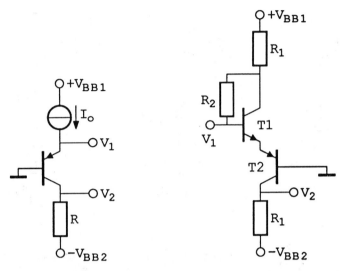

Fig. 5.36(a) *Using a pnp-transistor for obtaining a level shift.* (b) *Driving the CB-level-shift stage from an emitter follower.*

In multistage amplifiers it is possible to design a biasing circuit that is common to more than one stage. In order to ascertain stability of the biasing quantities, d.c.-feedback has to be provided. This feedback will not usually be optimal from the point of view of signal transfer. The d.c.-feedback can be made ineffective for signal frequencies by the judicious employment of bypassing capacitors. In monolithic implementations such methods are in most cases not attractive. However, here compensation methods can provide excellent solutions for biasing arrangements. This point will be taken up in the next chapter.

5.15 Drift effects

The second problem that is evoked by d.c.-coupling of subsequent stages is the suppression of drift effects. The pre-eminent method for achieving this is the use of compensation techniques. The general idea of such methods is the intentional generation of a drift effect of the same magnitude but of opposite polarity with respect to the effect to be suppressed. By judicious arrangement of the circuitry, both effects are made to cancel each other. Of course, the arrangement should be such that regular signals do not take part in the compensation. The most consequent realization of this idea involves the application of two identical circuits in balance. Elaboration of this concept leads to a basic circuit of great versatility and one that is highly suited to monolithic implementation: the differential pair. In the next chapter various embodiments of this approach will be discussed.

Further reading

Most books on electronics contain material concerning basic transistor configurations and on biasing methods and circuits. Here, only a few modern textbooks are mentioned.

Colclaser, R. A., Neamen, D. A and Hawkins, C. F. (1984) *Electronic Circuit Analysis*, Wiley, New York.
Gray, P. R. and Meyer, R. G. (1984) *Analysis and Design of Analog Integrated Circuits*, Wiley, New York.
Grebene, A. B. (1984) *Bipolar and MOS Analog Integrated Circuit Design*, Wiley, New York.
Smith, R. J. (1980) *Electronics: Circuits and Devices*, Wiley, New York.

6 Differential amplifiers and operational amplifiers

6.1 Introduction: differential-mode and common-mode signals

In the preceding chapter, in the context of d.c.-coupling between stages, the point was raised as to how to cope with the effects of drift. First, let a simple example illustrate the seriousness of the problem. Figure 6.1 shows a combination of a CE-stage and an emitter follower. The CE-stage ($T1$) is assumed to be biased at $I_C = 2.5$ mA, while the emitter voltage is assumed to be 1 V. Since $V_{BB} = 9$ V, the collector voltage V_C of $T1$ is 4 V. Neglecting the base current of $T2$, and assuming $V_{BE(T2)} = 0.7$ V, the emitter of $T2$ is at 3.3 V and $I_{C(T2)} = 3.3$ mA. Let us now assume that the temperature changes by $\Delta T = 10\,°C$. As was found previously, V_{BE} of a transistor changes by an amount of $2\,\text{mV}\,°C^{-1}$. At $I_C = 2.5$ mA, the transconductance of $T1$ is $100\,\text{mA V}^{-1}$, hence, the change in I_C is $\Delta I_C = 2$ mA, so that $\Delta V_C = 4$ V and $V_C = 0$ V. Hence, $T1$ is in saturation and $T2$ is in cut-off. Obviously, even a moderate temperature shift causes total disruption of the circuit function!

It will be clear that something has to be done to prevent this sensitivity to an ordinary effect such as a slight change in temperature.

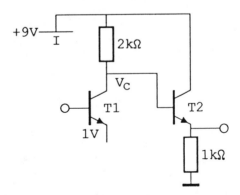

Fig. 6.1 *A simple example illustrating the effect of temperature drift.*

An obvious method is to use d.c-feedback. This could be accomplished by inserting an emitter resistor R_E in $T1$. For convenience we will assume that the resistor is connected to a negative supply voltage such that the emitter voltage remains at 1 V. If, for instance, the resistor is 2 kΩ, the effective g_m of $T1$ is 0.5 mA V^{-1}, hence with $\Delta T = 10\,°\mathrm{C}$, $\Delta I_C = 10\,\mu\mathrm{A}$ and $\Delta V_C = 20$ mV. However, to maintain the a.c.-transconductance at its previous value, R_E must be capacitively bypassed by a large capacitor, and such a measure is alien to monolithic circuit design.

As an alternative to feedback we can choose a compensation method. Figure 6.2 shows a balanced circuit consisting of two identical CE-stages. For simplicity, we assume that the transistors are biased by fixing their base currents with the aid of the resistors R_B. This is a very poor biasing method that will never be used in practice, but for the moment this is not considered. The input signal is applied *between* the bases of $T1$ and $T2$. For the time being we assume that no other signals are present at the bases of the transistors. If we neglect the base resistances r_b, the signal current in the emitters of $T1$ and $T2$ is $v_i/2r_e$, since both emitter junctions are in series. But of course $i_{e2} = -i_{e1}$, and hence $v_{c1} = -i_e R_L$, whereas $v_{c2} = +i_e R_L$. If we now take $v_{c1} - v_{c2}$ as the output signal, we have $v_o = -g_e R_L v_i$. The voltage transfer of the pair equals that of a single CE-stage. If both transistors undergo the same change in temperature, both collector voltages change by the same amount, letting $v_o = v_{c1} - v_{c2}$ undisturbed. However, the change of both collector voltages still disturbs the d.c.-level of the output.

The signal that is applied between both bases is called a *differential-mode signal* (DM-signal). The internally generated voltage due to the shift ΔT of the ambient temperature is equally present on both sides of the circuit. It is therefore called a *common-mode signal* (CM-signal). It may be noted also that an externally applied common-mode signal does not affect the DM-output signal.

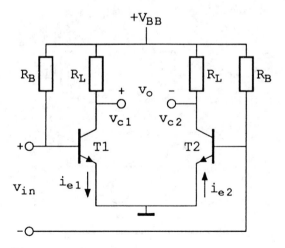

Fig. 6.2 *Balanced circuit.*

At this stage, the remaining problem is that CM-signals still disturb the output d.c.-level. To combat this very undesirable effect we must look for a method that prevents the transfer of CM-signals to the output terminals. The unimpeded transfer of these signals is caused by the fact that they still give rise to signal currents in the individual transistors. However, if we isolate the emitters from the common ground, then no CM-current can flow because for CM-signals both bases are at the same potential. The isolation can be accomplished in several ways but by far the most simple method is to use a current source for this purpose. A true current source has infinite (differential) impedance and hence is an ideal isolator. Figure 6.3 shows the circuit obtained in this way.

Due to the isolation provided by the current source, CM-signals are not transferred. Their only effect is that they raise the base and emitter potentials. Hence, if large CM-signals must be taken into account, the collector d.c.-voltages should be sufficiently high to prevent V_C from reaching the saturation level. The circuit we have devised is called a *differential amplifier*, because it transfers only differential signals (DM-signals) while rejecting CM-signals, irrespective of whether internally generated or externally applied.

In practice, CM-signals will not be completely suppressed. Practical current sources have finite internal impedance. Moreover, perfect symmetry cannot be achieved. Let, for instance, the only deviation from perfect symmetry be a difference ΔR in the load resistors R_L. If, due to the finite internal impedance of I_o, common-mode signal currents $i_{c1} = i_{c2} = i_c$ occur, then a DM-output voltage $v_o = i_c \Delta R$ will show up between the collectors. This is called CM-to-DM conversion. When this occurs, a DM-signal is generated which in subsequent stages will be handled as a DM-signal and hence will further enjoy the privileged transfer of such signals!

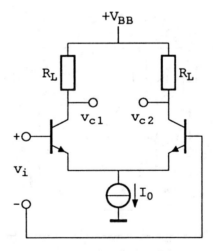

Fig. 6.3 *Differential amplifier.*

DIFFERENTIAL AMPLIFIERS AND OPERATIONAL AMPLIFIERS 157

To determine the properties of a differential amplifier it makes sense to introduce two quantities. What we are primarily interested in is the amount of rejection of CM-signals. The *common-mode rejection ratio* (CMRR) is defined as the ratio of a DM-voltage and a CM-voltage at the input producing an equal (DM) output signal. It will be clear that with ideal isolation the CMRR is bound to be infinite, even if there is no perfect symmetry. Likewise, if the isolation is imperfect, but there is perfect symmetry, the CMRR will also be infinite. Since the quality of the isolation is very important, it is worthy of qualification by a special quality parameter. The effect of the isolation is discrimination of the CM-signal. The discrimination factor is defined as

$$F = \frac{DM\text{-}amplification}{Single\text{-}sided\ CM\text{-}amplification} = \frac{(v_o/v_i)DM}{(v_c/v_i)CM} \tag{6.1}$$

Let us now see how the various amplification factors can be conveniently calculated. The differential pair consists of two identical halves. We can therefore split the circuit in a left half-circuit and a right half-circuit. Since the DM-signal does not produce a signal between the emitters and the common ground, the impedance of the isolating current source is inconsequential for the DM-signal. Hence, the half-circuit for the DM-signal reduces to that shown in Figure 6.4(a). Neglecting r_{ce}, we can write for the low-frequency transfer

$$A_v = \frac{v_c}{v_{dm}} = \frac{-1}{2} \frac{R_L}{r_e + r_b/\beta_0}$$

Taking into account that for the other half-circuit the driving signal is $-\frac{1}{2}v_{dm}$, we obtain for the voltage gain of the full differential pair

$$A_{v\ dm} = -\frac{R_L}{r_e + r_b/\beta_0} \tag{6.2}$$

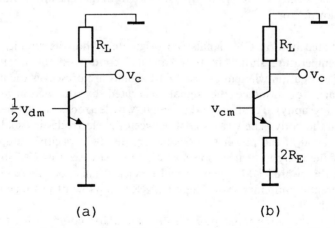

Fig. 6.4(a) *DM-signal half-circuit.* (b) *CM-signal half-circuit.*

For the transfer of the CM-signal, the internal impedance R_E of the isolating current source is, of course, absolutely essential. If we split the circuit, R_E splits up into two resistances $2R_E$ in parallel. The relevant half-circuit is shown in Figure 6.4(b). Hence,

$$A_{v\,cm} = -\frac{R_L}{2R_E + r_e + r_b/\beta_0} \approx -\frac{R_L}{2R_E} \qquad (6.3)$$

Substituting eqs. (6.2) and (6.3) in eq. (6.1) we find

$$F = \frac{2R_E}{r_e + r_b/\beta_0} \qquad (6.4)$$

For the CMRR to be finite some asymmetry must be present. As an example, let us assume that the load resistors differ by an amount ΔR_L. The CM-signal will then give rise to a DM output signal $v_o = v_{cm}\,\Delta R_L/2R_E$. Hence, $v_o/v_{cm} = \Delta R_L/2R_E$. Since for the DM-signal $v_o/v_{dm} = R_L/(r_e + r_b/\beta_0)$, we have

$$CMRR = \frac{R_L}{\Delta R_L} \cdot \frac{2R_E}{r_e + r_b/\beta_0} \qquad (6.5)$$

As should be the case, CMRR $\to \infty$ if $\Delta R_L \to 0$ and if $R_E \to \infty$.

EXAMPLE

Let $R_E = 10\,\mathrm{M}\Omega$ and $\Delta R_L/R_L = 0.05$. Further, let $r_b = 100\,\Omega$, $\beta_0 = 200$ and $I_C = 0.1\,\mathrm{mA}$, hence $r_e = 250\,\Omega$.

Substituting these values in eq. (6.5) we find $CMRR = 16.10^4$. Usually CMRR is given in dB. Doing this in our example we find $CMRR = 20\log 16.10^4 = 110\,\mathrm{dB}$. In practice such a large value of CMRR is not easily obtained since, apart from the assumed asymmetry in the values of R_L, some asymmetry in the transistors will also be present. And, of course, at frequencies where the parasitic capacitances across R_E cannot be neglected, CMRR will suffer from the ensuing reduction in the discrimination factor.

As was stated before, CM-signals can originate within the circuit, for instance due to a temperature shift, but they can also come with the input signal. An example of a situation wherein external CM-signals are present is the measurement of biosignals, for instance the signal generated by the action of the heart (electrocardiography). To measure this signal two electrodes are applied at suitable locations on the body. The weak signal between the electrodes is usually corrupted by a large signal at the power-line frequency, due to capacitive and/or inductive coupling of the body to nearby power lines. In most cases this CM-signal is much larger than the desired DM-signal. It will be clear that under these circumstances only a differential amplifier with a high CMRR is capable of producing the correct output signal.

Not only is the differential pair ideally suited for coping with the effects of unwanted CM-signals, it also offers an extremely simple possibility for biasing. In the absence of a DM-signal, both bases are at the same potential, hence $V_{BE1} = V_{BE2}$ and $I_{E1} = I_{E2} = \frac{1}{2}I_o$, irrespective of any CM-signals. Hence, for the biasing current to be well-defined we have only to take care that I_o is well-defined. Methods for the generation of accurate currents will be dealt with in a later section. With regard to the collector bias voltage V_C: if I_o is given, then $V_C = V_{BB} - \frac{1}{2}I_o R_L$, hence accuracy of V_C comes along with that of I_o.

The attractive properties of the differential pair have long been recognized. However, in the era of discrete-component electronics, these properties could not be satisfactorily exploited. Essential for the proper operation of the circuit is perfect, or at least nearly perfect, symmetry and this can hardly be accomplished in a discrete-component implementation. In a batch of discrete components no two individual specimens will be identical, let alone samples taken from different batches. Moreover, ideal thermal coupling of the two transistors cannot be accomplished. On the contrary, in a monolithic implementation all components are made in the same substrate at the same time, and by choosing a judicious layout of the circuit, thermal matching is easily achieved. The differential pair is the backbone of analogue monolithic electronics. Its great versatility will become apparent in subsequent sections and chapters. While in the foregoing the bipolar version has served as a vehicle for the introduction of the circuit, it will be clear that MOS-versions are equally practicable.

6.2 Current sources

In the preceding section it was found that both for adequate CM-suppression and for correct biasing of a differential pair a current source is indispensable. For rendering both services the current source should possess high impedance and provide accurate bias current. As a matter of fact, current sources play an important part in analogue integrated circuitry and therefore there is ample reason to dwell upon their realization. The simplest implementation is a high-valued resistor. Needless to say, a large resistor makes a very poor current source. The d.c. operating voltage is $I_o R$, thus depending on I_o as well as on R. In the era of discrete-component electronics resort was sometimes taken to this implementation for economic reasons. Much better properties are obtained by using a transistor operating in its active region (Figure 6.5).

For the collector impedance of the bipolar transistor we find by straightforward calculation, in the frequency region where transistor capacitances can be neglected,

$$R_o = r_{ce}\left(1 + \frac{\beta_0 R_E}{r_b + r_\pi + R_E}\right) + R_E\left(1 - \frac{R_E}{r_b + r_\pi + R_E}\right) \quad (6.6)$$

Hence if $R_E \to 0$, $R_o = r_{ce}$ and if $R_E \gg r_b + r_\pi$, $R_o \to r_{ce}(1 + \beta_0)$. Further, $r_{ce} = V_A/I_C$, where V_A denotes the Early-voltage. For example, if $V_A = 100$ V and $I_C = 0.1$ mA, then $r_{ce} = 1$ MΩ. And with $\beta_0 = 100$, then $\beta_0 r_{ce} = 100$ MΩ. Doing the

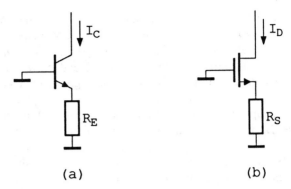

Fig. 6.5 *Generalized one-transistor current source with BJT (a) and with MOST (b).*

same exercise for a MOST (Figure 6.5b), we find

$$R_o = (1 + g_m R_S) r_d \qquad (6.7)$$

With $R_S \to 0$, $R_o = r_d$ and with $R_S \to \infty$, $R_o \to \infty$. Though in general r_d for a MOST will be much smaller than for a BJT, except for unduly large MOSTs, source degeneration can yield higher low-frequency output impedance than does emitter degeneration. But in general the value of R_o obtained with a BJT with sufficient emitter degeneration will be large enough to be useful practically.

In a monolithic implementation the use of large resistors (R_E, R_S) is impractical. Cascoding is a useful, and even preferable, alternative. In the circuit of Figure 6.6(a), the output resistance of $T1$ (point A) is at least r_{ce}, hence R_o of $T2$ (point B) approaches $\beta_0 r_{ce}$. The desired current is obtained by choosing R_1 and R_2 properly. The diode D is inserted for compensation of the temperature dependence of V_{BE}. The voltage V_{RE} across R_E equals the voltage across R_2, hence $V_{RE} = V_{BB2} R_2 / (R_1 + R_2)$ and $I_E = V_{RE}/R_E$. In the MOST-version of Figure 6.6(b), the output resistance of $T1$ is $r_d(1 + g_m R_S)$ and at point B it amounts to $r_d[1 + g_m r_d(1 + g_m R_S)]$. If, for instance, $r_d = 100$ kΩ, $g_m = 0.2$ mA V^{-1} and $R_S = 5$ kΩ, then $R_o \approx 4$ MΩ. However, it should be noted that this value is only correct for low frequencies, where the effect of capacitances, notably the substrate capacitance, can be neglected.

Though the use of resistors of moderate value, such as R_E and R_S is not precluded in monolithic circuits, their use can be avoided. Figure 6.7 shows the prototype of a current source that has found widespread use in analogue IC-electronics. The configuration operates as a current-driven current source. The driving current is fed to $T1$, which is diode-connected. From $V_{BE(T1)} = V_{BE(T2)}$

$$V_T \ln \frac{I_{C1}}{I_{s1}} = V_T \ln \frac{I_{C2}}{I_{s2}} \qquad (6.8)$$

DIFFERENTIAL AMPLIFIERS AND OPERATIONAL AMPLIFIERS 161

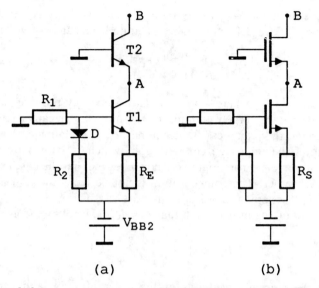

Fig. 6.6 *Cascoded current sources* (a) *with BJTs;* (b) *with MOSTs.*

follows, where V_T is the thermal voltage kT/q and I_s is the emitter junction saturation current. Remember that I_s is proportional to the emitter junction area.

If we neglect base currents, $I_2 = I_1(I_{s2}/I_{s1})$ and if $T1$ and $T2$ are identical transistors, it follows that $I_1 = I_2$. If base currents I_B are taken into account, then $I_1 = I_{C1} + 2I_B$ and with $I_B = I_C/\beta_F$,

$$I_2 = I_1\left(1 - \frac{2}{\beta_F + 2}\right) \tag{6.9}$$

To a first approximation, the current I_1 is 'copied' by the circuit, hence its commonly used name of 'current mirror'.

In its basic form, the circuit is afflicted with two imperfections. First, copying is incorrect by the factor between brackets in the right-hand side of eq. (6.9). And, second, the output resistance is moderate, since $R_o = r_{ce}$. Moreover, if the

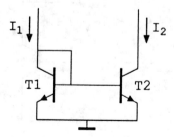

Fig. 6.7 *Basic current mirror.*

transistors have different V_{CB}, the Early-effect causes a further deviation of true mirror operation. If we delete the effect of finite base current, we find

$$\frac{I_2}{I_1} = \frac{1 + V_{CE(T2)}/V_A}{1 + V_{CE(T1)}/V_A} \qquad (6.10)$$

Of course $V_{CB(T1)} = 0$ V, whereas $V_{CE(T2)}$ can assume any positive value.

These effects can be largely diminished by applying negative feedback. In a monolithic circuit this can best be accomplished by adding a third transistor. Figure 6.8 shows the improved configuration, which is known by the name of 'Wilson current mirror'. From Figure 6.8 it can be seen that the base currents nearly compensate each other. The compensation is not perfect, since I_{B3} is slightly less than I_{B1} and I_{B2}, assuming identical transistors. Assuming equal values of β_F for all transistors and taking into account that $I_B = I_C/\beta_F$, it follows from $I_{C1} = I_{C2}$ that

$$I_2 = I_1 \left(1 - \frac{2}{\beta_F^2 + 2\beta_F + 2}\right) \qquad (6.11)$$

The error in the current transfer is of the order of $1/\beta_F^2$, and hence has been improved by the factor β_F. To eliminate the effect of unequal values of V_{CB} a diode can be added, as shown in Figure 6.9.

In order to calculate the input resistance of the circuit reference is made to Figure 6.10. Since T2 and T3 carry equal currents, a voltage v at the base of T3 produces $\frac{1}{2}v$ at its emitter. Hence $i_{c(T1)} = \frac{1}{2}g_e v$ and $R_i = v/i = 2r_e$. The output resistance is found to be $\approx \beta_0 r_{ce}$, which could be expected in view of the small-signal loop gain β_0.

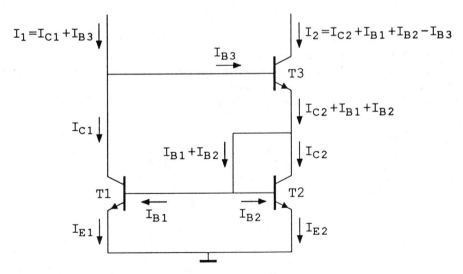

Fig. 6.8 *Improved current mirror.*

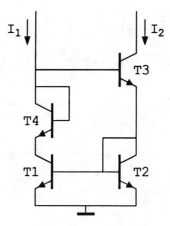

Fig. 6.9 *Matching $V_{CB(T1)}$ and $V_{CB(T2)}$.*

It is frequently desired that the output current I_2 of a current mirror is different from the input current. Such current rating is easily obtained by using different emitter-junction areas in $T1$ and $T2$. This is called 'emitter scaling'. If $I_{s2} = cI_{s1}$, then $I_2 = cI_1$ also. However, if emitter scaling is applied to the circuit of Figure 6.8, the compensation of base currents is no longer effective. By considering that base currents scale along with collector currents, it can easily be seen that eq. (6.11) has to be replaced by

$$\frac{I_2}{I_1} = c \left[1 - \frac{\beta_F(c - 1/c) + 1 + c}{\beta_F^2 + (\beta_F + 1)(1 + c)} \right] \tag{6.12}$$

If the value of c differs significantly from 1, the transfer error resumes to be of the order of $1/\beta_F$. Also, significant differences in the values of β_F cause the error term to be of the order $1/\beta_F$ instead of $1/\beta_F^2$.

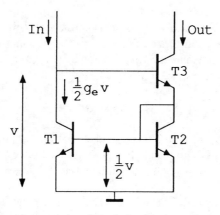

Fig. 6.10 *Schematic for determining R_i of the improved current mirror.*

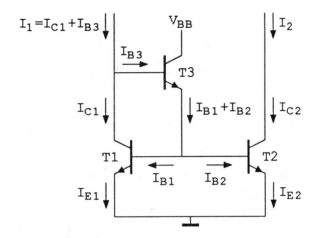

Fig. 6.11 *Current mirror with external supply of the base currents.*

An alternative method for compensation of the base current is to supply these currents externally, as shown in Figure 6.11. In this configuration the base currents I_{B1} and I_{B2} are supplied by the emitter follower $T3$. Inspecting Figure 6.11 we find, assuming again that $I_{s2} = cI_{s1}$,

$$\frac{I_2}{I_1} = c \left[1 - \frac{1+c}{1 + \beta_F \beta_{F3} + c} \right] \qquad (6.13)$$

where β_{F3} denotes I_{C3}/I_{B3}. Since the current carried by $T3$ is much smaller than that of $T1$ and $T2$, usually $\beta_{F3} \neq \beta_{F1,2}$. By virtue of the external supply of base currents, the value of c now has only a minor effect on the error term. However, this configuration also has a disadvantage in that its output resistance is still r_{ce}. This is, of course, a consequence of $T2$ not being part of the feedback loop. The output resistance can be improved by inserting emitter resistors in $T1$ and $T2$ or by cascoding, as shown in Figure 6.12.

If a large value of the ratio I_2/I_1, or alternatively I_1/I_2, is desired, the application of emitter scaling can be impractical in view of the silicon area involved. In this case an alternative approach may be taken, shown in Figure 6.13. Assuming identical transistors and putting $V_{BE1} - V_{BE2} = \Delta V_{BE}$ we have

$$I_2 R = \Delta V_{BE} = V_T \ln \frac{I_1}{I_2} \qquad (6.14)$$

By way of example: If $I_1 = 100\ \mu\text{A}$, and $I_2 = 0.1\ I_1$, then $R = 6\ \text{k}\Omega$. Thanks to R the output resistance is enhanced. But it should be noted that current transfer in this configuration is temperature-dependent due to the asymmetry introduced by the insertion of R.

Current mirrors using FETs can be devised in much the same way as their bipolar counterparts. Of course, only FETs of the 'normally-off' (enhancement) variety

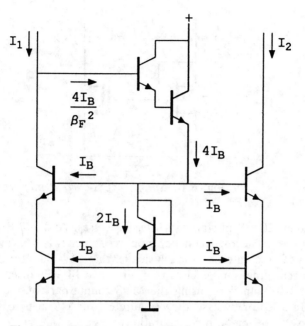

Fig. 6.12 *Enhancement of R_o by cascoding.*

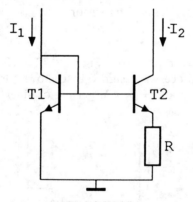

Fig. 6.13 *Current-transforming current mirror.*

qualify for the job, since 'normally-on' types cannot be biased properly in this application. Hence JFETs are ruled out. With 'normally-off' MOSTs, all configurations discussed before can be realized. Figure 6.14 shows a simple two-transistor current mirror. A distinct advantage is the absence of gate currents and the ensuing transfer error. On the other hand, it should be noted that in general the deviations from perfect symmetry in MOS-circuits are considerably larger than in bipolar circuits. The difference ΔV_t in the threshold voltage of an MOS-pair can

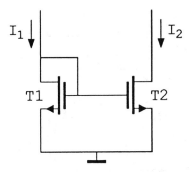

Fig. 6.14 *Simple MOS-current mirror.*

amount to about 20 mV in standard MOS-processes. And $\Delta\beta/\beta$ values can range from 0.5 up to 5%. The low-frequency input resistance is $1/g_m$, while $R_o = r_d$. Since r_d for MOSTs of moderate size is much smaller than r_{ce}-values for BJTs conducting the same current, the simple circuit of Figure 6.14 will often be inadequate. Improvement is obtained by using the MOS-counterpart of the Wilson current mirror (Figure 6.15) or by cascoding (Figure 6.16). As can be easily derived, the impedance levels are for both configurations

$$R_i = 2/g_m \text{ and } R_o = r_d(2 + g_m r_d)$$

under the assumption of identical transistors.

EXAMPLE

Design the current mirror circuit of Figure 6.15 such that $I_1 = I_2 = 50\ \mu A$ and $R_o \approx 25\ M\Omega$. A MOS-process is assumed that allows for 3 μm minimum features. Assume $\beta_{sq} = 40\ \mu A\ V^{-2}$ and $V_A = 4$ V per μm channel length.

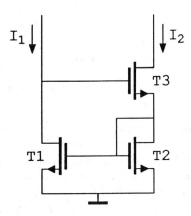

Fig. 6.15 *MOS-Wilson current mirror.*

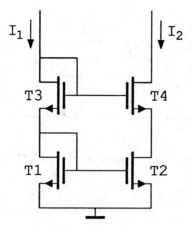

Fig. 6.16 Cascoded MOS-current mirror.

SOLUTION
Using $g_m = \sqrt{2\beta I_D} = \sqrt{2\beta_{sq} I_D (W/L)}$ and $r_d = V_A/I_D = 4L/I_D$, we find

$$R_o = r_d(2 + g_m r_d) \approx g_m r_d^2 = \left(\frac{4L}{I_D}\right)^2 \left(2\beta_{sq}\frac{W}{L} I_D\right)^{1/2} = 25 \text{ M}\Omega$$

Substituting the known values of β_{sq} and I_D and choosing $W = 3$ μm for minimum silicon area consumption, we find $L \approx 11$ μm. Hence, the channel dimensions of the MOSTs are $W = 3$ μm and $L = 11$ μm.

6.3 Signal transfer of current mirrors

In several applications current mirrors are used to transfer signal currents. In such cases the high-frequency behaviour is of interest. Let us first consider the simple bipolar mirror of Figure 6.7. As can easily be seen, for the transfer of signal currents the small-signal counterpart of eq. (6.9) applies, hence

$$\frac{i_2}{i_1} = 1 - \frac{2}{\beta + 2} = \frac{1}{1 + 2/\beta} \tag{6.15}$$

Using the hybrid-π equivalent circuit of a transistor (Figure 6.17), we find for the complex current transfer

$$\beta = \frac{I_c}{I_b} = \frac{\beta_0}{1 + s\tau_\pi} \tag{6.16}$$

with $\tau_\pi = r_\pi C_\pi$.

Replacing i_1 and i_2 by their complex counterparts I_1 and I_2 and substituting eq.

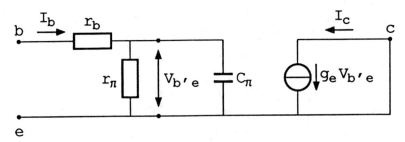

Fig. 6.17 *Equivalent circuit for the calculation of complex current transfer.*

(6.16) into eq. (6.15) we find

$$\frac{I_2}{I_1} = \frac{1}{1 + 2/\beta_0} \cdot \frac{1}{1 + 2s\tau_\pi/\beta_0} \quad (6.17)$$

The transfer is characterized by one pole $p = -1/2C_\pi r_e$. Applying the same routine to the Wilson current mirror (Figure 6.8), we find

$$\frac{I_2}{I_1} = \frac{\beta_0(\beta_0 + 2)}{\beta_0^2 + 2\beta_0 + 2} \cdot \frac{1 + 2sr_e C_\pi}{1 + 2sr_e C_\pi + 2s^2(r_e C_\pi)^2} \quad (6.18)$$

The transfer is characterized by two poles and a zero (Figure 6.18).

For the MOS-versions of these configurations similar expressions are found. To calculate the transfer of the simple circuit of Figure 6.14 we use the simplified MOS-model according to Figure 6.19. Figure 6.20 shows the equivalent circuit of the current mirror, where Z denotes the parallel connection of $1/g_m$ and $2C_{gs}$.

Current transfer is given by

$$\frac{I_2}{I_1} = \frac{1}{1 + 2sC_{gs}/g_m} \quad (6.19)$$

which shows that the MOS-circuit behaves like its bipolar counterpart.

Using the same simple MOS-model, Figure 6.21 shows the equivalent circuit for the Wilson current mirror. Assuming identical transistors, we find for the current

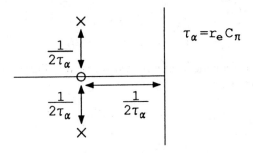

Fig. 6.18 *Pole–zero plot of the Wilson mirror current transfer.*

transfer

$$\frac{I_2}{I_1} = \frac{1 + 2s\tau}{1 + 2s\tau + 2s^2\tau^2} \tag{6.20}$$

with $\tau = C_{gs}/g_m$.

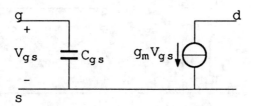

Fig. 6.19 *Simplified MOS-model.*

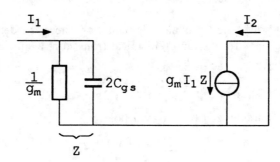

Fig. 6.20 *Simplified equivalent circuit for the two-transistor mirror.*

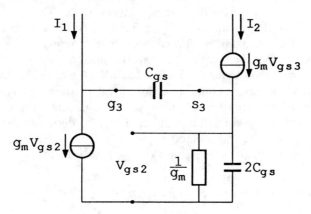

Fig. 6.21 *Equivalent circuit for the Wilson mirror.*

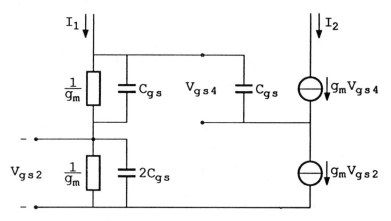

Fig. 6.22 *Equivalent circuit for the cascoded mirror.*

Comparison with eq. (6.18) shows that the pole–zero pattern is similar to that of the BJT-version.

Figure 6.22 shows the equivalent circuit for the cascoded current mirror according to Figure 6.16. Assuming identical transistors again, current transfer is found to be given by

$$\frac{I_2}{I_1} = \frac{1}{1 + 4s\tau + 2s^2\tau^2} \qquad (6.21)$$

The transfer is characterized by two real poles

$$p_{1,2} = -\frac{1}{\tau}(1 \pm \tfrac{1}{2}\sqrt{2}) \qquad (6.22)$$

If we compare this transfer with that of the Wilson version we note that the latter has somewhat larger bandwidth. On the other hand its complex poles indicate proneness to causing overshoot in the transient response. The difference in transfer properties is quite in accord with what could be expected. The complex poles of the Wilson mirror are due to the internal feedback loop. Since such a loop is absent in the cascoded mirror, this version cannot exhibit complex poles.

It may be noted that the given expressions for high-frequency current transfer are approximations due to the use of a simplified MOS-model. The advantage of this approach is that it simplifies the calculations greatly, while preserving the essential characteristics of the transfer. If more precise data are wanted it is far more appropriate to resort to computer simulation than to derive analytical expressions based on an extended MOS-model. Such exercises tend to become quite cumbersome, with circuits containing three or four transistors.

6.4 Reference voltage and current sources

In a monolithic circuit built up of many elementary circuits a large number of current sources is needed. The driving currents for these sources can be derived

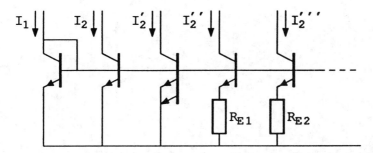

Fig. 6.23 *Formation of secondary currents I_2, I_2', ... from a primary current I_1.*

from one reference current source by multiple mirroring, as is indicated in Figure 6.23. By employing emitter scaling and/or adding emitter resistors, different values of the generated currents can be obtained. The accuracy of all the current values depends on the accuracy of the primary current that feeds the slave sources. Hence, the formation of this primary current is of crucial importance.

Nature has not granted us a mechanism to provide a reference current that is directly related to a well-defined physical quantity. On the other hand many possibilities are available for deriving a reference voltage. The formation of a reference current is therefore bound to rely on a reference voltage, which is converted into a current by means of a resistor. Obviously, we first have to list useful methods for deriving a reference voltage. In monolithic technology the following voltage references are possible candidates for fruitful application:

(a) the power-supply voltage;
(b) the avalanche breakdown voltage;
(c) the thermal voltage $V_T = kT/q$;
(d) V_{BE} of a forward-biased junction;
(e) the silicon bandgap voltage.

(a) Though use of the power-supply voltage is obvious and quite simple, it is rarely capable of providing a good solution. Power-supply voltage is often corrupted with spurious signals and in many cases it is not accurately defined. Figure 6.24 shows a simple configuration for the formation of positive and negative current sources by using the power-supply voltage.

(b) Avalanche breakdown of a reverse-biased junction occurs at a well-defined voltage. It exhibits a positive temperature coefficient that is almost equal to the negative temperature coefficient of the voltage V_{BE} of a forward-biased junction. Hence, by using a series combination of both devices an almost temperature-independent voltage reference is obtained. However, there are two disadvantages that frequently preclude the use of this method. First, the avalanche mechanism produces a lot of noise and in cases where very small voltages must be handled the signal-to-noise ratio may be intolerably spoiled by the additional noise. Second, the avalanche breakdown voltage depends on the doping levels and these are specified by the process technology. In current bipolar processes the avalanche voltage of the

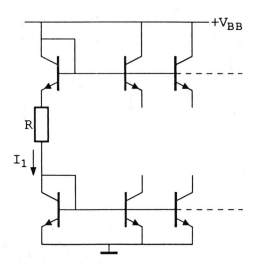

Fig. 6.24 *Formation of a reference current from the ps-voltage*
$I_1 = V_{BB} - 2V_{BE}/R$.

emitter junction is of the order of 5–6 V, which is frequently too high to be practicable. Figure 6.25(a) shows a Zener reference current source. The biasing current can be derived from the generated current, as shown in Figure 6.25(b).

(c) The use of V_T as a voltage reference is widespread. This voltage can easily be obtained from the difference of the V_{BE} of two transistors with different emitter junction areas and/or biased at different currents (Figure 6.26). Assuming that the emitter area of $T1$ is m times that of $T2$, and assuming that $T2$ carries a current that is n times that of $T1$, we have

$$RI = \Delta V_{BE} = V_T \ln \frac{nI}{I_s} - V_T \ln \frac{I}{mI_s} = V_T \ln mn \qquad (6.23)$$

Hence,

$$I = \frac{\Delta V_{BE}}{R} = \frac{V_T}{R} \ln mn \qquad (6.24)$$

The reference current can be extracted from the circuit by mirroring, as indicated in the diagram.

The circuit can be made to be self-biasing by coupling the current of $T1$ and $T2$ with the aid of a (pnp) current mirror, as indicated in Figure 6.27. In most bipolar processes the pnp-transistors are of poor quality. Improved current transfer can then be obtained by using an improved current mirror, for instance of the type shown in Figure 6.8 or in Figure 6.11. A further improvement is obtained by also replacing the npn-current mirror $T1$–$T2$ by a three-transistor mirror. Since the reference voltage (V_T) used in these current sources is proportional to absolute temperature (PTAT), the same is true for the current generated. In several

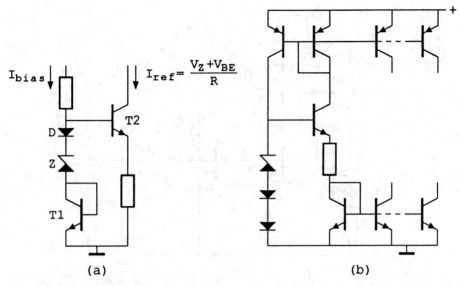

Fig. 6.25(a) *An avalanche breakdown reference source.* (b) *A reference source generating its own bias current.*

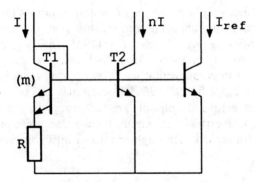

Fig. 6.26 *Formation of a voltage that is proportional to T (PTAT-voltage).*

applications this is a desirable feature. For instance, in an amplifier device transconductances are usually preferred to be constant.

For a bipolar transistor $g_e = I/V_T$. Hence, if I is PTAT, g_e is independent of T.

EXAMPLE
Design the reference circuit source of Figure 6.27 such that $I_{ref} = 100 \ \mu A$. In view of convenience of circuit lay-out, use integer values of m and n.

174 ANALOGUE ELECTRONIC CIRCUIT DESIGN

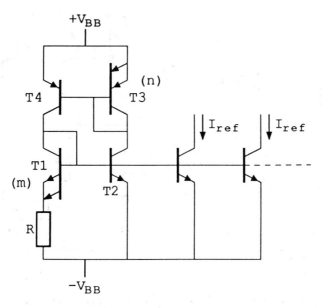

Fig. 6.27 *Self-biased PTAT-reference source.*

SOLUTION

$I_{ref} = V_T/R \ln mn$. Choosing large values of m and n implies large junction areas. Silicon area is minimized by choosing a low value of mn. The minimum possible value of mn is 2. Choosing $mn = 2$ yields $R = 179 \, \Omega$. This is a rather low value. In view of the uncertainty of the values of the parasitic resistances of the base and emitter regions and of interconnection leads, $mn = 3$ or 4 might be preferable. If we choose $m = 2$, $n = 2$, we find $R = 360 \, \Omega$, which may be considered to be a good compromise between area consumption and accuracy. In passing, note that an analysis of the noise properties of the circuit reveals that noise on the output signal is proportional to $1/\ln(mn)$, which might be taken into consideration in choosing mn.

The method of self-biasing shown in the circuits in Figure 6.25(b) and in Figure 6.27 requires a provision for starting up the generation of the reference current. In fact, such circuits have two stable states, the one that is desired and the state $I = 0$. When the circuit is switched on, it may tend to remain in the zero current state. This can be prevented by feeding a small leakage current to the circuit. Any finite starting current will cause the circuit to reach the desired stable state.

(d) The voltage across a forward-biased junction is well determined, given the doping levels of the technological process. But it also depends on the junction current, be it only slightly. Doubling of the current involves a change $\Delta V_{BE} = V_T \ln 2 \approx 18$ mV. Moreover, V_{BE} is temperature-dependent with a

DIFFERENTIAL AMPLIFIERS AND OPERATIONAL AMPLIFIERS 175

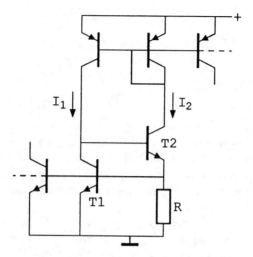

Fig. 6.28 V_{BE} reference current source.

coefficient of about 2 mV $°C^{-1}$. When these dependencies can be tolerated, the use of V_{BE} as a reference can be attractive in view of its simplicity. Figure 6.28 shows a self-biased version of a V_{BE}-related current source. From $I_1 = I_2$ and $I_2 R = V_{BE1}$ the implicit relation for I_1 follows:

$$I_1 = \frac{kT}{qR} \ln \frac{I_1}{I_s} \qquad (6.25)$$

(e) The bandgap voltage of silicon is a quantity that is firmly anchored in the atomic properties of silicon. It is by far the most accurate reference quantity to exploit in monolithic circuitry. However, it is not directly available. In order to make it available we have to resort to an indirect method, using V_{BE} as an intermediary quantity. The voltage across a forward-biased junction depends on the bandgap voltage V_G, but it is also temperature-dependent with a temperature coefficient of -2 mV $°C^{-1}$. If a voltage that is proportional to absolute temperature (PTAT) is added to V_{BE}, the sum of both voltages can be made to be equal to V_G. Figure 6.29 shows the principle of the bandgap reference. The current I is PTAT and hence the same applies to IR. The reverse current I_s of a junction obeys the equation

$$I_s = BT^\eta \exp(-qV_G/kT) \qquad (6.26)$$

where B and η are constants. Hence, referring to Figure 6.29

$$V_{ref} = V_{BE} + ATR = V_T(\ln A - \ln B + RAq/k) - (\eta - 1)V_T \ln T + V_G \qquad (6.27)$$

By choosing appropriate values for A and R we can arrange that at a certain temperature T_0, $dV_{ref}/dT = 0$ holds. Hence, from (6.27)

$$(\eta - 1)(1 + \ln T_0) = \frac{q}{k}RA + \ln \frac{A}{B} \qquad (6.28)$$

Substitution of eq. (6.28) into eq. (6.27) yields

$$V_{ref} = V_G + (\eta - 1)V_T - (\eta - 1)V_T \ln (T/T_0) \qquad (6.29)$$

The values of V_G and η depend slightly on the constructive details of the junction. For a given manufacturing process these quantities can be found by measurement. Representative values are $V_G = 1.166$ V and $\eta = 3.72$. With these values eq. (6.29) yields $V_{ref} = 1.236$ V at $T = T_0$. The reference temperature T_0 can be freely chosen. For obvious reasons it is customary to choose the ambient temperature for T_0, hence $T_0 \approx 300$ K.

Proper adjustment of the circuit is a simple matter. If care is taken that the chip is at temperature T_0, only A or R have to be adjusted so that the actual value of V_{ref} corresponds to its (calculated) nominal value.

If the actual value of T deviates from T_0 by an amount $\Delta T \ll T_0$, eq (6.29) can be approximated by

$$V_{ref} = V_G + (\eta - 1)\frac{kT_0}{q} - (\eta - 1)\frac{kT_0}{2q}\left(\frac{\Delta T}{T_0}\right)^2 \qquad (6.30)$$

The deviation of V_{ref} from its value at T_0 depends quadratically on ΔT. However, the error is very small. For instance, if $|\Delta T| = 20$ K, then $\Delta V_{ref} = -156 \mu V$.

As to the practical realization of the circuit, the formation of a PTAT current has already been discussed. Using the method indicated in Figure 6.26, the circuit shown in Figure 6.30 is capable of producing the bandgap reference voltage. Assuming that $T1$ and $T2$ are identical transistors and that the emitters of $T5$ and

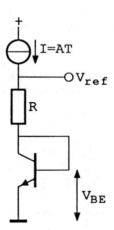

Fig. 6.29 *Principle of a bandgap voltage reference source.*

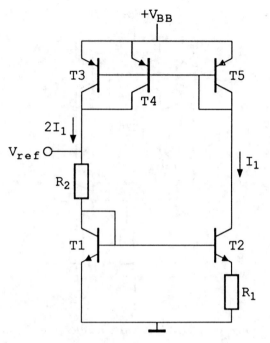

Fig. 6.30 *Simple bandgap voltage reference circuit.*

$T3,4$ scale by a factor of n (the figure suggests $n = 2$), then

$$V_{ref} = V_{BE1} + nV_T \frac{R_2}{R_1} \ln n \tag{6.31}$$

A possible temperature dependency of R_1 and R_2 does not cause problems, provided both resistors are at the same temperature and have the same temperature coefficient. The circuit shown in Figure 6.30 can be improved in many ways. In particular the Early-effect, the finite base currents and the poor quality of the pnp-transistors cause aberrations from the ideal operation. All these effects can be compensated for, be it at the cost of considerable extensions of the circuit (see, for instance, Meijer and Verhoeff, 1976).

For the formation of a bandgap-related reference *current*, the current produced by V_{ref} in a resistor can be used. Since resistors are already present in the circuit, no additional resistor is needed. Figure 6.31 shows a simple modification of the circuit of Figure 6.30, producing a reference current $I_{ref} = V_{ref}/R_2$, which is easily verified by considering that $I_{ref} = V_{BE1}/R_2 + (n+1)V_T/R_1 \ln n$. And in view of eq. (6.31) this equals

$$\frac{V_{ref}}{R_2} + \frac{V_T}{R_1} \ln n \text{ if } \frac{R_2}{R_1} = \frac{V_{ref} - V_{BE}}{nV_T \ln n}$$

178   ANALOGUE ELECTRONIC CIRCUIT DESIGN

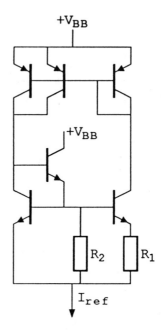

Fig. 6.31 *Bandgap current reference circuit.*

6.5 MOS reference sources

The inventory of possible sources for the formation of a reference voltage, made up at the beginning of the preceding section, reveals that designers of analogue MOS-circuits are far less well off than their bipolar colleagues. Nevertheless, MOS-technology can be harnessed to provide satisfactory tools for the implementation of a reference voltage source. The counterpart of the bipolar V_{BE} is the MOS-threshold voltage V_t. However, threshold voltage is much more process-dependent than V_{BE}. Most NMOS-processes are capable of providing depletion devices as well as enhancement devices. The difference $\Delta V_t = V_{t\,enh} - V_{t\,depl}$ is much less process-dependent than the individual threshold voltages. Figure 6.32 shows a circuit for the formation of a reference voltage based on this observation. Transistor $T1$ is an enhancement (normally-off) type, while $T2$ is a depletion (normally-on) type. The feedback through the differential amplifier A forces the voltage across both resistors to be equal, so that both drain currents must also be equal. Using the formula

$$V_{GS} = \sqrt{\frac{2I_D}{\beta}} + V_t \qquad (6.32)$$

and assuming equal values of β, we obtain

$$V_{ref} = |V_{t1}| - |V_{t2}| \qquad (6.33)$$

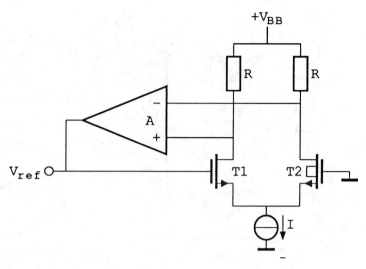

Fig. 6.32 *Formation of a reference voltage using the difference of two threshold voltages.*

In practice, β-values for enhancement and depletion types are slightly different due to different values of electron mobility μ. This can be compensated for by choosing slightly dissimilar W/L ratios, so that

$$\mu_{enh}\left(\frac{W}{L}\right)_{enh} = \mu_{depl}\left(\frac{W}{L}\right)_{depl}$$

In CMOS-technology the generation of a bandgap related voltage is within reach. Its realization requires the availability of a PTAT-current and a forward-biased junction voltage (V_{BE}). In any CMOS-process a bipolar transistor can be formed in the way indicated in Figure 6.33 for a p-well process. The n-region is formed along with the regular drain and source regions. Since the substrate constitutes the collector, the transistor has to be used accordingly. The formation of a PTAT current can be based on the exponential voltage-to-current characteristics of MOSTs operating in subthreshold mode. Figure 6.34 shows a bandgap reference circuit based on this approach. Transistors $T1$ through $T4$ operate in the

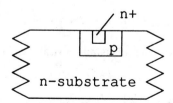

Fig. 6.33 *npn-transistor in a CMOS-process.*

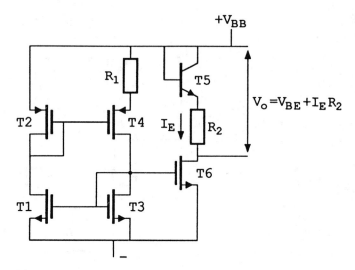

Fig. 6.34 *Bandgap reference in CMOS-technology.*

subthreshold mode. The circuit is essentially equivalent to its bipolar counterpart of Figure 6.27. The current generated in the circuit is mirrored into $T6$. Straightforward calculation yields

$$I_E R_1 = V_T \frac{S_6}{S_3} \cdot \ln \left(\frac{S_4}{S_2} \frac{S_1}{S_3} \right) \tag{6.34}$$

where S_n denotes the 'shape factor' W/L of transistor T_n.

Using MOSTs in the subthreshold mode is not without problems. Transistor parameters tend to spread widely, even within the confines of a chip, and the inherently very low drain currents can easily be spoiled by leakage currents. Using the MOS-transistors in the quadratic region of operation is therefore to be preferred. Figure 6.35 shows a circuit using two bipolar transistors for the formation of the PTAT-current. In this example an n-well process is assumed so that the transistors are of the pnp variety. The schematically indicated differential amplifier in the feedback path is a conventional MOS-amplifier. As in the circuit of Figure 6.32 the feedback brings about that $I_1 = I_2$. If the emitter area of $T1$ is n times that of $T2$, then

$$I_1 = I_2 = I = \frac{V_T}{R_2} \ln n$$

Because I is PTAT, R_2 can be adjusted so that

$$V_o = V_{BE1} + IR_2 = V_{gap}$$

A disadvantage is that a possible offset-voltage V_{os} of the differential amplifier causes an error $V_{os}(1 + R/R_2)$ in V_o.

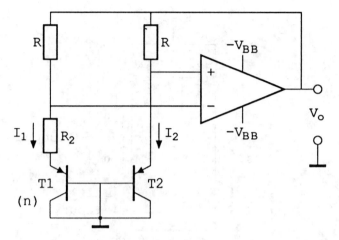

Fig. 6.35 *CMOS-bandgap reference.*

6.6 Active loads

At this stage we must return to the discussion of monolithically integratable amplifier stages. As we saw earlier, the differential pair as shown in its basic form in Figure 6.3 is a building block that is pre-eminently suited for monolithic integration. Having discussed the implementation of the emitter current source fully, we must now turn our attention to the collector load, indicated in Figure 6.3 by R_L. To obtain maximum signal-current transfer to the next stage, R_L should be maximized. This implies maximum amplification of the DM-signals at the same time, since $A_{vdm} = g_e R_L$. Here, too, replacement by an active impedance is the obvious approach. One possibility is the use of (negative) current sources for loading the collectors. But, since the collector itself acts as a current source, a current source load cannot be applied right away. After all, a series connection of two current sources is only possible if the two currents are exactly equal. Even the slightest inequality will cause one of the transistors constituting these sources to saturate. Coupling of the sources to be connected in series can be accomplished by a feedback arrangement. However, in many cases a simpler solution is possible. Frequently, it is preferable to obtain the amplified DM-signal in a single-sided fashion, i.e. between one of the collectors and ground. This can be simply achieved by using a current mirror as the loading device. Figure 6.36 shows the basic circuit, where $T3$ and $T4$ form a pnp-current mirror. Driving the circuit with a DM-signal causes signal currents of opposite polarity to flow in both halves of the differential pair. The loading pair mirrors the signal current i, so the signal current in the output circuit is $2i$, implying that the total differential current is transferred to the output circuit.

If the transfer of the pnp-mirror is inaccurate, the d.c.-currents of $T2$ and $T4$ will be different and consequently an offset-current ΔI will flow in the output circuit.

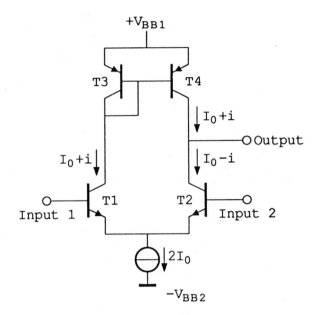

Fig. 6.36 *pnp-current mirror as active load.*

In most cases this is unwanted and therefore the transfer of $T3$–$T4$ should be as accurate as is possible. If the pnp-transistors are of the lateral variety, the error can be quite large. Henceforth, in most cases an improved current mirror will be needed. Even then, the circuit is not ideal because of the poor high-frequency transfer of lateral transistors.

The impedance level at the output node is determined by the parallel connection of three internal impedances: the output impedance of $T2$, the output impedance of $T4$ and the input impedance of the next stage. The last-mentioned impedance can be made high, for instance by using a Darlington configuration as the next stage. The low-frequency output impedance of $T2$ is $r_{ce(npn)}$. If the loading current mirror is of the Wilson type, its output impedance is $\beta_0 r_{ce(pnp)}$, which is usually much higher than the output impedance of $T2$. Hence, to enhance the impedance level of the output node, the output impedance of $T2$ should be enhanced. This can be done by cascoding the differential pair, as shown in Figure 6.37.

EXAMPLE
For estimating the low-frequency voltage gain of this configuration, let us assume the following data: $V_A(\text{npn}) = 100$ V, $\beta_0 = 200$ and for the pnp $V_A = 50$ V, $\beta_0 = 30$. Further, let R_L (the input impedance of the next stage) be 20 MΩ and let $I_o = 100\ \mu\text{A}$, hence $g_e = 2$ mA V^{-1}. With these data and writing $r_c = \beta_0 r_{ce}$ we find $r_c(\text{npn}) = 200$ MΩ, $r_c(\text{pnp}) = 15$ MΩ. Hence $r_c(\text{npn}) \| r_c(\text{pnp}) \| R_L \approx 8.2$ MΩ and $A_v = 16{,}400$.

This example makes it clear that further measures to increase the impedance level

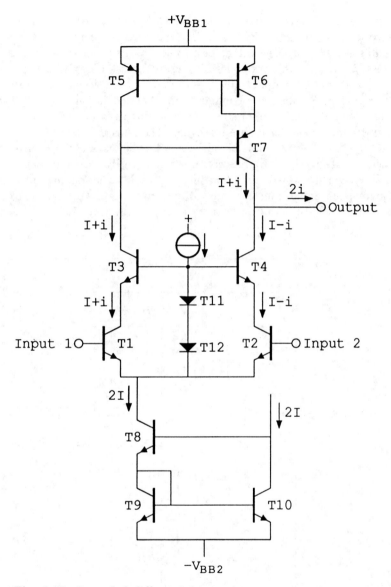

Fig. 6.37 *Cascoded differential input stage.*

of the output node will rarely make sense, since for all but signals of very low frequency the capacitive component in the load will tend to dominate the impedance level. After all, the output node is loaded with the substrate capacitances of two collectors together with the input capacitance of the next stage.

In spite of its good qualification as an amplifier of low-frequency DM-signals, the circuit of Figure 6.37 has several disadvantages. First, the circuit exhibits a very unfavourable level shift. Output d.c-level is close to the positive supply voltage. And second, the pnp-load badly affects the high-frequency behaviour. Both disadvantages can to a large extent be avoided by using pnps for the cascoding transistors. Of course, this precludes biasing of the cascoding transistors through the biasing current of the input transistors, and hence additional current sources are needed. Figure 6.38 shows the configuration obtained in this way. It should be noted that the additional current sources are not involved in signal transfer, hence their high-frequency properties are irrelevant. The pnps that are involved in signal transfer are CB-connected and this alleviates their effect on the transfer at high frequencies considerably. Moreover, they can easily be bypassed by small integratable capacitors, as is indicated in Figure 6.38. The loading current mirror that is involved in signal transfer is now of the npn variety. And, thanks to the level shift produced by the pnp-transistors, the d.c.-outut level is near the negative

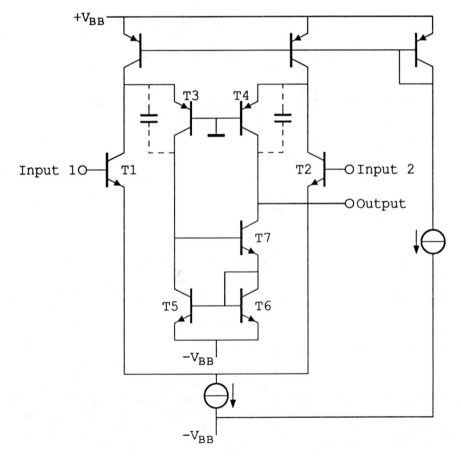

Fig. 6.38 *Alternative cascoded input stage.*

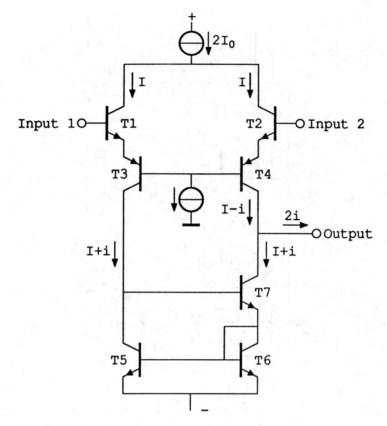

Fig. 6.39 *An alternative solution for obtaining level shift.*

supply voltage, leaving room for a level shift towards ground potential in subsequent stages.

As an alternative, the input transistors can be connected as emitter followers (Figure 6.39). This arrangement for creating a level shift has already been mentioned in Chapter 5 (Figure 5.36b). It maintains the possibility of biasing the CB-stages through the biasing current of the input transistors. For convenience, in Figure 6.39 the biasing current source has been symbolically indicated in the collector leads of $T1$ and $T2$. But in practice, matching the collector currents to the biasing current source can only be accomplished through feedback. This can be done as indicated in Figure 6.40. The sum of the collector currents is measured by $T8$ and through $T9$ compared with the current of the biasing source $2I_o$. Any inequality is fed back to the bases of $T3$–$T4$. An alternative way to comprehend this biasing arrangement is to look upon it as a Wilson current mirror consisting of $T8$, $T9$ and the cluster $T1$–$T4$ operating as one single transistor.

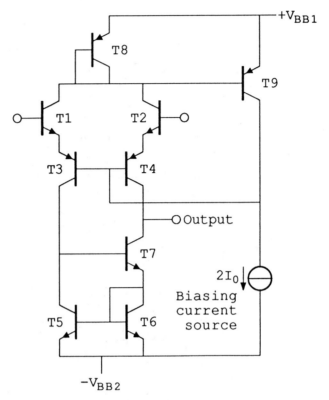

Fig. 6.40 *Biasing method for the configuration of Figure 6.39.*

6.7 Active loads in MOS-technology

In the preceding section we discussed several bipolar amplifier configurations with active loads. These configurations can, of course, also be implemented in CMOS-technology. Figure 6.41 shows the MOS-version of the simple circuit of Figure 6.37. It should, however, be noted that in general the gain of an MOS-stage is much lower than that of its bipolar counterpart. The maximum voltage gain of a common source amplifier stage is $g_m r_d$. Since $g_m = \sqrt{2\beta I_D}$ and $r_d = V_A/I_D$, where V_A is the Early-voltage (for these relations see Chapter 2), $g_m r_d = V_A \sqrt{(2\beta/I_D)}$. Since V_A is approximately proportional to the channel length L and β is proportional to W/L, where W is the channel width, the value of $g_m r_d$ depends on channel geometry as well as on biasing current. To acquire some feeling for quantitative aspects, let us assume that V_A is 5 V μm^{-1} and that $\beta = 40(W/L)\mu$A V^{-2}. With $W = 25$ μm and $L = 5$ μm, we have $V_A = 25$ V and $\beta = 200$ μA V^{-2}. Assuming $I = 100$ μA, we find $g_m r_d = 50$. Compare this with a bipolar transistor. Here maximum voltage gain is $g_e r_{ce} = V_A/V_T$. With $V_A = 100$ V and $V_T = 25$ mV, $g_e r_{ce} = 4000$, independent of geometry and current level.

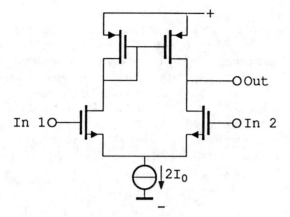

Fig. 6.41 *Basic MOS-differential pair with active load.*

Much improvement is obtained by cascoding through its ensuing enhancement of the output resistance. Maximum gain is then $r_d(1 + g_m r_d)g_m \approx (g_m r_d)^2$, which is still below the gain provided by a non-cascoded bipolar transistor! Figure 6.42 shows a cascoded MOS differential pair with active load. The load consists of a cascoded current mirror since, of course, cascoding the input transistors is of no

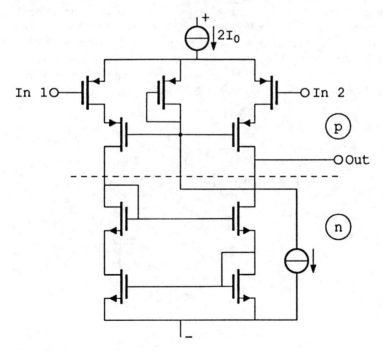

Fig. 6.42 *Cascoded MOS-differential pair.*

avail with a low impedance load. A definite advantage of CMOS-technology is that the quality of the p-channel devices is not significantly below that of the n-channel devices. Hence, the decision as to whether to use p-channel input devices with n-channel loads, or the complementary counterpart, can be taken on the grounds of convenience with regard to the relevant application. By way of illustrating this point Figures 6.41 and 6.42 show complementary configurations.

If an MOS-process must be used that provides n-channel MOSTs only, the possibilities of designing an actively loaded amplifier stage are very much restricted. If the process is only capable of providing enhancement MOSTs, resort must be made to the configuration shown in Figure 6.43. The load impedance offered by $T3$ and $T4$ is $1/g_m$, hence

$$A_v = \frac{g_{m1}}{g_{m3}} = \sqrt{\frac{(W/L)_1}{(W/L)_3}} \qquad (6.35)$$

under the assumption that the body-effect can be neglected. In practice, this will rarely apply. The body-effect can be accounted for in eq. (6.35) by introducing a factor α_B in the right-hand side. As has been explained in Chapter 2, α_B depends on substrate doping and on the bias voltage V_{SB} between source and body of the load MOSTs. In practice, α_B will usually assume a value between 0.5 and 1.

Moreover, if the output signal is taken single-sidedly, a further attenuation factor of 0.5 applies. From these considerations it is clear that even a voltage gain of the order of 10 requires load transistors with excessive channel length. Some improvement is possible by replacing $T3$ and $T4$ with two or more transistors in series, but this requires a larger supply voltage and it enhances the body-effect.

If the NMOS-process is capable of providing MOSTs of both the depletion and the enhancement variety, a somewhat better result can be obtained. With the gate connected to the source, the transistor behaves as a current source (Figure 6.44a).

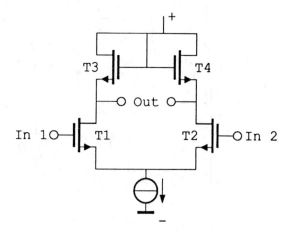

Fig. 6.43 *All-enhancement NMOS-differential pair.*

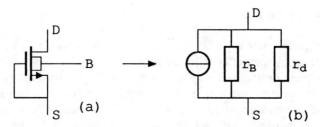

Fig. 6.44 *Depletion MOST used as current source (a) and equivalent circuit (b).*

The impedance of this source consists of the parallel connection of r_d and the impedance seen into the source of the 'back-gate' controlled transistor. If the transconductance of this transistor, whose existence is due to the body-effect, is given by g_{mB}, the impedance seen into the source is $r_B = 1/g_{mB}$ (Figure 6.44b). As a rule $r_d \gg r_B$ will apply and hence $R_o \approx r_B$. Hence, the gain of the depletion load differential pair shown in Figure 6.45 is given approximately by

$$A_v = \frac{g_m(T1)}{g_{mB}(T3)} = \frac{1}{\lambda_B}\sqrt{\frac{(W/L)_1}{(W/L)_3}} \qquad (6.36)$$

where λ_B denotes g_{mB}/g_m. Practical values of λ_B are commonly of the order of 0.2 to 0.3. This shows that the improvement that can be obtained by using depletion loads is not impressive. Several methods for further improvement have been devised, none of them being very effective. Since NMOS analogue circuitry is of minor and even shrinking interest due to the growing preference for CMOS, we will not dwell upon the subject any longer.

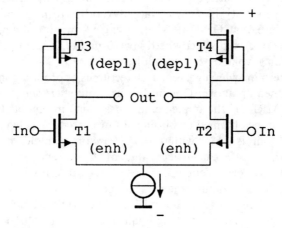

Fig. 6.45 *NMOS-differential pair with depletion load.*

6.8 Operational amplifiers

Amplification is the most central electronic signal-processing function. Amplifiers thus constitute an important class of electronic circuits. But the requirements to be met depend largely on the nature of the electronic system they are part of, and of their particular function therein. In many cases these requirements are so special that only dedicated, tailored to measure, amplifier circuitry can offer optimum performance. In particular, if high dynamic range and/or large bandwidth are at a premium, careful matching to the signal source and optimized circuit design are imperative. However, there are also design situations where requirements are sufficiently lax to justify the use of a general purpose standard configuration with the ensuing advantage of fast design by the use of standard building blocks. The standard building block that has found the most widespread application is named the 'operational amplifier' (OPAMP, or OPA). The name derives from the days of analogue computers, when this type of amplifier was used as the general device for performing mathematical operations. The general requirements to be met by these amplifiers come down to the following main features: differential input, high CMRR, often single-sided output, high-gain, high-input impedance (voltage input), low-output impedance (voltage output), no overall level shift. In recent times modifications with different output properties have gained some interest. The best known of these is the variety with a high-impedance (current source) output. This variant is known under the name of operational transconductance amplifier (OTA).

The general idea of the OPAMP is to provide so much gain that the transfer properties can be adapted to the relevant application by using an appropriate feedback arrangement. At first sight, it might be thought that almost any transfer function can be optimally realized in this way. But this is not true: the internal structure of the amplifier block is fixed and hence also many important parameters such as its noise behaviour, bandwidth, large-signal response, and so on. No doubt, the OPAMP is a very useful and versatile building block, but time and again it appears that its versatility is overestimated. Another form of misuse that is not uncommon is the use of an OPAMP where a single transistor or a pair of transistors can offer the same or even distinctly better performance. It seems that the versatility of the concept can invoke a kind of addiction.

Let us now determine which general architecture is appropriate in order to realize the features summed up above. Obviously, the input stage should be a differential pair with high CMRR. If the implementation is done in bipolar technology, high input resistance can be obtained by biasing at low current. The DM-input resistance of a differential pair is $2\beta_0 r_e$ with $r_e = V_T/I_C$. For example: if $I_C = 10\,\mu\text{A}$ and $\beta_0 = 200$, then $R_i = 1\,\text{M}\Omega$. A higher value of R_i can be obtained by using a Darlington input stage (see Section 5.11). With regard to input resistance an FET, either JFET or MOST, input stage is, of course, the best choice. However, technological processes providing both JFETs and BJTs are more expensive than bipolar-only processes. To achieve high gain the input stage is best to be actively loaded. In view of the requirement of low-output impedance the output stage is best

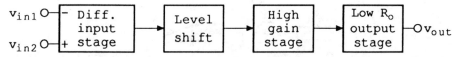

Fig. 6.46 *General architecture of OPAMP.*

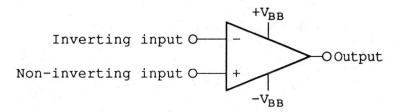

Fig. 6.47 *General symbol for the OPAMP.*

to be configured as an emitter follower. However, this implies that no voltage gain is contributed by this stage. To obtain satisfactory overall voltage gain an additional stage cannot usually be dispensed with. This leads to the general architecture indicated in Figure 6.46. Figure 6.47 shows the symbol for the OPAMP in general use. The two input terminals are commonly designated as the *inverting* ($-$) and the *non-inverting* ($+$) input.

6.9 Offset and offset-compensation

Before discussing signal-processing circuits using OPAMPs as building blocks, some general problems that are inherent to the underlying concept must be faced. The first concerns the transfer of the input d.c.-level. For proper operation no level shift is allowed to occur between the input and output terminals. In a practical circuit the occurrence of some d.c.-offset cannot be avoided. The input-related offset-voltage is defined as the input voltage needed to obtain zero output. The difference between *static* and *dynamic* offset must be made. By 'static offset' is meant the offset occurring at the nominal operating temperature. Offset alteration due to temperature changes is designated as dynamic offset. The main source of offset is the input stage, since any offset-signal originating in this stage is transferred to the output fully amplified. Offset originating in subsequent portions of the circuitry can be accounted for by an equivalent input offset. Figure 6.48 shows a schematic representation of the input offset by an offset-voltage source V_{off}. In a bipolar implementation of the input stage some d.c.-current must flow in the external driving circuit. Unequal currents in both input leads can be accounted for by an offset-current source $I_{\text{off}} = I_{B1} - I_{B2}$. If the d.c.-resistances of these input leads are denoted by R_g, the total offset can be represented by one voltage source

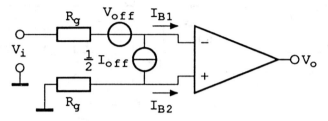

Fig. 6.48 *Representation of offset in an OPAMP.*

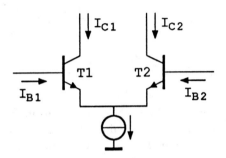

Fig. 6.49 *Bipolar pair.*

$V_{\text{off}} + R_g I_{\text{off}}$. A dominant cause of offset in a bipolar input stage is dissimilar values of V_{BE}. Referring to Figure 6.49,

$$\Delta V_{BE} = V_T \ln \frac{I_{C1}}{I_{s1}} \cdot \frac{I_{s2}}{I_{C2}} \tag{6.37}$$

According to transistor theory (Chapter 2)

$$I_s = \frac{q^2 n_i^2 D_n A}{\int_0^W (N_A - N_D) d_x} \tag{6.38}$$

where n_i denotes intrinsic mobile charge concentration, D_n is the diffusion coefficient for electrons, A is the junction area, W the base width and $N_A - N_D$ the net doping concentration of the base region.

If an input stage contains several transistor pairs, all of these contribute to the total offset. Of course, the magnitude and the polarity of the offset, being statistical parameters, cannot be predicted. In all applications where offset cannot be tolerated it has to be compensated. This is usually made possible by making a suitably chosen point in the OPAMP circuit externally accessible through a pin of the IC package. Figure 6.50 shows an example. At the compensation point a resistor can be connected or a compensation current can be applied.

Compensation of dynamic offset is much more difficult. Fortunately, by judicious chip layout thermal coupling between paired transistors can be made very

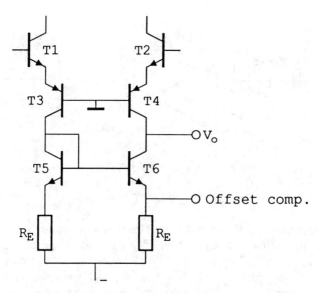

Fig. 6.50 *Common method of offset compensation.*

effective and in this way dynamic offset can be kept at a minimum (a typical value may be $1\ \mu V\ °C^{-1}$, with on-chip trimming $0.1\ \mu V\ °C^{-1}$). A rigorous measure to make sure that no temperature changes occur is to thermally stabilize the entire chip. This can be done by incorporating a temperature measuring element in the chip – which can be a forward-biased junction – and a heating element, usually a power transistor.

In MOST input stages the offset is caused by dissimilar values of β and V_t of the input transistors. If we denote the relative error in β by $\varepsilon_\beta = \Delta\beta/\beta$, we can write for the total offset

$$V_{\text{off}} = \Delta V_{\text{GS}} = \Delta V_t + I_D \varepsilon_\beta / g_m \tag{6.39}$$

MOS-stages show relatively large values of the offset-voltage (typically 1–20 mV). It is not to be expected that the offset-characteristics of MOS-circuits will substantially improve in the near future. They root to a large extent in basic attributes of MOS-technology. Since g_m-values are much smaller than for bipolar transistors, the voltage needed for compensation of a current imbalance is also much larger. Moreover, threshold voltage is a parameter that is very sensitive to doping aberrations. Finally, geometrical dissimilarities affect electrical parameters in MOS-structures more than they do in bipolar structures.

Fortunately, MOS-technology offers an elegant method for suppressing offset effects. This method is called 'auto-zeroing'. Its application is restricted to situations where short interruptions of the continuous signal flow can be tolerated. The method comes down to periodic measurement of the offset error. The measured value is stored in a memory that is implemented in the form of a

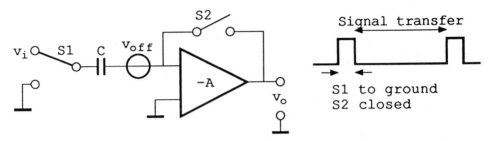

Fig. 6.51 *Principle of autozero method of offset compensation.*

capacitor. Subsequently, the stored error signal is subtracted from the signal that is corrupted with offset. The application of this method requires the availability of good switches and of leakage-free memory capacitors. MOS-technology provides both these facilities. Figure 6.51 shows a schematic representation of an offset compensation configuration. The switch connects the amplifier input with ground repetitively, thereby allowing the capacitor to store the offset-voltage. If during the subsequent signal-processing time interval, leakage through the switch or due to capacitor losses is negligible, offset is compensated by the capacitor voltage. As an illustration: if $C = 1$ pF and leakage current is $0.05\ \mu A$, then $\Delta V = 50\ \mu V$ in 1 ms. Hence, if a residual offset of 50 μV can be tolerated, the repetition frequency need not exceed 1 kHz. If 'floating' capacitances must be avoided, the configuration indicated in Figure 6.52 may be used (Klein and Engl, 1984), in which $A2$ is an auxiliary amplifier that is only involved in the handling of the offset signal. As can be easily derived the remaining offset is $V'_{\text{off}} = V_{\text{off2}}/A_1$.

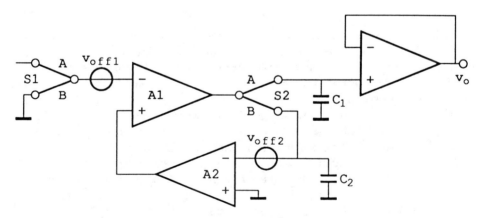

Fig. 6.52 *Autozero offset correction using an auxiliary amplifier.*

6.10 Slew rate

The second problem arising from the general concept of the OPAMP is its limited slew rate. By definition, slew rate is the maximum rate of change of the output voltage. Slew-rate limitation is not to be confused with finite rise time. Rise time is a phenomenon pertaining to linear operation, whereas slew-rate limitation is a non-linear effect occurring when driving with large signals. For signals with very low amplitude only, slew rate is determined by the rise time of the amplifier.

The cause of slew limitation lies in the measures taken for ascertaining the closed-loop stability of the OPAMP. An OPAMP circuit should be stable even if the output signal is fully fed back to the input. However, since the OPAMP architecture involves a structure containing at least three stages, at least three dominant poles must be present. Hence, a total phase shift exceeding $180°$ is quite possible at a frequency where the loop gain exceeds 0 dB, causing instability of the feedback configuration. Figure 6.53 depicts the Bode plot of the open amplifier (solid line). The amount by which the gain at the frequency where the phase shift is $180°$, is below 0 dB is called the *gain margin* M_{Ao}. In Figure 6.53 this frequency is indicated by f_{co}. According to the diagram M_{Ao} is negative, so that the amplifier is bound to be unstable when feedback is applied. The usual method of ensuring unconditional stability is to increase one of the time constants to such an extent that it fully dominates the frequency response of the open amplifier. The corresponding

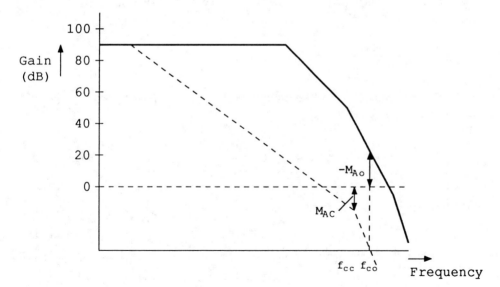

Fig. 6.53 *Bode plots of uncompensated (solid line) and of phase compensated OPAMP (broken line). M_{Ao} denotes gain margin of uncompensated amplifier, M_{Ac} denotes gain margin of compensated amplifier.*

196 ANALOGUE ELECTRONIC CIRCUIT DESIGN

Bode plot is represented by the broken line in Figure 6.53. In its modified form the amplifier exhibits a positive-gain margin M_{AC} at the critical frequency f_{cc}, so that stability under feedback operation is warranted. The simplest method of realizing the large time constant is to apply capacitive local feedback in the second stage. This measure introduces excessive pole splitting (Section 5.4). Thanks to the high gain of the second stage, the capacitor needed can be small enough to be implemented on-chip.

Figure 6.54 shows the make-up of the phase compensated OPAMP schematically. In order to analyze the slewing effect (Figure 6.55a) we can simplify the circuit indicated in Figure 6.54 by way of the model shown in Figure 6.55(b). The first stage behaves as a voltage-to-current converter feeding a signal current $g_e v_i$ to the second stage. If we assume the gain of this stage ($T5$–$T6$) to be large, then

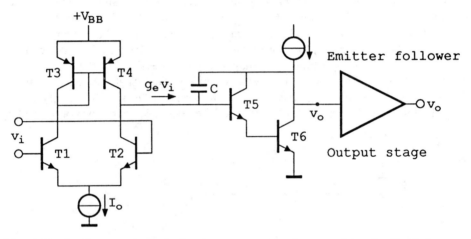

Fig. 6.54 *Implementation of phase compensation by capacitive feedback in high-gain second stage.*

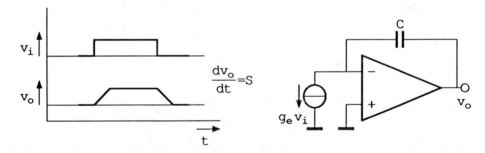

Fig. 6.55(a) *Response of output to a step-type input signal. (b) Simple model for the analysis of slew-rate limitation.*

we can write, using complex notation,

$$\frac{V_o}{V_i} = \frac{g_e}{sC} \qquad (6.40)$$

Hence, the frequency by which gain assumes unity value (0 dB) is given by $\omega_0 = 2\pi f_0 = g_e/C$.

Since $g_e = \frac{1}{2}I_0/V_T$, we find

$$\omega_0 = \frac{1}{2}I_0/CV_T \qquad (6.41)$$

Let us now assume that an external feedback loop is connected to the OPAMP, so that the closed-loop gain G_c is very much below the open-loop gain G_o. When an input signal v_i is applied, the output voltage cannot immediately respond with the steady-stage output signal $G_c v_i$ since the rate of change of the charge on C is limited by the maximum current that can be supplied to it. Because G_o is much larger than G_c, even a very small signal v_i will drive the input stage into a state wherein one of the transistors conducts all the current provided by I_o. Clearly, this is the maximum charging current of C and hence the maximum rate of change of v_o amounts to $S = dv_o/dt = I_o/C$. Substituting this into eq. (6.41) we find for the slew rate

$$S = 2\omega_0 V_T \qquad (6.42)$$

This is an important conclusion, since it reveals that S is uniquely concatenated with ω_0. However, the value of ω_0 is dictated by the requirement of a positive-gain margin. Obviously, increasing I_o does not improve slew rate. This is because g_e is proportional to I_o, and hence to maintain the value of ω_0, C should increase along with I_o.

From our analysis it is clear what can be done to improve S. One approach is to increase I_o without a proportional increase of transconductance. This can be done by adding emitter resistors R_E. But the improvement obtained by such 'emitter degeneration' is bound to be small. Effective transconductance is $g_e/(1 + g_e R_L)$, so that for substantial improvement $R_E \gg 1/g_e$ is required. If $r_e = 1/g_e = 2$ kΩ, chip area economy dictates that R_E be not much larger than $3r_e$. Moreover, R_E is a considerable additional source of thermal noise (Chapter 8) and for this reason alone this approach is unacceptable in many applications. A better method is to use FETs in the input stage since g_m/I for an FET is much lower than for a BJT. This generally unfavourable feature is an advantage in this particular application. A second possibility is to increase ω_0. This can be achieved by shifting upward the pole that is responsible for the first corner frequency in the Bode plot of the open loop. In bipolar designs this frequency is usually determined by the pnp level-shift circuit. If the use of pnps can be circumvented, ω_0 can be chosen higher without risking instability. If p-channel FETs (either JFETs or MOSTs) are available, these can be used as a preferred substitute for lateral pnps. Finally, capacitive bypassing of the pnp-transistors can be practised, but careful design is needed in view of the additional phase shift it can introduce.

As an alternative to specifying slew limiting by the slew rate, the *full-power bandwidth* f_{fp} can be used. This is the highest frequency that can be handled at maximum output signal without slew distortion being present. Writing the output signal as $v_o = \hat{v}_o \sin \omega t$, the maximum value of dv_o/dt is given by $\omega \hat{v}_o$. Hence, the maximum angular frequency ω_{fp} at which v_o can be obtained without slew distortion is given by $\omega_{fp} = S/\hat{v}_o$, hence $f_{fp} = S/2\pi\hat{v}_o$.

EXAMPLE

For a bipolar OPAMP let the requirement for positive gain and phase margins dictate $f_0 = 2.1$ MHz, hence $\omega_0 = 13.2 \times 10^6$ rad s^{-1} and let $I_o = 20$ μA. Then $g_e = 400$ μA V^{-1} and $C = g_e/\omega_0 \approx 30$ pF, so that $S = I_o/C = 0.67$ V μs^{-1}. Let us now assume that p-type FETs are available with $g_m = 40$ μA V^{-1} at $I_o = 20$ μA. Assuming again $f_0 = 2.1$ MHz, then $C = 3$ pF and $S = 6.7$ V μs^{-1}. However, if, realistically, we assume that replacing the pnp level shift by one using p-FETs justifies a fivefold increase of f_0, hence, $f_0 = 10.5$ MHz, then $C = 0.6$ pF and $S = 33$ V μs^{-1}. This clearly illustrates the benefits of using FETs in the input- and level-shift stages with regard to slew rate. Moreover, it must be noted that the compensation capacitances needed are smaller than in bipolar implementations, with the ensuing advantage of requiring less chip area.

6.11 Dedicated OPAMPS

Though the very concept of the OPAMP implies versatility and broad applicability, they come in a great variety of actual implementations. A first distinction can be made between true general purpose types and dedicated types. The OPAMPs of the first class are usually packed separately or in groups of four at most. Their design is directed towards very high open-loop gain, high R_i low R_o and phase compensation for unconditional stability. The class of dedicated OPAMPs, strictly speaking, does not merit being so called. In several system architectures elementary circuits abound that can be modelled as OPAMP-like elements with known loads (mostly of a capacitive nature) and with known feedback paths. Notable examples can be found in the area of on-chip filter techniques. Their implementation nearly always relies on MOS-technology, and large numbers of elementary OPAMP-like structures are usually needed. Because the operating conditions are fixed, the requirements with regard to the electrical parameters can be relaxed considerably. In particular, the requirement of low output impedance can often be given up, with the ensuing advantage that the final stage can be omitted. Figure 6.56 shows an example of such a simple OPAMP. Its very modest claims with regard to chip area allow for accommodating hundreds of such elementary structures in a single chip.

The configuration shown in Figure 6.56 contains an input stage loaded with a current mirror, followed by a complementary final stage ($T8$–$T9$) in CS configuration. The source follower $T6$ provides the level shift needed for correctly biasing the gates of $T8$ and $T9$, whereas $T5$ and $T7$ operate as current sources. Pole

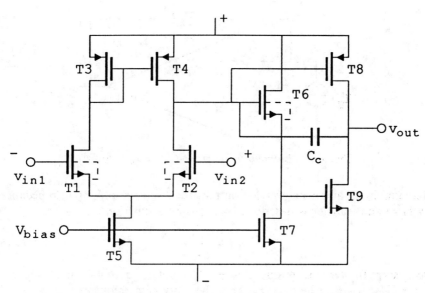

Fig. 6.56 *Simple OPAMP in CMOS-technology.*

splitting is effected by the compensation capacitor C_c. In passing it may be noted that in MOS-implementations the additional right half-plane zero in the transfer function that is concomitant with the pole splitting capacitance (see Section 5.4) can be so close to the origin that its existence cannot be neglected. As was seen in Section 5.4, the location of this zero is given by $z = (g_{m8} + g_{m9})/C_c$. In bipolar configurations g_m is usually so high that the zero has little effect on the phase characteristic. The zero can be shifted to a safe location by replacing C_c by a series connection of a capacitor and a resistor R_c. In that case,

$$z = \frac{1}{[R_c - 1/(g_{m8} + g_{m9})] C_c} \quad (6.43)$$

By choosing $R_c = 1/(g_{m8} + g_{m9})$, the zero is even shifted to infinity. The resistor can simply be implemented by a MOST operating in the triode region.

6.12 OPAMP circuits: the inverting amplifier

Standard operational amplifiers are designed for application in a variety of signal-processing functions. Manufacturers of such OPAMPs provide data sheets with a wealth of application suggestions. In this section only the basic principles underlying design with OPAMPs as building blocks for electronic circuits will be dealt with.

To a first approximation an OPAMP can be modelled as an amplifier with a

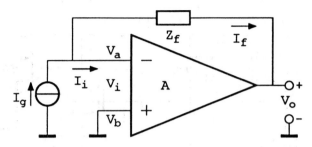

Fig. 6.57 *Elementary current-to-voltage converter.*

differential input and a single-sided output. Ideally, gain and input impedance are very large and transfer is described by a one-pole function

$$A = \frac{A_0}{1 + s\tau} \qquad (6.44)$$

Since, except for very low frequencies, $\tau \gg 1/\omega$, holds, eq. (6.44) can also be written as $A = A_0/s\tau = \omega_1/s$, where $\omega_1 = A_0/\tau$ is the unity-gain frequency.

For low-frequency signals the OPAMP approximates the ideal nullor. Figure 6.57 shows a frequently used basic OPAMP circuit operating as a current-to-voltage converter.

Using the nullor concept ($A_0 \to \infty$) we note that $V_i = V_a - V_b \to 0$ and $I_i = 0$, hence $I_f = I_g$ and $V_o = -I_g Z_f$. The input impedance Z_{if} as seen at the inverting input is given by $Z_{if} \approx Z_f/A \to 0$. The inverting input is said to be 'virtually grounded'. In the circuit of Figure 6.58, $I_g = V_g/Z_g$, since $V_a \to 0$ (virtual ground). Hence,

$$\frac{V_o}{V_g} \approx -\frac{Z_f}{Z_g} \qquad (6.45)$$

In this configuration Z_g converts voltage into current and the circuit operates as an inverting voltage amplifier.

If it is desired to account for the non-idealities of the OPAMP, a more complete

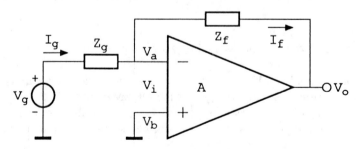

Fig. 6.58 *Inverting voltage amplifier.*

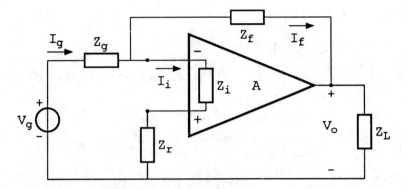

Fig. 6.59 *Non-ideal inverting amplifier.*

schematic must be used. In Figure 6.59 the finite input impedance and the impedance between the non-inverting input and ground Z_r is accounted for.
The following equations apply:

$$V_o = -AV_i = -AI_iZ_i \tag{6.46}$$

$$I_g = I_f + I_i \tag{6.47}$$

$$V_g = I_gZ_g + I_i(Z_i + Z_r) \tag{6.48}$$

$$I_i(Z_i + Z_r) = I_fZ_f + V_o \tag{6.49}$$

Using these equations we find

$$\frac{V_o}{V_g} = \frac{-AZ_f}{Z_g(1+A)} \left[1 + \frac{Z_r}{Z_i(1+A)} + \frac{Z_f}{Z_i(1+A)} + \frac{Z_f(Z_i + Z_r)}{Z_iZ_g(1+A)}\right]^{-1} \tag{6.50}$$

In any OPAMP Z_i and A are so large that

$$\frac{Z_r}{Z_i(1+A)} + \frac{Z_f}{Z_i(1+A)} + \frac{Z_f(Z_i + Z_r)}{Z_iZ_g(1+A)} \ll 1 \tag{6.51}$$

Using this relation, eq. (6.50) simplifies to

$$\frac{V_o}{V_g} = \frac{-AZ_f}{Z_g(1+A)} \left[1 - \frac{Z_r + Z_f}{Z_i(1+A)} - \frac{Z_f(Z_i + Z_r)}{Z_iZ_g(1+A)}\right] \tag{6.52}$$

This equation can be used for the calculation of V_o/V_g in cases where the finite values of A, Z_i and Z_r cannot be neglected. With $A \gg 1$, eq. (6.52) simplifies to

$$\frac{V_o}{V_g} \approx -\frac{Z_f}{Z_g} \tag{6.45}$$

which relation was found previously by using the nullor concept.
In many practical applications, the most important deviation from ideal nullor

behaviour is brought about by the presence of the pole in A. Writing

$$A = \frac{A_0}{1 + s\tau} \quad (6.44)$$

we find for the voltage transfer A_f of the inverting amplifier, assuming $Z_i \to \infty$,

$$A_f = \frac{V_o}{V_g} = \frac{-Z_f}{Z_g + (Z_g + Z_f)/A} \quad (6.53)$$

If $Z_f = R_f$ and $Z_g = R_g$, eq. (6.53) can be written in the form

$$A_f = -\frac{A_0 R_f}{A_0 R_g + R_g + R_f} \cdot \frac{1}{1 + s\tau_f}$$

with

$$\tau_f = \frac{\tau(R_g + R_f)}{A_0 R_g + (R_g + R_f)}$$

Hence, the 3-dB bandwidth ω_{0f} is given by

$$\omega_{0f} = \frac{A_0 R_g + (R_g + R_f)}{R_g + R_f} \cdot \frac{1}{\tau} \quad (6.54)$$

EXAMPLE
Let $\tau = 2.50 \cdot 10^{-3}$ s, $R_g = 1$ kΩ, $R_f = 10$ kΩ, and $A_0 = 10^5$. Then:

$$A_{f0} = \left(\frac{V_o}{V_g}\right)_{(\omega=0)} = -10 \text{ and } \omega_{0f} = 3.6 \cdot 10^6, \text{ so that } f_{3db} = 0.58 \text{ MHz}.$$

This demonstrates that even in the case of strong feedback ($A_{f0} \ll A_0$), the pole in A still affects the transfer of high-frequency signals significantly. This is, of course, the price that must be paid for the unconditional stability brought forth by the phase compensation.

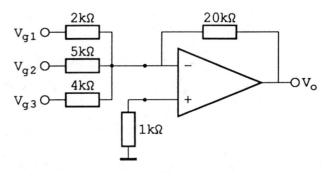

Fig. 6.60 *Summing amplifier.*

If the input stage of the OPAMP is equipped with bipolar transistors, their biasing base currents will flow in the input circuits. In order to prevent offset due to these currents, Z_r should exhibit the same d.c.-resistance as the parallel connection of Z_g and Z_f.

By being virtually grounded the inverting input is an ideal summing node. Several input signal circuits can be connected to this terminal without mutual interaction. As an example, Figure 6.60 shows a summing amplifier whose output signal is given by $V_o = -(10\,V_{g1} + 4\,V_{g2} + 5\,V_{g3})$.

6.13 Integrating amplifier

In analogue computers and in instrumentation systems a frequently occurring signal-processing function is *time integration*. An obvious method is to charge a capacitor with a signal current (Figure 6.61).

If i_g is a constant current, and $v_c = 0$ at $t = 0$, then

$$v_c = i_g R_0 \{1 - \exp(-t/\tau)\} \text{ with } \tau = R_0 C$$

As long as $t \ll \tau$, $\exp(-t/\tau)$ can be approximated by $1 - t/\tau$, hence $v_c = i_g t/C$, representing true integration. If i_g is an arbitrary signal and $R_0 \to \infty$, then

$$v_c = \frac{q_c}{C} = \frac{\int_0^t i_g\,dt}{C} \tag{6.55}$$

The time interval wherein accurate integration is accomplished can be greatly expanded by using C as the feedback element in an OPAMP circuit (Figure 6.62a).

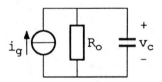

Fig. 6.61 *A simple integrating circuit.*

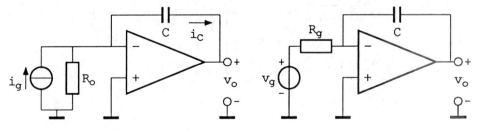

Fig. 6.62(a) *Improved integrator with OPAMP.* (b) *Voltage integrator.*

Thanks to the virtual grounding of the inverting input, $i_g = i_c$ and hence $v_o = -v_c$, while eq. (6.55) applies for an infinite time interval. In practice, the interval of accurate integration is limited by the finite gain of the OPAMP. The input impedance of the inverting input node is $1/sAC$, which is equivalent to a capacitance AC. Hence, in comparison with the circuit of Figure 6.61, the interval of accurate integration has been expanded by a factor A. This result can also be stated in terms of pole positions. True integration corresponds to a pole in $s = 0$. The actual pole position in the integrator of Figure 6.61 is $s = -1/\tau$ and in that of Figure 6.62(a) it is $s = -1/A\tau$. If the signal is obtained from a voltage source (Figure 6.62b), a resistor R_g can be used to convert the signal into a current. The output signal is given by

$$v_o = -\frac{1}{\tau} \int_0^t v_g \, dt \text{ with } \tau = R_g C$$

6.14 Non-inverting amplifier

Subsequently, we consider the case of driving at the non-inverting input. It goes without saying that the feedback impedance has still to be connected to the inverting input for accomplishing *negative* feedback. Figure 6.63 shows the configuration for this case.

Using the nullor concept, we can write $V_i = 0$, $I_i = 0$, hence $V_a = V_b$ and $I_r = I_f$, so that

$$-\frac{V_g}{Z_r} = \frac{V_g - V_o}{Z_f}$$

yielding

$$\frac{V_o}{V_g} = 1 + \frac{Z_f}{Z_r} \tag{6.56}$$

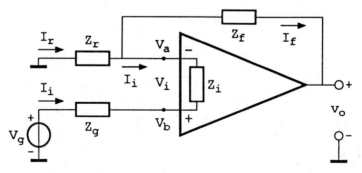

Fig. 6.63 *Non-inverting amplifier.*

The input impedance of the non-inverting input is very high. Straightforward calculation yields

$$Z_{if} = \frac{V_b}{I_i} = Z_i \left(1 + \frac{AZ_r}{Z_r + Z_f}\right) \qquad (6.57)$$

Thanks to its high input impedance the non-inverting amplifier makes an excellent buffer amplifier.

If Z_r is omitted from the circuit ($Z_r \to \infty$), the configuration exhibits unity gain. This has given birth to the commonly used name of *voltage follower*. The circuit possesses a very high input impedance and a very low output impedance. The latter is found to be given by

$$Z_{of} = \frac{Z_o}{1 + A} \qquad (6.58)$$

where Z_o denotes the output impedance of the open amplifier. Obviously, the voltage follower behaves as a nearly ideal voltage source.

By driving an OPAMP at both input terminals, subtraction of the input signals is accomplished. If, in the circuit shown in Figure 6.64, $R_{g1} = R_{g2} = R_g$, then

$$V_o = \frac{R_f}{R_g}(V_{g2} - V_{g1}) \qquad (6.59)$$

A disadvantage of this circuit is its lack of symmetry. The impedance levels of both input terminals differ largely. Even if R_{g1} and R_{g2} are equal, the unavoidable parasitic capacitances cause aberrations of true subtraction for high-frequency components of the input signals. More important: the internal impedances of the signal sources are part of R_{g1} and R_{g2}. If these internal impedances are unequal or not well-defined, which is frequently the case in instrumentation systems, the subtractive action is corrupted. A better approach in such cases is to apply both input signals to a non-inverting input, using two OPAMPs. The outputs of these OPAMPs behave as voltage sources with well-defined low internal impedances that can be safely used for driving a circuit such as that of Figure 6.64. Figure 6.65

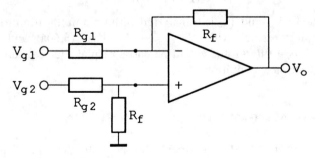

Fig. 6.64 *Subtracting amplifier.*

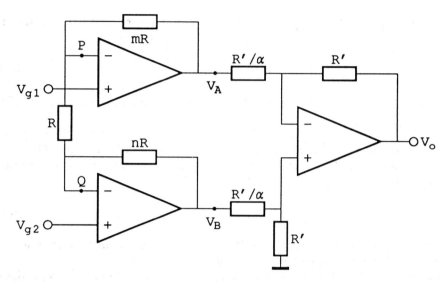

Fig. 6.65 *Improved subtracting amplifier.*

shows the configuration obtained in this way. For the calculation of signal transfer we apply the superposition principle. If only V_{g1} is present, then point Q is virtually grounded. Hence, $V_A = (1+m)V_{g1}$ and $V_B = -nV_{g1}$. And if only V_{g2} is present, then P is virtually grounded, so that $V_A = -mV_{g2}$ and $V_B = (1+n)V_{g2}$. Adding both results yields

$$V_A = (1+m)V_{g1} - mV_{g2} \qquad (6.60a)$$

$$V_B = (1+n)V_{g2} - nV_{g1} \qquad (6.60b)$$

so that

$$V_A - V_B = (1+m+n)(V_{g1} - V_{g2})$$

If the gain of the third OPAMP is α, then

$$V_o = \alpha(1+m+n)(V_{g1} - V_{g2}) \qquad (6.61)$$

A CM-signal in V_{g1} and V_{g2} is transferred without amplification to the inputs of the final amplifier, since with $V_{g1} = V_{g2} = V_g$, eqs. (6.60a) and (6.60b) yield $V_A = V_B = V_g$. As a rule m and n will be chosen equal, but eq. (6.61) shows that this is not a necessary condition.

6.15 Effects of offset-voltages

The effects of offsets in OPAMP circuits can be easily determined by using again the nullor concept. In Figure 6.66 offset-voltages are indicated at each of the three

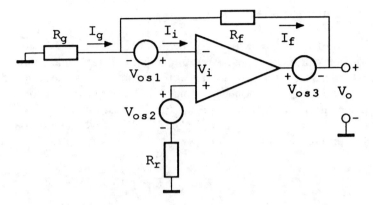

Fig. 6.66 *Offset in an OPAMP.*

terminals. By virtue of the superposition principle the effect of each offset-voltage can be calculated separately. Assume first that only $V_{os\,1}$ is present. From $V_i = 0$ and $I_i = 0$ (nullor concept):

$$\left.\begin{array}{l} V_o = -I_f R_f - V_{os\,1} \\ I_g R_g - V_{os\,1} = 0 \\ I_g = I_f \end{array}\right\} \rightarrow V_o = -V_{os\,1}\left(1 + \frac{R_f}{R_g}\right) \quad (6.62)$$

Proceeding in the same way for $V_{os\,2}$ yields

$$V_o = V_{os\,2}\left(1 + \frac{R_f}{R_g}\right) \quad (6.63)$$

and for $V_{os\,3}$:

$$V_o = 0 \quad (6.64)$$

The result of eq. (6.64) is not surprising. It is a consequence of the infinite loop gain in the feedback loop. If A is finite, we find

$$V_o = \frac{V_{os\,3}}{A}\left(1 + \frac{R_f}{R_g}\right) \quad (6.65)$$

From eqs. (6.62) and (6.63) we conclude that input offset-voltages are transferred like signals at the non-inverting input. Hence, in a given application the allowable signal-to-offset ratio determines the amount of offset that can be tolerated. By the way, note that in the integrator configuration even a very small offset voltage will ultimately provoke saturation of the OPAMP.

6.16 Logarithmic amplifier

By employing a non-linear element in the feedback loop, a defined non-linear transfer function can be obtained. An example is the logarithmic amplifier shown in Figure 6.67. This circuit uses to good advantage the fact that the relation between I_C and V_{BE} of a BJT is an exponential function. For a good transistor the validity range of this relation can encompass even nine decades. For positive values of v_g, by virtue of the virtual grounding of point A: $i_{C1} = i_g = v_g/R_g$. The second OPAMP amplifies the difference $\Delta V_{BE} = V_{BE1} - V_{BE2}$. With $i_{C1} = v_g/R_g$ and $i_{C2} = I_0$:

$$v_o = \left(1 + \frac{R_f}{R_1}\right) \Delta V_{BE} = \left(1 + \frac{R_f}{R_1}\right) V_T \ln \frac{v_g}{R_g I_o} \qquad (6.66)$$

The resistor in the emitter leads of $T1$ and $T2$ reduces load fluctuations of the first OPAMP.

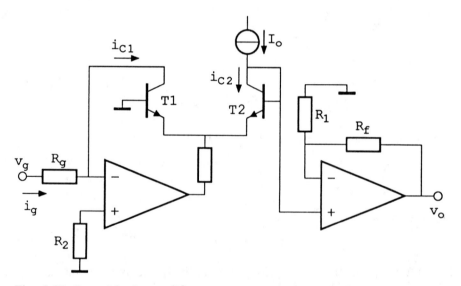

Fig. 6.67 *Logarithmic amplifier.*

References

Klein, H. W. and Engl, W. L. (1984) 'Design technique for low noise CMOS operational amplifiers', *Proceedings European Solid-State Circuits Conference*, Edinburgh, 1984, pp. 27–30.

Meijer, G. C. M. and Verhoeff, J. B. (1976) 'An integrated bandgap reference', *IEEE Journal of Solid-State Circuits*, **SC 11**, (3), 403–6.

Further reading

Allen, P. A. and Holberg, D. R. (1987) *CMOS Analog Circuit Design*, Holt, Rinehart and Winston, New York.

Gray, P. R. and Meyer, R. G. (1984) *Analysis and Design of Analog Integrated Circuits*, Wiley, New York.

Grebene, A. B. (1984) *Bipolar and MOS Analog Integrated Circuit Design*, Wiley, New York.

7 Power amplifiers

7.1 Introduction

Though amplification of signal power is a feature of virtually any amplifier, the term 'power amplifier' is usually reserved for designating an amplifier whose primary function is to drive a load that absorbs a significant amount of power. Consequently, the amplifier must be capable of coping with large voltage and current swings at its output. In view of this requirement due attention has to be paid to certain properties that are of restricted importance in small-signal amplifiers. The most notable among these are as follows:

- efficiency η of the conversion of d.c.-power into signal power;
- prevention of damage by overdriving;
- distortion due to the non-linearity of transfer characteristics;
- output-impedance characteristics.

7.2 Driving considerations

Figure 7.1 shows the output characteristics of a bipolar transistor. We assume a resistive load R_L. Further, for convenience of the discussion we assume that the driving signal is sinusoidal. For any transistor, either discrete or integrated, the maximum tolerable values of the collector current I_C, the collector voltage V_{CE} and of the collector dissipation P_C are specified by the manufacturer. Together these data determine what is called the *safe operating area*. In Figure 7.1 the maximum values of V_{CE} and I_C are denoted V_{max} and I_{max}. The minimum value of V_{CE} is the 'knee-voltage' ($V_{CE\,sat}$), the minimum value of I_C is, of course, $I_C = 0$.

The locus of the points where $P_C = I_C V_{CE} = P_{max}$ is indicated by the hyperbolic line $P_C = P_{max}$. Operation above this line is prohibited at the penalty of irreversible damage to the transistor.

The equation relating I_C and V_{CE} at a load R_L is

$$V_{BB} = I_C R_L + V_{CE} \tag{7.1}$$

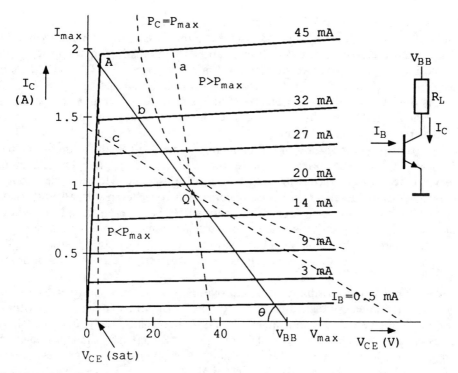

Fig. 7.1 I_C-vs-V_{CE} characteristics of a bipolar power transistor with load lines. Cotan $\theta = R_L$. Load line b represents the case of optimum power matching.

The graphical representation of this equation is a straight line in the I_C–V_{CE} plane. In Figure 7.1 three load lines are drawn under the assumption that $V_{BB} < V_{max}$. Line a represents a load that is too small to provide full use of the available voltage swing capability of the transistor; line c represents a load that is too large for full profit to be taken of the available current capability. Simultaneous use of the current and voltage capabilities is obtained by the load R_{Lopt} represented by line b that connects the points $(V_{BB}, 0)$ and (V_{min}, I_{max}). Obviously, this is the load permitting maximum output power without violation of the safe operating area condition. If positive as well as negative voltage swings of equal magnitude must be accommodated, the quiescent point Q, i.e. the point indicating biasing voltage and current, must be chosen halfway down the loadline. In the case of the optimum load resistance, then $V_{CE} = \approx \frac{1}{2} V_{BB}$ and $I_C = \frac{1}{2} I_{max}$.

The power taken from the power supply is $P_{dc} = I_C V_{BB}$. The power content of a sinusoidal output signal with voltage $v = v_{max} \cos \omega t$ and current $i = i_{max} \cos \omega t$ is $\frac{1}{2} i_{max} v_{max}$. Hence, the power efficiency of the amplifier is

$$\eta = \frac{\frac{1}{2} i_{max} v_{max}}{I_C V_{BB}} = \frac{i_{max} v_{max}}{V_{BB} I_{max}} \quad (7.2)$$

From Figure 7.1 we take it that under maximum driving conditions $v_{max} = \frac{1}{2}V_{BB}$ and $i_{max} = \frac{1}{2}I_{max}$, so that $\eta_{max} = 0.25$ or 25%. In the absence of a driving signal, P_{dc} is divided equally between collector dissipation P_c and dissipation P_R in R_L. With increasing driving, a.c.-power is developed in R_L at the cost of P_C. In the case of maximum driving, $P_C = 0.25\ P_{dc}$, $P_{ac} = 0.25\ P_{dc}$ and $P_R = 0.50\ P_{dc}$.

It is obvious that the power efficiency of the transistor operating in this way is low. One method to improve η is to connect the load via a transformer (Figure 7.2), so that no d.c.-power is lost in R_L.

The transformer is also useful for matching a given load R_L to the optimum load $R_{Lopt} = V_{BB}/I_{max}$. The turns ratio n follows from $n = \sqrt{R_{Lopt}/R_L}$. The maximum power efficiency obtained with this arrangement is 50%. However, in modern electronics transformers are often unwelcome. Moreover, even with a transformer, power efficiency remains unfavourable, particularly because P_{dc} is independent of the a.c.-output, implying that η decreases dramatically at low driving levels.

With regard to JFETs or MOSTs in power stages, similar considerations apply. Figure 7.3 shows the output characteristics of a power MOST with the optimum load line.

Before discussing possibilities for improving power efficiency further let us first deal with the aspect of distortion due to driving with large signals. It goes without saying that small-signal equivalent circuits cannot adequately describe signal transfer under large-signal driving conditions. Either large-signal models or graphical methods have to be used for analyzing signal transfer. However, the use of small-signal models need not be rejected altogether. The values taken by small-signal parameters depend on the actual biasing point. Large-signal operation can be looked upon as operation under biasing that depends on momentary signal value. Distortion occurs if in the small-signal transfer function parameters show up that are bias-dependent. For instance, I_C versus V_{BE} of a BJT is governed by a highly non-linear relation. This manifests itself in the small-signal parameter g_e in that g_e is proportional to I_C.

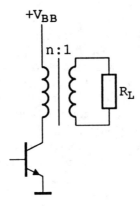

Fig. 7.2 *Loading via a transformer.*

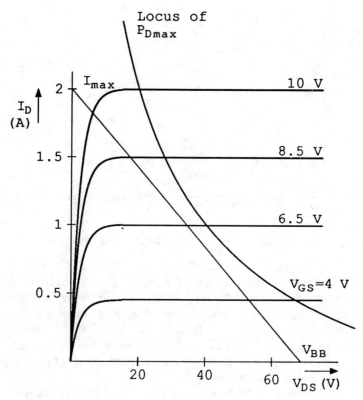

Fig. 7.3 I_D-vs-V_{DS} characteristics of a power-MOST with load lines for optimum power matching.

What are the dominant non-linear effects in a BJT? The most dominant one is the just-mentioned exponential relation between V_{BE} and I_C. In a small-signal transistor the current-transfer function I_C vs I_B is usually rather linear. But in a power transistor several effects are responsible for considerable deviations from linear behaviour. At large current densities high-level injection phenomena occur. The concentration of minorities is no longer far below that of the majorities. As a consequence, reverse injection from the base into the emitter enhances, with the ensuing decrease of emitter efficiency. Moreover recombination in the base region increases. Also emitter crowding (Section 2.5) can affect current transfer to a fairly large extent. In the low-current region current transfer is also affected, here mainly by recombination in the emitter junction. Due to all these effects, in a power transistor maximum β occurs at a certain 'medium' value of bias current, sloping down both at lower and at higher currents. Figure 7.4 shows the general shape of a β versus I_C plot for a power transistor. But, all in all, it remains true that non-linearity of current transfer is much less pronounced than that of voltage-to-current transfer.

214   ANALOGUE ELECTRONIC CIRCUIT DESIGN

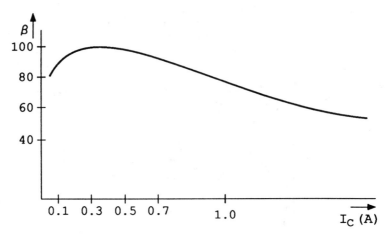

Fig. 7.4 *Dependency of β on I_C for a power transistor.*

With this in mind we conclude that for restricting the effects of non-linearity under large-signal driving conditions, current driving is to be preferred above voltage driving. Figure 7.5(a) and (b) shows the basic driving arrangements. In terms of small-signal transfer we can write $v_o = -g_e R_L v_i$ for Figure 7.5(a) and $v_o = -\beta i_i R_L$ for Figure 7.5(b). Since β depends much less on momentary current than does g_e, these relations confirm that, with regard to distortion, current driving is the preferred mode of operation.

As has been discussed in Section 4.2, the nature of the load dictates whether it should be driven from a voltage source, i.e. a low-impedance source, or from a

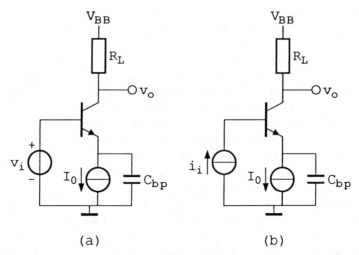

Fig. 7.5 *Basic CE-configurations: (a) voltage driving; (b) current driving.*

current source, i.e. a high-impedance source. With many common loads, voltage driving is preferred. A familiar example of such a load is a loudspeaker. The required low output impedance can be obtained by using overall feedback. But this is not always implemented without problems in a power amplifier. An alternative is to apply local feedback. An obvious way to implement this is to use the emitter as the output terminal. This means using the transistor as an emitter follower. Figure 7.6(a) and (b) gives the basic configurations. In terms of small-signal transfer, neglecting r_b, we can write $v_o = v_i g_e R_L/(1 + g_e R_L)$ for the configuration of Figure 7.6(a), and $v_o = i_i \beta R_L$, for the configuration of Figure 7.6(b). Again, it turns out that current driving is advantageous with regard to distortion, except for the unlikely case that $R_L \gg 1/g_e$ over the entire range of values that g_e can assume.

It may be argued that the output impedance of a current-driven emitter follower is still high, since $R_o = r_e + R_g/\beta$, R_g being the impedance of the current source. In practice this is usually the collector impedance r_{ce} of the driving source. This impedance is in any event much smaller than the collector impedance r_{ce} of the CE-configuration. And, if desired, lowering R_o of the emitter follower can be accomplished by replacing the power transistor by a Darlington pair.

A further advantage of the emitter-follower option is that biasing is much easier. It can be accomplished by a current source in the emitter lead, as is indicated in Figure 7.6(a, b). Biasing the CE-stage is more difficult, unless a large bypass capacitor can be used, as is indicated in Figure 7.5(a, b).

The emitter-follower approach has one disadvantage. Since its voltage gain assumes unity value at maximum, the driving voltage at the base of the transistor is much larger than in the CE-case. This also true in the case of current driving, since then $v_i = i_i R_i = i_i \beta (r_e + R_L)$. As a consequence, the voltage swing that must be accommodated by the driver transistor is large, and thus care must be taken to avoid distortion effects in this driving stage. However, this stage can be designed

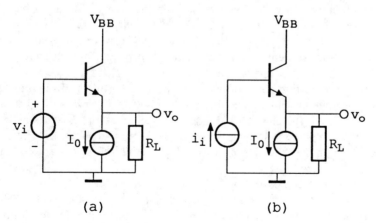

Fig. 7.6 *Basic emitter-follower configurations:* (a) *voltage driving;* (b) *current driving.*

for producing voltage at high-impedance level, so only little signal power is involved.

Turning our attention now to FET-power stages, we must note that in this case there is no choice between voltage and current driving. Because of the infinite input impedance, only voltage driving can be considered. But this is not at all a disadvantage, since voltage-to-current transfer is much less non-linear than in a BJT. Particularly in power-MOSFETs, g_m tends to lose its dependence on instantaneous bias current because of the reduction of electron mobility in high fields. Power-MOSFETs exhibit very attractive properties for power amplification. They can possess very large values of g_m, for instance 1 A V^{-1} at $I_o \approx$ 200 mA. A further advantage is that the input capacitance is much lower than in a BJT operating at the same current level. Preserving adequate high-frequency response is therefore much easier than in a bipolar design. An additional advantage is the negative temperature coefficient of g_m. With increasing temperature, current therefore tends to decrease with the ensuing decrease in dissipation, which prevents self-heating. In a BJT the temperature coefficient is positive so that here self-heating can occur with the associated risk of self-destruction due to 'thermal runaway'. Therefore, taking provisions for adequate cooling is a matter of serious concern in bipolar power-amplifier design.

Although a judicious choice of driving conditions can effectively contribute to keeping distortion within bounds, in many applications further reduction will be imperative. The obvious way to achieve this is the application of overall feedback over the loop consisting of the driver amplifier and the power stage. But in the case of a bipolar power stage in particular, the design of such a feedback loop requires considerable care. Power transistors exhibit large diffusion and junction capacitances. The diffusion capacitance is heavily current-dependent and, therefore, the time constants in the feedback loop are signal-dependent. The feedback loop must remain stable and well-behaved under all circumstances. Moreover, the possibility of slew-rate limiting (Section 6.10) has to be taken into consideration. Using a MOSFET-power stage largely alleviates the risks of undesirable effects of the application of feedback. However, for economic reasons BJTs will often be the elements of choice, and it turns out that with adequate design efforts quite satisfactory results can be achieved.

Note, by the way, that some designers of power amplifiers still favour the use of electron tubes, which can be classified as field-effect devices with vacuum as the 'material' of which the channel is made. Their characteristics are similar to those of FETs. Though there is some ground for using these devices, their technological incompatibility makes their application unattractive unless extremely high power must be delivered.

An obvious further possibility for reducing distortion is the use of a balanced output stage. Figure 7.7 shows the principle of this approach. When both transistors operate in their active region, all even harmonics in the drain currents represent common-mode signals. Consequently, they do not show up in the differential output signal. In Figure 7.7 the subtraction of both drain currents is

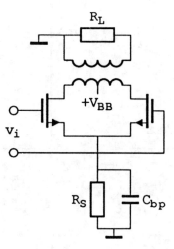

Fig. 7.7 *A balanced output stage.*

accomplished by means of an output transformer that can also be used for load-impedance transformation. In the past the use of output transformers was commonplace in audio output stages. Nowadays, complementary transistor configurations are usually used. Examples will be given in Section 7.4 in the context of the implementation of class-AB amplifiers. These configurations can also be biased for continuous operation in the active region (so called class-A operation, see next section). Balancing as a tool for distortion reduction is particularly helpful when field-effect devices are used. Because of their approximately quadratic characteristics, distortion in these devices is predominantly of even order. In order to preserve the quadratic nature of voltage-to-current transfer, no source degeneration should be applied. Hence, source biasing can best be accomplished by a capacitively bypassed resistor or current source.

7.3 Classes of operation

We must now face the problem of improving the efficiency of power amplification. Under the operating conditions we have assumed up to now, a large amount of d.c.-power is wasted due to the high quiescent current level. This level is chosen such that accommodation is provided for full output current and voltage swing of both polarities. As a consequence, d.c.-power delivered by the power supply is independent of the magnitude of the driving signal. This is, of course, an inefficient way of converting d.c.-power into signal power. Figure 7.8(a and b) indicates graphically how a signal is transferred in a BJT and in an FET using the familiar biasing arrangement. If Θ denotes the fraction of a full period the transistor is conducting, in this case $\Theta = 1$. This type of operation is commonly called 'operation in class-A'.

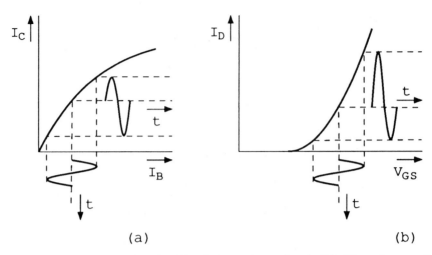

Fig. 7.8(a) *Class-A operation for bipolar transistor; $\theta = 1$. (b) Class-A operation for enhancement FET; $\theta = 1$.*

Figure 7.9(a and b) depicts transfer if the quiescent point is chosen in the origin, i.e. $I_C = 0$, $I_D = 0$ respectively, so that only signal parts of positive polarity are transferred ($\Theta = 0.5$). It goes without saying that in this case distortion is excessively large. But this can be corrected for by using a second transistor that is driven with a signal of opposite polarity so that it conducts during the half-period in which the first transistor is cut off. Combining both output signals re-establishes the situation of full signal transfer. This type of operation is called 'operation in class-B'.

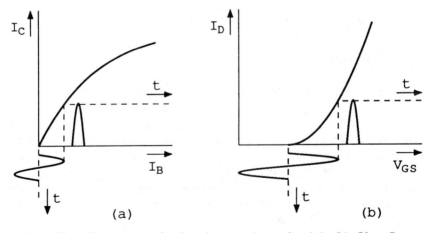

Fig. 7.9(a) *Class-B operation for bipolar transistor; $\theta = 0.5$. (b) Class-B operation for enhancement FET; $\theta = 0.5$.*

POWER AMPLIFIERS 219

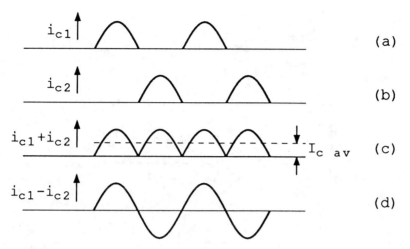

Fig. 7.10 *Waveforms in the class-B output stage: (a) and (b) collector currents; (c) sum of both collector currents; (d) difference of both collector currents.*

Figure 7.10 depicts the waveforms of both output signals, together with the sum and the difference of these signals. As Figure 7.10(d) shows, subtraction of both signals yields the proper output signal.

Figure 7.11 shows the output characteristics of a BJT under the assumption of class-B operation. The quiescent point is located on the V_{CE}-axis at $V_{CE} = V_{BB}$, since with $I_C = 0$, $V_{CE} = V_{BB}$.

To find the efficiency of a class-B stage, reference is made to Figure 7.10. The d.c.-power P_{dc} delivered by the power supply is given by $V_{BB} \cdot I_{C\,av}$, where $I_{C\,av}$ is the average value of the collector current. From Figure 7.9(c) we take $I_{C\,av} = 2/\pi\, i_{max}$, where i_{max} is the amplitude of the output signal. Signal power is given by $P_{ac} = \frac{1}{2} v_{max} i_{max}$, so that

$$\eta = \frac{P_{ac}}{P_{dc}} = \frac{\pi}{4} \frac{v_{max}}{V_{BB}} \tag{7.3}$$

From Figure 7.11 we see that the maximum output signal v_{max} that can be accommodated is $v_{max} = V_{BB}$. Hence, under peak driving conditions, efficiency is $\frac{1}{4}\pi$ or 78.5%. This is certainly much better than the 50% of the class-A stage. But even more important is that P_{dc} decreases with decreasing driving signal. As is obvious from eq. (7.3), η decreases with decreasing v_{max}, but since P_{dc} also decreases, the low value of η with small driving signals does not imply much wastage of power. This can be seen convincingly by considering the case of no driving. With $v = 0$, $\eta = 0$, but also $P_{dc} = 0$.

It must be noted that our calculation of power efficiency is for driving with a sinusoidal signal. If the class-B stage is driven with an ideal square wave signal of maximum allowable magnitude, η can reach close to the 100% mark, since then the

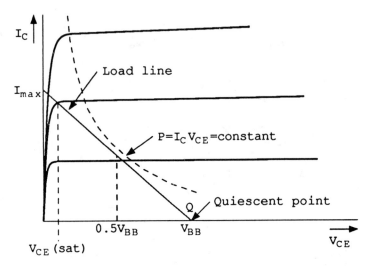

Fig. 7.11 *Quiescent point and load line for a power transistor operating in class-B.*

signal causes the stage to switch between two extreme states, viz. $I_C = 0$, $V_{CE} = V_{BB}$ and $I_C = I_{max}$, $V_{CE} = V_{CE(sat)} \to 0$. Thus in both states $P_C \to 0$.

In practice, full class-B operation is not propitious. Close to the cut-off point, the transconductance is so low that hardly any signal transfer occurs. This leads to a 'dead zone' in the output signal, as indicated in Figure 7.12. As a consequence, the output signal is seriously distorted. This type of distortion is called 'crossover distortion'. It can be avoided by allowing some biasing current to flow in the no-driving state. In practice the quiescent current can be a small fraction of the average current under full driving, so that efficiency is not essentially affected. This class of operation is indicated by the name 'class-AB operation'.

An additional advantage of class-AB operation is the reduction of distortion that it brings about. In class-B or class-AB operation the transfer characteristics are identical for signal parts of opposite polarity. Hence, the effective transfer characteristic is symmetrical and therefore, in principle, no even harmonics occur.

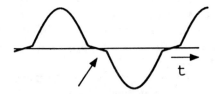

Fig. 7.12 *Cross-over distortion at zero crossings of a sinusoidal signal.*

7.4 Implementation of class-AB output stages

In the foregoing section we found that for optimum power efficiency a power amplifier, often alternatively named an 'output amplifier', should consist of two transistors, each handling one polarity of the driving signal. Presently, we must turn our attention to the realization of such amplifier stages. We will concentrate first on bipolar implementations.

If two transistors of the same type (both npn or pnp) are to be used in the balanced output stage, they must be driven with signals of opposite polarity. Such driving signals can be obtained from both collectors of a differential pair. According to the observations made in Section 7.3, the output currents of both power transistors must be subtracted. This can be achieved by using a balanced transformer. In the past this was the usual implementation. By way of illustration, Figure 7.13 shows an example with the output transistors in CE-configuration. The power transistors $T5-T6$ and the associated driver transistors $T3-T4$ form Darlington pairs, driven in turn from the differential pair $T1-T2$. In the resistive biasing circuitry a diode is incorporated for thermal stabilization. Note that the emitter resistors needed for obtaining AB-operation are not capacitively bypassed.

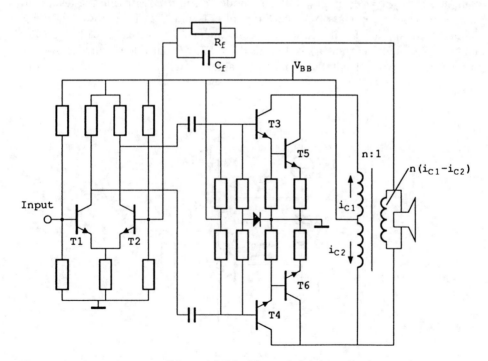

Fig. 7.13 *An output amplifier with biasing and driving circuitry, using an output transformer.*

Such bypassing is not possible in this case since the average current in the transistors depends on the amplitude of the driving signals, and capacitors would constitute a memory for past signal levels. In order to prevent too much transconductance degeneration, the emitter resistors should be kept small.

Reduction of distortion and of output impedance is accomplished by the application of overall feedback through the feedback network $R_f - C_f$. A detailed discussion of the circuit is omitted here, since with the present state of the art the use of transformers is restricted to very special cases.

A transformer for high-power audio-frequency signals is an expensive and bulky item and it is no surprise that circuit configurations have been devised that can do without this component. Though it is possible to devise transformerless circuits with transistors of the same type, this approach has been largely abandoned in view of several disadvantages inherent in this type of circuit. This was made possible by the advent of high-quality complementary transistor pairs, i.e. pairs of npn- and pnp-transistors with nearly identical characteristics, except, of course, polarity.

Figure 7.14 shows a basic complementary transistor configuration. The output transistors $T1-T2$ are operating as emitter followers. The circuit is fed from a symmetrical power supply delivering $+V_{BB}$ and $-V_{BB}$. The circuit is adjusted for $I_{C3} = I_o$ in the absence of a driving signal v_i. Due to the symmetry of the configuration, $I_{C1} = I_{C2}$ and $I_L = 0$. The diodes $D1$ and $D2$ provide the level shift that is necessary for class-AB operation of $T1$ and $T2$. If a positive driving voltage v_i is applied i_{C3} increases and consequently i_{B2} increases with the same amount. The signal current i_{b2} is amplified by $T2$, so that a load current $\beta_{T2} i_{b2}$ is produced. Likewise, if v_i takes a negative value, i_{C3} decreases, leading to a signal

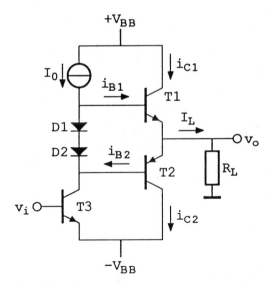

Fig. 7.14 *Basic complementary output stage.*

current i_{b1}, which is amplified by $T1$ and transferred to the load. Hence, R_L carries no d.c.-current, and no d.c.-power is dissipated in R_L. This is the basic mode of operation of the circuit.

Several remarks may be made concerning details of the operation of the circuit. Let us first consider the way in which AB-operation is established. If, for simplicity, we assume that the base currents of $T1$ and $T2$ in the quiescent state can be neglected, so that $D1$ and $D2$ both carry the biasing current I_o, we can write

$$V_{D1} + V_{D2} = V_{BE1} + V_{EB2}$$

hence

$$2V_T \ln \frac{I_o}{I_{sd}} = V_T \ln \frac{I_{C1}}{I_{s1}} + V_T \ln \frac{I_{C2}}{I_{s2}} \tag{7.4}$$

where I_{sd} denotes the reverse current of $D1$ and $D2$, and I_{s1}, I_{s2} the reverse currents of $T1$ and $T2$. With true complementary transistors $I_{s1} = I_{s2} = I_s$. From eq. (7.4) we then obtain

$$I_{C1} = I_{C2} = I_o \frac{I_s}{I_{sd}} \tag{7.5}$$

This implies that the quiescent current in $T1$–$T2$ is determined by I_o and the ratio of the junction areas of the diodes and the transistors. Assuming good thermal coupling of all components, I_{C1} and I_{C2} do not depend on temperature. As long as the base currents of $T1$ and $T2$ are small in comparison with I_o, relation (7.4) remains valid, so that under driving conditions

$$i_{C1} i_{C2} = I_o^2 \frac{I_s}{I_{sd}} \tag{7.6}$$

still holds. This implies that no hard changeover of signal current between $T1$ and $T2$ occurs. If, for instance i_{C1}, increases by a factor of 10, then i_{C2} will still flow, albeit decreased by a factor of 0.1. As a consequence a smooth changeover takes place which is a beneficial situation with regard to the occurrence of crossover distortion.

The current I_o cannot be freely chosen, since it is the biasing current for $T3$ and hence has to accommodate the driving requirements for this transistor. This implies that, given I_o, the operating point of $T1$ and $T2$ must be adjusted through a judicious choice of the ratio I_s/I_{sd}. However, since $D1$ and $D2$ never have to conduct a current in excess of the small current I_o, it is propitious to use small-area diodes. But then the ratio I_s/I_{sd} could prove to be too large for efficient class-AB operation. This problem can easily be coped with by replacing the diodes with an alternative level shift arrangement, such as that shown in Figure 7.15(a) which has been discussed previously in Section 5.14. Another alternative is shown in Figure 7.15(b), where T_A only conducts the base current of T_B plus the current through the (large) resistor R.

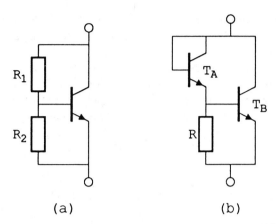

Fig. 7.15(a) *Alternative level shift circuit.* (b) *Another level shift circuit.*

Several modifications of the basic circuit, as shown in Figure 7.14, have found their way into practical applications. If true complementary transistors are available, the basic circuit can provide satisfactory results. However, in most current bipolar IC-processes, only low-quality pnp-transistors can be produced. Lateral transistors are not useable as power transistors since their generally poor quality worsens still more at high currents. Fortunately, an alternative is available. Since in the configuration of Figure 7.14 the pnp-collector is at $-V_{BB}$, a substrate-pnp can be employed. Though substrate pnp-transistors are also of poor quality, they can still be laid out for large-current operation. However, they remain poor counterparts for the npns in complementary configurations. Improvements can be obtained in various ways. Figure 7.16 shows a modification in which $T3$ of Figure 7.14 has been replaced by a pnp-transistor operating as an emitter follower. Consequently, v_o/v_i assumes approximately unity value, largely independent of

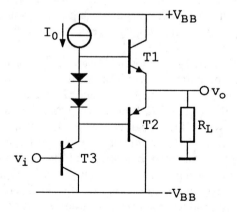

Fig. 7.16 *An alternative output stage.*

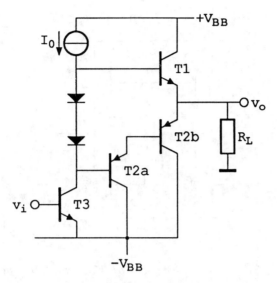

Fig. 7.17 *Replacement of T2 by a Darlington pair.*

β-values of $T1$ and $T2$ and hence of dissimilarities between them. The disadvantage of this approach is that $T1$ and $T2$ are no longer current-driven; an advantage is the lower value of R_o. In the circuit of Figure 7.17 improvement is obtained by replacing $T2$ by a Darlington pair. A disadvantage is the occurrence of an additional pole in the transfer function with its associated extra phase shift. Moreover, the maximum negative-output voltage swing is lowered by one V_{BE}. A slightly better alternative is replacement of $T2$ by a combination transistor (Section 5.11), as indicated in Figure 7.18. The combination behaves approximately as an improved pnp. The extra phase shift is less, thanks to the better high-frequency behaviour of the added npn. The best approach is, of course, the use of a technology that is capable of providing good vertical pnp-transistors. Such processes have been developed, but they require more masking steps.

A matter of concern in the design of output stages is their current- and voltage-handling capability. For a brief discussion of this point we consider the circuit of Figure 7.14. The maximum positive current in R_L, supplied by $T1$, is $\beta_{T1} I_o$. Negative load current has to be supplied through $T2$. Since with v_i positive, the current for $T2$ can be supplied from the negative power supply and base current of $T2$ is sinked through $T3$, essentially no current limiting takes place. In order to prevent the occurrence of a current in $T2$ beyond its safe operating area a provision can be made restricting the current that can be conducted by $T3$.

With regard to the maximum-output voltage that can be accommodated, the positive value of v_o is not allowed to rise above the level where $T1$ and the transistor constituting the current source I_o are no longer in their active region. Hence

$$v_{o\,max}^+ = V_{BB} - V_{BE} - V_{CE\,sat(I_o)}$$

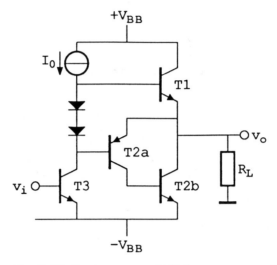

Fig. 7.18 *Replacement of T2 by a compound transistor.*

On the negative side, a similar reasoning yields

$$v_{o\,max}^{-} = V_{BB} - V_{BE} - V_{CE\,sat(T3)}$$

7.5 MOST-output stages

With availability of MOS-transistors, the design of output stages follows the same pattern as with bipolar transistors. The basic configurations are quite similar. However, there are less side-effects due to the absence of gate current and to the smoother transfer characteristics. Figure 7.19 shows a basic complementary output stage. It may be observed that it is topologically identical to the bipolar circuit of Figure 7.14. Level shift is provided by the diode-connected transistors $T4$–$T5$, while $T6$ operates as a current source. Since the output transistors $T1$–$T2$ operate as source followers, they do not provide voltage gain. Their driving signals v_1 and v_2 must therefore equal v_o. The driving signals are provided by the driving transistor $T3$. Since in a good CMOS-process, p-channel and n-channel transistors are well-matched, problems due to lack of complementary symmetry are much less prominent than in the bipolar counterpart.

The level shift circuit consisting of $T4$ and $T5$ should ideally provide the desired level shift without introducing appreciable signal attenuation. The small-signal impedance of $T4$ and $T5$ is $1/g_{m4} + 1/g_{m5}$. Hence,

$$v_1 = -g_{m3}v_i\,[r_{d3} \parallel (r_{d6} + 1/g_{m4} + 1/g_{m5})]$$

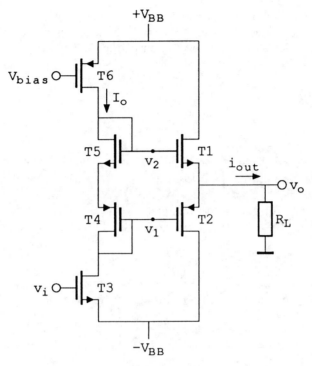

Fig. 7.19 *Complementary MOS-output stage.*

and

$$v_2 = \frac{r_{d6}}{1/g_{m4} + 1/g_{m5} + r_{d6}} v_1$$

For $v_1 \approx v_2$, $1/g_{m4} + 1/g_{m5} \ll r_{d6}$ must hold.

Using the simple transfer formula for a MOS-transistor (Section 2.8),

$$I_D = \tfrac{1}{2}\beta(V_{GS} - V_t)^2 \tag{7.7}$$

with $\beta = W/L$, β_{sq}, g_m is given by $g_m = \sqrt{2\beta I_D}$

Further, $r_d = V_A/I_D$, where V_A denotes Early-voltage. Since V_A is approximately proportional to the channel length L (Section 2.8), we can write $V_A = L V_{Ao}$, with V_{Ao} the Early-voltage per μm. We can now compile the following formulas for designing the output stage:

$$r_{d6} = \frac{L_6 V_{Ao}}{I_o}, \quad g_{m4,5} = \sqrt{2\left(\frac{W}{L}\right)_{4,5} \beta_{sq} I_o}$$

$$g_{m1,2} = \sqrt{2\left(\frac{W}{L}\right)_{1,2} \beta_{sq} I_{1,2}}, \quad g_{m3} = \sqrt{2\left(\frac{W}{L}\right)_3 \beta_{sq} I_o} \tag{7.8}$$

We note that V_{Ao} and β_{sq} are parameters that are determined by the technological process. Typical values are $V_{Ao} \approx 5$ V μm^{-1} and β_{sq} (n-channel) ≈ 40 μA V^{-2}, β_{sq}(p-channel) ≈ 20 μA V^{-2}. By inspection of Figure 7.19, we find

$$V_{GS5} + V_{SG4} = V_{GS1} + V_{SG2} \tag{7.9}$$

From eq. (7.7)

$$V_{GS} = V_t + \sqrt{\frac{2I_D}{\beta}} \tag{7.10}$$

follows where V_t denotes threshold voltage, which is determined by the technological process; usually $|V_t|$ is different for the p-channel and the n-channel devices.

If we denote V_t for both transistor types by V_{tn} and V_{tp}, and β_{sq} for both types by β_{sqn} and β_{sqp}, from eqs. (7.9) and (7.10) it follows, when $i_{out} = 0$,

$$V_{tn} - V_{tp} + \sqrt{\frac{2I_o}{\beta_5}} + \sqrt{\frac{2I_o}{\beta_4}} = v_{tn} - V_{tp} + \sqrt{\frac{2I_1}{\beta_1}} + \sqrt{\frac{2I_2}{\beta_2}}$$

where $I_1 = I_2$ is the quiescent current in $T1-T2$. If we choose

$$\frac{W_4}{L_4} = \frac{W_5}{L_5} \frac{\beta_{sqn}}{\beta_{sqp}} \quad \text{and} \quad \frac{W_2}{L_2} = \frac{W_1}{L_1} \frac{\beta_{sqn}}{\beta_{sqp}}$$

we find

$$I_1 = I_o \left(\frac{W_2}{W_4} \frac{L_4}{L_2} \right) \tag{7.11}$$

For $r_{d6} \gg 1/g_{m4} + 1 g_{m5}$ to be valid, L_6 should be taken as large as is compatible with bandwidth requirements. Further, g_{m4} and g_{m5} can be maximized by designing $T4$ and $T5$ as minimum-length transistors.

The minimum useable value of I_o is mainly dictated by bandwidth requirements. When I_o is given, I_1/I_o can be assigned the desired value with the aid of the ratio $W_2/W_4 = W_1/W_5$, assuming that $L_4/L_2 = 1$. In general I_1 will be chosen considerably higher than I_o, so that $W_2/W_4 \gg 1$. A large value of W_1 and W_2 is desirable in any event in order to prevent excessive current density in $T1-T2$.

If no satisfactory design can be accomplished along these lines, the level shift produced by $T4$ and $T5$ can be reduced by replacing these transistors by an alternative level shift, for instance one that is similar to that of the bipolar configuration of Figure 7.15(a).

With regard to the required magnitude of the driving signal: since $T1$ and $T2$ operate as source followers, v_1 and v_2 are approximately equal to v_o. If $1/g_{m4} + 1/g_{m5} \ll r_{d6}$, $v_1 \approx v_2 = g_{m3} v_i (r_{d6} \| r_{d3})$. Hence,

$$v_i \approx \frac{v_o}{g_{m3}(r_{d6} \| r_{d3})} \tag{7.12}$$

In order to calculate the bandwidth of the stage, the various capacitances must be taken into account. Such calculations tend to be cumbersome, and in many cases computer simulations will prove to be the fastest way to do the job. Using computer simulation as a design tool implies that design is done by repeated analysis. In order to restrict the number of iterations necessary to obtain a useful result, it is advisable to start with a reasonable approximation, found by design on paper. The preceding discussion has indicated how this can be done.

If large output power is required, on-chip realization of the power amplifier cannot be accomplished in view of the inherent large dissipation. Discrete MOS-transistors are available with quite large power capabilities. In such cases it is often propitious to use bipolar transistors in the low-power part of the amplifier. This option is also available for on-chip medium power designs if manufacturing is planned in a BIMOS-process. Such processes for mixed bipolar–MOS-circuits are becoming within reach with the progress of the state of the art in IC-technology.

Further reading

Gray, P. R. and Meyer, R. G. (1984) *Analysis and Design of Analog Integrated Circuits*, Wiley, New York.

Grebene, A. B. (1984) *Bipolar and MOS Analog Integrated Circuit Design*, Wiley, New York.

8 Noise in electronic circuits

8.1 Introduction

Any physical system is afflicted with stochastical variations caused by random processes. The occurrence of such variations is a fundamental phenomenon that is a consequence of the second law of thermodynamics. An example is the well-known 'Brownian motion' of particles in an environment at non-zero absolute temperature. As a consequence, the pressure which a gas exerts on the surface of a container cannot be measured with absolute accuracy. The accuracy can be improved by averaging many individual measurements, since the long-term average of the random fluctuations tends to zero. In electronic systems for the processing of audio signals, these fluctuations manifest themselves as noisy sounds when the signals are converted into sound. The term 'noise' has found general acceptance for indicating such stochastical fluctuations, irrespective of the nature of the signals afflicted with them. Therefore, it is common use to speak about, for instance, noise in television signals or in binary signals.

Being a fundamental phenomenon, noise cannot be avoided. But its effects can be minimized. The art of noise minimization can therefore be characterized as 'learning to live with noise'. The amount of noise that a signal is afflicted with determines the minimum signal power that can be accurately detected. Therefore, the noise threshold is a fundamental determinant with regard to the dynamic range of a signal. If noise was entirely absent, even the smallest signal power would encompass an infinite dynamic range. The larger the noise threshold is, the larger the maximum undistorted signal power that must be accommodated by the circuitry handling the signal. Minimization of the noise level is therefore a key objective in any design situation.

8.2 Description of noise in electronic systems

Let v_n be a stochastically fluctuating electrical signal. The long-term average value $\overline{v_n}$ of v_n is assumed to be zero. This implies that positive and negative excursions of v_n have equal probability. From now on, we will indicate the long-term average of a quantity by a bar over the symbol for that quantity. Since positive and negative values of v_n are assumed to have equal probability, a convenient measure of the randomness of v_n is the average σ^2 of the square of $v_n - \bar{v}_n$, hence

$$\sigma^2 = \overline{(v_n - \bar{v}_n)^2} \tag{8.1}$$

and since we have assumed $\bar{v}_n = 0$, $\sigma^2 = \overline{v_n^2}$. The quantity σ^2 is called the variance of v_n, while σ is called the 'standard deviation'. If v_n has the dimension of a voltage (or, alternatively, a current), σ^2 is related to the power content of the 'noise signal'. Because of this physical connotation, the squared average is a useful measure for the magnitude of a noise signal. A quantity with the same dimension as v_n is obtained by taking the square root of $\overline{v_n^2}$. This is called the root mean square (rms) value of the noise signal, hence

$$v_{n(rms)} = \sqrt{\overline{v_n^2}} \tag{8.2}$$

Let us now consider a series connection of two noise voltage sources (Figure 8.1). The total noise voltage is given by

$$\overline{v_n^2} = \overline{v_{n1}^2} + \overline{v_{n2}^2} + \overline{2v_{n1}v_{n2}} = \overline{v_{n1}^2} + \overline{v_{n2}^2} \tag{8.3}$$

since $\overline{v_{n1}v_{n2}}$ is bound to be zero if v_{n1} and v_{n2} are mutually independent noise voltages. This is easily seen by taking into account that positive and negative momentary values have equal probability. If the noise voltages are interdependent, $\overline{v_{n1}v_{n2}}$ can take a finite value. Such noise sources are called *correlated* noise sources. In many practical situations the relevant noise sources are uncorrelated or only weakly correlated and therefore in most cases the additivity rule, as expressed by eq. (8.3), is valid. Generalizing, in the case of n uncorrelated voltage sources in series:

$$\overline{v_n^2} = \sum_{j=1}^{j=n} \overline{v_{nj}^2}$$

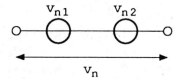

Fig. 8.1 *Series connection of noise sources.*

8.3 Autocorrelation and spectral power density

If a noise signal is displayed on an oscilloscope screen a certain coherence between neighbouring signal values is observed. This is understandable because the maximum slope dv/dt of signal transients is limited by the rise time of the signal amplifier. In terms of a spectral description of the signal this implies that correlation between nearby signal values is related to bandwidth. The internal coherence of a signal can be mathematically described by the *autocorrelation function* $R(\tau)$ of the signal. By definition

$$R(\tau) = \overline{v_n(t) \cdot v_n(t+\tau)} = \lim_{T \to \infty} \frac{1}{T} \int_{-1/2T}^{+1/2T} v_n(t) v_n(t+\tau) \, dt \qquad (8.4)$$

This formula has a simple interpretation: to find the correlation between signal values separated by a time interval τ, a great many determinations of $v(t)$ and $v(t+\tau)$ are performed, and subsequently their product is averaged. If there is no correlation at all between values at t and $t+\tau$, the averaged product must be zero by virtue of the equal probability of positive and negative momentary values. If there is some correlation, a finite value of $R(\tau)$ is found. In general, $R(\tau)$ can be expected to decrease with increasing τ.

The relation between autocorrelation and bandwidth, as suggested above, can be formulated in a mathematically exact way by using Fourier analysis. As is assumed to be known to the reader a spectral function $F(\omega)$ can be assigned to any time function $v(t)$ according to

$$F(\omega) = \int_{-\infty}^{+\infty} v(t) \exp(-j\omega t) \, dt \qquad (8.5a)$$

The reverse relation can be written

$$v(t) = \frac{1}{2\pi} \int_{-\infty}^{+\infty} F(\omega) \exp(j\omega t) \, d\omega \qquad (8.5b)$$

$F(\omega)$ and $v(t)$ are said to constitute a 'Fourier pair'. Let us now consider a signal bounded within a time interval T. In order to characterize the spectral distribution of signal power we define the *spectral power density* $S(\omega)$ according to

$$S(\omega) = \lim_{T \to \infty} \left[\frac{1}{T} F(\omega)^2 \right] \qquad (8.6a)$$

Through the Fourier transform a function of time is assigned to $S(\omega)$:

$$f_p(\tau) = \frac{1}{2\pi} \int_{-\infty}^{+\infty} S(\omega) \exp(j\omega\tau) \, d\omega \qquad (8.6b)$$

After some mathematical manipulating, eq. (8.6b) can be shown to be equivalent to

$$f_p(\tau) = \lim_{T \to \infty} \int_{-\infty}^{+\infty} v(t) v(t+\tau) \, dt = R(\tau) \qquad (8.7)$$

Hence, it appears that $R(\tau)$ and $S(\omega)$ constitute a Fourier pair, which is the mathematical formulation sought.

8.4 Physical aspects of noise: thermal noise

Noise in electronic circuits is caused by several physical fluctuation mechanisms. An important source of noise is the random motion of electrons in a conductor due to thermal agitation. Random motion of electric charge implies the occurrence of random currents and hence of a random voltage between the terminals of the conductor with a resistance R. Nyquist has shown that the noise voltage generated by this mechanism obeys the relation

$$\overline{v_n^2} = 4kTR\Delta f \qquad (8.8)$$

where k is Boltzmann's constant, T denotes absolute temperature and Δf is the bandwidth of the system under consideration. According to eq (8.8), thermal noise has a flat spectral power density. Such noise is called *white noise*. On the basis of eq. (8.8) a resistor can be modelled as a hypothetical noise-free resistor and a noise source, as indicated in Figure 8.2(a). Alternatively, the noisy resistor can be modelled as a conductor $G = 1/R$ and a noise current source $\overline{i_n^2} = 4kTG\Delta f$ (Figure 8.2b). If the noisy resistor is connected to a load R_L that is in thermal equilibrium with R (Figure 8.2c), the maximum power that can be delivered to R_L amounts to

$$P_{n(max)} = \frac{\overline{v_n^2}}{4R} = kT\Delta f \qquad (8.9)$$

occurring when $R = R_L$ (power matching). $P_{n(max)}$ is called the *available noise power*. If $T = 300$ K, $P_{n(max)} = 4 \times 10^{-21}$ W Hz^{-1}.

If eq. (8.9) were valid in an infinitely vast frequency band, $P_{n(max)}$ would tend to infinity. This is, of course, physically incomprehensible. However, the derivation of eq. (8.8) does not take quantum effects into account. This implies that the validity range of eq. (8.8) is restricted to the frequency band where $hf \ll kT$ (with h being Planck's constant). Since $hf = kT$ at 6000 GHz, in practice thermal noise can be considered as white noise. By the way, note that in the configuration of Figure 8.2(c) no net exchange of power occurs, since R_L delivers the same amount of power to R as R does to R_L.

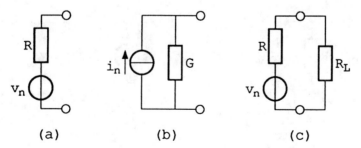

Fig. 8.2(a) *Equivalent circuit of noisy resistor.* (b) *Current source representation.* (c) *Noise source with load resistor R_L.*

234   ANALOGUE ELECTRONIC CIRCUIT DESIGN

If a network contains reactive components (L and C), no noise sources have to be added related to these components. However, this does not imply that these components are free of noise. As an illustration we consider the simple RC-combination of Figure 8.3(a). For the calculation of the noise energy in the capacitor we use the equivalent circuit of Figure 8.3(b).

For any spectral component of v_n, the noise voltage v_{nc} is given by

$$|v_{nc}(f)| = \frac{|v_n|}{|Z|\omega C}$$

with

$$|Z| = \left(R^2 + \frac{1}{\omega^2 C^2}\right)^{1/2} \tag{8.10}$$

Hence, the noise voltage per Hz is

$$\overline{v_{nc}^2}(f) = \frac{\overline{v_n^2}}{|Z|^2 \omega^2 C^2} = \frac{4kTR}{1 + R^2 \omega^2 C^2} \tag{8.11}$$

The total noise voltage is given by

$$\overline{v_{nc}^2} = \int_0^\infty \frac{4kTR}{1 + 4\pi^2 f^2 R^2 C^2}\,df = \frac{kT}{C} \tag{8.12}$$

Obviously, v_{nc} depends solely on C, irrespective of R. The noise energy in the capacitor is

$$\tfrac{1}{2} C \overline{v_{nc}^2} = \tfrac{1}{2} kT$$

The RC-combination acts as a filter on the noise generated by R. If the noise of R were filtered by an ideal low-pass filter with bandwidth B, then according to eq. (8.8) the noise voltage is given by

$$\overline{v_n^2} = 4kTRB \tag{8.13}$$

Since $\overline{v_{nc}^2} = kT/C$ with respect to the filtering of the noise the bandwidth is

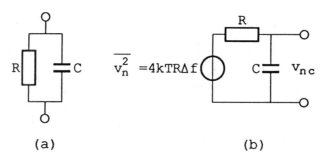

Fig. 8.3(a) *RC-circuit.* (b) *Equivalent circuit for the calculation of noise voltage.*

$1/(4RC) = \frac{1}{2}\pi B_3$, where B_3 is the 3-dB bandwidth of the RC circuit, which amounts to $1/(2\pi RC)$. This effective bandwidth is called the *noise bandwidth* B_n of the circuit. Formulated in a more general way, B_n is the bandwidth of an ideal low-pass filter whose area under the square of the amplitude characteristic equals that of the filter under consideration.

8.5 Shot noise

A second important source of noise in electronic circuits is related to the stochastic nature of the flow of charge carriers that pass a potential barrier, for instance emitted by a cathode or a pn-junction. Because of the discrete nature of the emitted charge carriers, mostly electrons, the current is quantized. As a consequence, the number of emitted carriers per given time interval Δt is a fluctuating quantity. As an example we consider a vacuum photocell. Due to the photoelectric effect, electrons are emitted by the cathode. If on average $\bar{N}$ electrons are emitted in the time interval Δt, then $\bar{I} = \bar{N}q/\Delta t$. Since N is fluctuating, the same applies to I. Putting $i_n = I - \bar{I} = (N - \bar{N})q/\Delta t$, we obtain

$$(\overline{i_n^2})_{\Delta t} = \frac{\overline{(N-\bar{N})^2}}{(\Delta t)^2} q^2 \qquad (8.14)$$

The quantity $\overline{(N-\bar{N})^2}$ depends on the statistical distribution function that describes the emission of electrons. The emission of electrons is a sequence of mutually independent events. And because the number of electrons participating in the emission process during even a very short time interval Δt is very large, the distribution function corresponds with the well-known Poisson distribution. Because of this $\overline{(N-\bar{N})^2} = \bar{N}$ applies. Substitution into eq. (8.14) yields

$$(\overline{i_n^2})_{\Delta t} = \frac{\bar{N}q^2}{(\Delta t)^2} = \frac{\bar{I}q}{\Delta t} \qquad (8.15)$$

The larger the averaging time Δt, the smaller the noise current, in agreement with what is expected.

Equation (8.15) is concerned with the time average of the noise current associated with the process of electron emission. In many cases a formulation in terms of a spectral distribution is preferred. The spectral equivalent of eq. (8.15) can be proved to be

$$\overline{i_n^2} = 2q\bar{I}\,\Delta f \qquad (8.16)$$

This type of noise is commonly called *shot noise*. According to eq. (8.16) it is, like thermal noise, characterized by a flat spectral power density or, in other words, it is white.

8.6 Other types of noise

Thermal noise and shot noise are the main types of noise in electronic circuits. However, in certain cases other noise-producing mechanisms have to be taken into account. In certain active devices a current that is generated by an emission process splits up along two or more paths. If there is interaction between the partition mechanism and the emission mechanism, the partition may act as an autonomous cause of additional noise. This type of noise is called *partition noise*. The best-known example is the partition of the cathode current in a pentode vacuum tube between the screen grid and the anode. This causes a pentode tube to be more 'noisy' than a triode tube. In a bipolar transistor partition also occurs: the emitter current splits up into a base current and a collector current. However, in this particular case there is no interaction whatsoever between the partition mechanism and the emission mechanism, and therefore the base current as well as the collector current are independently burdened with shot noise, and no additional partition noise occurs.

Another cause of noise is current pulses induced by moving electrons in nearby electrodes. The best-known example is the current induced in the gate electrode of a FET by electrons travelling from source to drain in the channel region. This type of noise is called *induced noise*. It is by no means white since its spectrum does not contain frequencies with a time period that is large compared with the transit time of the noise-inducing charge carriers. Induced noise is to a certain degree correlated with the noise associated with the primary flow of charge carriers causing the induced charges. Cases wherein this type of noise is significant are rare.

A final type of noise that must be mentioned is the noise caused by stochastical variations in electron conduction mechanisms. An example is provided by the fluctuations of the contacts between carbon particles in a carbon microphone. In semiconductors, fluctuations of surface effects and of recombination and generation mechanisms are responsible for this type of noise. A characteristic property is that its spectral power density is approximately inversely proportional to frequency. A general expression for the spectral density is

$$\overline{i_{nf}^2} = KI^m \, \Delta f / f^\alpha \tag{8.17}$$

where K, m and α are constants. In most practical cases m takes values between 1 and 2, while α is usually close to unity. It is commonly denoted as $1/f$ noise, although 'flicker noise', 'excess noise' and 'current noise' can be used. It is only of interest at low frequencies. Of practical interest is the frequency f_l where it is equal to the white noise that is concurrently present. When the source of white noise that it is compared with is shot noise, it is convenient to rewrite eq. (8.17) in the form

$$\overline{i_{nf}^2} = 2qI\Delta f \cdot (f_l/f) \tag{8.18}$$

where I denotes the current producing the concurrent shot noise. When the concurrent white noise is characterized as thermal noise produced by a conductance

G, a convenient notation is

$$\overline{i_{nf}^2} = 4kTG\Delta f \cdot (f_1/f) \tag{8.19}$$

The value of f_1 depends on the type of component that is afflicted with $1/f$ noise. For good bipolar transistors f_1 ranges from a few Hz up to several tens of Hz; for MOSTs f_1 is much higher, ranging from, say, 100 kHz to a few MHz. This is a consequence of the predominance of surface effects in MOSTs. Because of the large amount of $1/f$ noise present in MOSTs, these components are in general not useable in low-noise amplifiers for low-frequency signals.

8.7 Noise in bipolar transistors

Having made an inventory of noise-generating physical mechanisms, we are now capable of mapping the noise effects in electronic devices. To start with, we consider the pn-junction. In a forward-biased pn-junction we have two mutually independent charge-transport mechanisms. The total current is composed of a diffusion current and a current consisting of thermally generated minorities. In a reverse-biased junction only the last-mentioned current is present. Both constituents are afflicted with shot noise. Since they are mutually independent, the associated shot noise currents i_{nd} and i_{ng} are additive. If we denote the diffusion current by I_d and the thermal generation current by I_s, the total junction current I is given by

$$I = I_d - I_s = I_s[\exp(V/V_T) - 1] \tag{8.20}$$

where V is the biasing voltage. The noise current is given by

$$\overline{i_n^2} = \overline{i_{nd}^2} + \overline{i_{ng}^2} \tag{8.21}$$

where $\overline{i_{nd}^2} = 2qI_d \Delta f$ and $\overline{i_{ng}^2} = 2qI_s \Delta f$.
Because $I_d = I + I_s$, $\overline{i_n^2}$ can also be written

$$\overline{i_n^2} = 2q(I + 2I_s) \Delta f \tag{8.22}$$

Under reverse biasing, $I = -I_s$, hence $\overline{i_n^2} = 2qI_s \Delta f$. If $V = 0$, the junction is in thermal equilibrium. It is to be expected that in this particular situation the incremental resistance $r_d = kT/qI_s$ acts as a source of thermal noise. Putting $I = 0$ into eq. (8.22) yields

$$\overline{i_n^2} = 4qI_s \Delta f = 4kTg_d \Delta f \tag{8.23}$$

where $g_d = 1/r_d$. This result is in agreement with our expectation. In the forward-biased region $I \approx I_d = I_s \exp(V/V_T)$, so that

$$\overline{i_n^2} = 2qI \Delta f \tag{8.24}$$

Where $g_d = I/V_T$, eq. (8.24) can be written

$$\overline{i_n^2} = 2kTg_d \Delta f \tag{8.25}$$

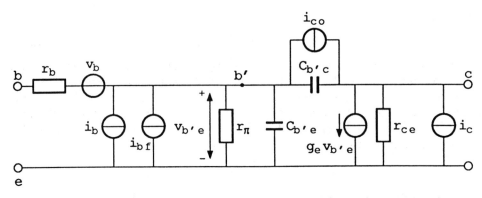

Fig. 8.4 *Hybrid-π equivalent circuit for BJT with noise sources.*

Under forward bias, the noise of the junction is equivalent to the thermal noise that $\frac{1}{2}g_d$ would generate under conditions of thermal equilibrium. This is sometimes expressed by stating that g_d is generating noise 'half thermally'. This is, of course, not in contradiction with eq. (8.8), since under forward bias there is no question of thermal equilibrium.

At this stage we can model the noise behaviour of the BJT. The current I_E injected by the emitter junction splits up into a base current I_B and a collector current I_C. Because the partition of I_E is independent of the injection mechanism, I_B and I_C are independently afflicted with shot noise. Further, the reverse-bias collector current I_s is afflicted with shot noise, while the base bulk resistance r_b is a source of thermal noise. And finally the $1/f$ noise must be taken into account. This noise current shows up as an additional noise current in the base circuit. We can now map all these noise contributions in the small-signal hybrid-π equivalent circuit, as shown in Figure 8.4. The following equations apply:

$$\overline{v_b^2} = 4kTr_b\,\Delta f \tag{8.26}$$

$$\overline{i_b^2} = 2qI_B\,\Delta f \tag{8.27}$$

$$\overline{i_{bf}^2} = 2qI_B\,\Delta f(f_1/f) \tag{8.28}$$

$$\overline{i_c^2} = 2qI_C\,\Delta f = 2kTg_e\,\Delta f \tag{8.29}$$

$$\overline{i_{co}^2} = 2qI_s\,\Delta f \tag{8.30}$$

As in good silicon transistors the reverse-bias current I_s is very small, the noise source i_{co} can nearly always be neglected.

8.8 Noise in field-effect transistors

A field-effect transistor consists of a channel region made up of semiconductor material, whose resistance is controlled by means of an electric field generated by the voltages between the terminal electrodes. The channel is, to a first approximation, in thermal equilibrium with the environment, and consequently it is a source of thermal noise. The channel region is capacitively coupled with the gate electrode. The carriers that are in motion in the channel induce charges in the gate electrode, thereby causing induced gate noise. This noise source exhibits, to a certain degree, correlation with the thermal noise in the channel. As will be shown later, the induced gate noise is mostly of minor importance in a wideband amplifier. It should, however, be kept in mind, since in tuned radio frequency amplifiers its effect cannot always be neglected. At low frequencies $1/f$ noise is present. This type of noise can be modelled as an additional noise source in the drain circuit. Finally, in a JFET, the shot noise due to the reverse-bias current I_G of the gate junction contributes a small amount of shot noise. This noise source can usually be neglected. The various noise sources can now be mapped into the small-signal equivalent circuit, as shown in Figure 8.5.

The thermal noise is accounted for by i_d. Because the resistance of the channel is closely related to the transconductance g_m of the FET, it stands to reason that the thermal noise is also related to g_m. Theoretical analysis yields the result that the equivalent resistance accounting for the thermal noise is $1/cg_m$, where c is a constant whose theoretical value is 0.66 for JFETs and about 1.33 for MOSTs. In practical FETs the actual value of c turns out to be somewhat larger, usually ranging from 1 to 2. Theoretical analysis predicts a value for the induced gate noise given by $\overline{i_{ig}^2} = 4kT\omega^2 C_{gs}^2 \Delta f / 3 g_m$, showing that this noise contribution is strongly frequency-dependent. Taking these results into account, the various noise sources are given by the following equations:

$$\overline{i_d^2} = 4kTcg_m \Delta f \tag{8.31}$$

$$\overline{i_{df}^2} = 4kTcg_m \Delta f (f_1/f) \tag{8.32}$$

$$\overline{i_g^2} = 2qI_G \Delta f \tag{8.33}$$

$$\overline{i_{ig}^2} = 4kT \frac{\omega^2 C_{gs}^2}{3 g_m} \Delta f \tag{8.34}$$

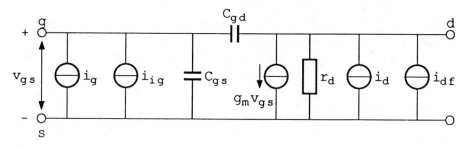

Fig. 8.5 FET equivalent circuit with noise sources.

8.9 A general model for noise in electronic circuits

In the preceding sections models concerning the noise aspects of passive and active devices have been developed. With the aid of these models noise spectra can be calculated at any point in an electronic circuit. Because signal transfer can also be calculated, the signal-to-noise ratio (S/N) can be determined as well. However, using the models as they are, one will frequently be confronted with cumbersome calculations because the noise sources are dispersed all over the circuit. It is therefore worth while looking for a method that is capable of simplifying such calculations. A very attractive model is obtained if all sources of noise are concentrated at the input of the circuit in question. This leaves the circuit noiseless as such, so that S/N at its output terminals is equal to S/N at the input terminals.

Rearranging all noise sources in an arbitrary circuit, no matter how complex, so that they concentrate at the input terminals, can be accomplished in two steps. In the first step the actual noise sources are shifted through the circuit until they reach its terminations. This stage is executed with the help of a simple network-theoretical method. At the end of this step, noise sources are only present at the input and output terminals. The last-mentioned group of noise sources should be transformed into equivalent noise sources at the input terminals. This can again be accomplished by using a network-theoretical procedure.

First we consider how a voltage source can be moved through a network. The appropriate tool is the *Blakesley transform*. Figure 8.6 illustrates its operation. The diagram shows an arbitrary node in a network. If in one of the branches that meet at the node a voltage source is present, this source can be moved through the node to all other branches. This is allowed since by doing this the mesh equations of meshes containing the node in question do not change. Of course, polarity has to be respected, even if the voltage source is a noise source, because by further shifts through the network, voltage sources coming from the same origin can meet again. The new sources can be shifted further through the network, until they finally reach the network terminals. If this procedure is applied to all sources contained in the network, the final result is a network that is free of voltage sources, except for such sources at the input and output terminals.

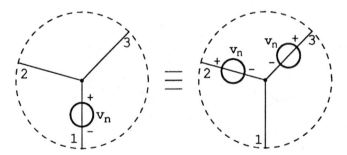

Fig. 8.6 *Blakesley transform of a (noise) voltage source.*

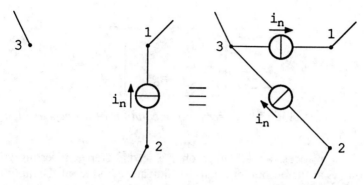

Fig. 8.7 *Shifting of current sources through a network.*

With regard to current sources a similar method is applicable. Figure 8.7 shows a number of nodes and a current source operating between nodes 1 and 2. The current source may be replaced by a number of identical current sources, provided that the sum of the currents that meet in each node of the network does not change. In the right-hand part of Figure 8.7 this is obviously the case: the algebraic sum of the currents in node 3 equals zero, in node 1 a current i_n is sinked, and node 2 sources a current i_n. This is equivalent to the original situation depicted in the left-hand part of Figure 8.7.

If a network contains transformers, voltage and current sources acting at the primary side can be shifted to the secondary side, and vice versa, taking into account the turns ratio (Figure 8.8).

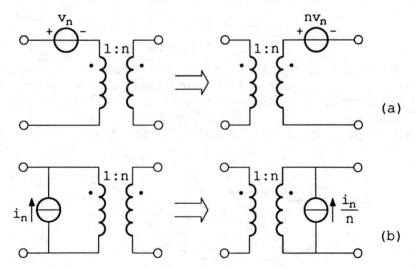

Fig. 8.8 *Transformation of voltage sources (a) and current sources (b) from the primary to the secondary side of a transformer.*

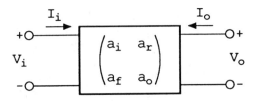

Fig. 8.9 *General representation of a linear twoport.*

We must now focus our attention on the second step: converting voltage and current sources at the output terminals to equivalent sources at the input terminals. Figure 8.9 shows once more the general symbol for a linear twoport. As has been explained in Section 4.5, the relations between the input and output quantities can be expressed by two linear equations in six different ways. Because we are interested in relating input quantities to output quantities, the set that is particularly useful for our purpose is the set governed by the *a*-matrix

$$V_i = a_i V_o + a_r I_o \tag{8.35a}$$

$$I_i = a_f V_o + a_o I_o \tag{8.35b}$$

If the currents and voltages are expressed as complex quantities, the *a*-parameters are also complex. They have different dimensions, as is clear from the definition formulas, eq. (8.35). For instance $a_i = (V_i/V_o)$ with $I_o = 0$ (open output), and so on.

Let us now consider a twoport consisting of the twoport of Figure 8.9, augmented with a voltage source and a current source at the output terminals (Figure 8.10). For this twoport eq. (8.35) can be written as

$$V_i = a_i(V_o + V_n) + a_r(I_o + I_n) \tag{8.36a}$$

$$I_i = a_f(V_o + V_n) + a_o(I_o + I_n) \tag{8.36b}$$

If V_n and I_n represent noise sources, we can describe these by their spectra, applying the complex notation to each spectral component. Based on eqs. (8.36) we can now transform the added sources towards the input terminals. Figure 8.11 shows the result.

Having performed the second step of our programme, we have achieved our

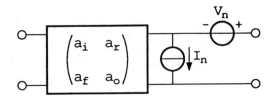

Fig. 8.10 *Representing a noisy twoport by a combination of a noise-free twoport and noise sources at the output.*

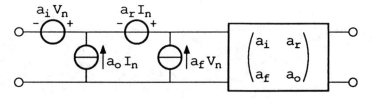

Fig. 8.11 *Transformation of output noise sources into input noise sources.*

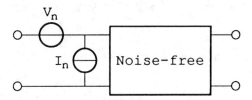

Fig. 8.12 *A general model for a noisy twoport.*

objective of configuring a model containing noise sources at the input terminals only. If all noise sources are uncorrelated, at least approximately, which is frequently the case, we can combine all voltage sources into one voltage source, and similarly with the current sources. The resulting model of striking simplicity is shown in Figure 8.12. It contains no more than one voltage source v_n and one current source i_n, while the twoport is noise-free.

8.10 Application of the generalized model for noisy twoports to the modelling of noise in active devices

The method developed in the preceding section is of a quite general nature. It can be applied to any linear twoport, no matter how complex. It may contain passive as well as active components. As an illustration we will now apply the method to the noise models of the BJT and the FET. At first we consider the BJT. For the sake of convenience, we will perform the operation on the *intrinsic transistor*, i.e. the transistor without the extrinsic base resistance r_b. In the equivalent circuit of the noisy BJT, according to Figure 8.4, the noise sources i_b and i_{bf} are already comfortably located at the input of the intrinsic transistor. The current source i_{co} can nearly always be neglected in comparison with the other noise sources. Hence, the only noise source that must be manipulated is the source i_c that acts on the output terminals. According to Figure 8.11, we must determine the parameters a_r and a_o for the intrinsic transistor. Straightforward calculation with the aid of the hybrid-π equivalent circuit for the intrinsic transistor yields

$$a_r = \frac{1}{g_e - j\omega C_{b'c}} \quad \text{and} \quad a_o = \frac{(1 + j\omega r_\pi C_\pi)/r_\pi + j\omega C_{b'c}}{g_e - j\omega C_{b'c}} \tag{8.37}$$

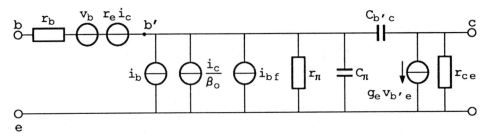

Fig. 8.13 *Noise equivalent circuit of BJT, wherein the noise source V_{i_c} is transformed into input noise sources.*

In the low-frequency region, where $\omega \ll 1/(r_e C_{b'c})$ and $\omega \ll 1/(r_\pi C_\pi)$, these expressions simplify into

$$a_r = 1/g_e = r_e \quad \text{and} \quad a_o = 1/(g_e r_\pi) = 1/\beta_0 \tag{8.38}$$

Using these results we find the transformed equivalent circuit for the noisy transistor represented in Figure 8.13.

The noise voltage source $r_e i_c$ and the noise current source i_c/β_0 both originate from the source i_c. Henceforth, they are correlated. However, the source i_c/β_0 can be specified by

$$\frac{\overline{i_c^2}}{\beta_0^2} = \frac{2qI_C\,\Delta f}{\beta_0^2}$$

while $\overline{i_b^2} = 2qI_B\,\Delta f$. Since $I_C \approx \beta_0 I_B$, and for low frequencies $\beta_0 \gg 1$, it is obvious that in all practical (low-frequency) cases the source i_c/β_0 can be removed from the equivalent circuit. Further, we can replace the voltage sources v_b and $r_e i_c$ by one voltage source. The result is the very simple equivalent circuit shown in Figure 8.14.

It is often convenient to describe noise sources, irrespective of the mechanism responsible for their occurrence, as if they were generated as thermal noise in a resistor. The magnitude of the resistor that would generate an amount of noise equal to that of the noise source under consideration, is called the *equivalent noise resistance*. Since

$$\overline{v_n^2} = \overline{v_b^2} + \overline{(r_e i_c)^2} = (4kTr_b + 2kTr_e)\,\Delta f = 4kT(r_b + \tfrac{1}{2}r_e)\,\Delta f \tag{8.39}$$

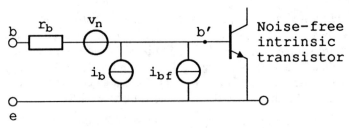

Fig. 8.14 *Simplified representation of the schematic of Figure 8.13.*

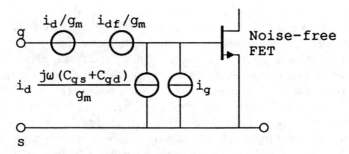

Fig. 8.15 *FET noise-equivalent circuit.*

the equivalent noise resistance assigned to v_n is $R_{nv} = r_b + \frac{1}{2}r_e$. Similarly, since $I_B \approx I_C/\beta_0$, the equivalent noise resistance assigned to i_b is $R_{ni} = 2\beta_0 r_e$.

Subsequently we turn our attention to the field-effect transistor. In the equivalent circuit of Figure 8.5, the current sources i_d and i_{df} have to be transformed to the input terminals. Form the small-signal equivalent circuit of the FET we find for the frequency region where $\omega \ll g_m/C_{dg}$

$$a_r = 1/g_m \text{ and } a_o = j\omega(C_{gs} + C_{dg})/g_m \tag{8.40}$$

Because $1/f$ noise is only significant at low frequencies, the transformed noise source $a_o i_{df}$ can be neglected. By inspecting and comparing eqs. (8.31) and (8.34) we can conclude that in general $i_{ig} \ll a_o i_d$ will be valid. Hence, the induced gate noise can be ignored. For all practical purposes the equivalent circuit of Figure 8.15 results. Because the sources $a_r i_d$ and $a_o i_d$ have a common origin, these sources exhibit a certain amount of correlation.

Applying the convention of representing noise sources by equivalent noise resistances again, we obtain the following values for these resistances:

$$R_{nv} = c/g_m \tag{8.41}$$

$$R_{ni} = \frac{g_m}{c\omega^2(C_{gs} + C_{dg})^2} \tag{8.42}$$

$$R'_{ni} = \frac{2kT}{qI_G} \tag{8.43}$$

Because in a good FET I_G is extremely small, only at very low frequencies may R'_{ni} be non-negligible in comparison with R_{ni}. In a MOST, no gate current is present, so that $R'_{ni} \to \infty$.

8.11 Qualification of noisy twoports

Before putting the question of how to optimize a circuit with regard to its noisiness, it is appropriate to discuss how noise behaviour can be specified. First of all we note that the quantity that is decisive for the quality of a signal is not the total amount of noise but the *signal-to-noise ratio* (SNR or S/N). Optimum SNR need not be concurrent with minimum noise. It is conceivable that a measure for minimizing noise degrades signal transfer to such an extent that, overall, SNR worsens.

How SNR should be specified is a matter of convenience. In many cases the logical choice is the ratio of signal power to noise power. However, this choice is impractical for signals whose power contents depend largely on their information content, such as video signals. A more suitable choice then is the ratio of two conveniently measured signals. In the case of a video signal this could be the ratio of the peak-to-peak signal voltage and the rms value of the noise voltage.

An alternative possibility is to define a quantity that qualifies the noise properties of a given twoport, for instance an amplifier or a transmission circuit. In the course of time, many proposals for such quantities have been made. Most of them have found little acceptance. One quantity has found wide acceptance: the *noise figure* of a twoport. The underlying idea is that a twoport always contains noise sources and thus is bound to add noise to the input signal. Hence, output SNR is lower than input SNR. The noise figure F is defined as the ratio of input SNR to output SNR. It is obvious that $F > 1$ must hold. For a (hypothetical) noiseless twoport, $F = 1$ and therefore $F - 1$ is called the *excess noise figure*. The noise figure is usually given in dB. It is more or less common practice to designate F as the *noise factor* instead of the noise figure if it is not expressed in dB.

Further, concerning the specification of the noise behaviour of a twoport by its noise figure, we consider a twoport together with its driving source and load (Figure 8.16).

In dealing with noise figure calculations, it is customary to assume that $Z_g = R_g + jX_g$ generates thermal noise at $T = 290$ K. The signal power delivered to the twoport is P_i. It is maximum if $Z_i = Z_g^*$, where Z_g^* denotes the conjugate complex value of Z_g. In this case the power delivered by V_g is called the *available input power* $P_{ia} = |V_g|^2/4R_g$. The rationale of the introduction of this quantity is that it depends solely on the properties of the source. The thermal noise supplied by R_g within a frequency band Δf is denoted P_{in}. If, again $Z_i = Z_g^*$, the *available noisepower* $P_{ina} = kT\Delta f$ is supplied. The ratio P_{ia}/P_{ina} is called the *available signal-to-noise ratio*.

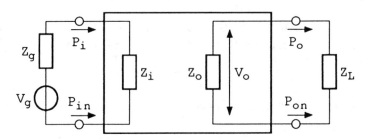

Fig. 8.16 *A twoport between signal source and load.*

The *power gain* of the twoport is $A_p = P_o/P_i$, where P_o is the signal power supplied to Z_L. The *available power gain* is defined as

$$A_{pa} = P_{oa}/P_{ia} \tag{8.44}$$

where P_{oa} is the power supplied to Z_L if $Z_L = Z_o^*$. By definition

$$F = \frac{SNR(input)}{SNR(output)} = \frac{P_i/P_{in}}{P_o/P_{on}} \tag{8.45}$$

As can be easily shown,

$$F = \frac{P_{ia}/P_{ina}}{P_{oa}/P_{ona}} \tag{8.46}$$

also holds.

An alternative definition of F is

$$F = \frac{P_{on}}{A_p P_{in}} \tag{8.47}$$

i.e. the ratio of the output noise power to the amplified input noise power. The definitions according to eqs. (8.45) and (8.47) are equivalent since, with $A_p = P_o/P_i$, eq. (8.46) can be rearranged in the form

$$F = \frac{P_{on}/P_o}{P_{in}/P_i} = \frac{P_i/P_{in}}{P_o/P_{on}}$$

In the foregoing, the noise power was tacitly considered to be restricted to a small frequency band Δf. Frequently F will be frequency-dependent. We can therefore specify P_{on} and P_{in} more appropriately as spectral power densities, i.e. as noise power per Hz bandwidth. Specified in this way F is commonly denoted as the *spot noise figure*. In wideband amplifiers F will usually vary over the bandwidth of the amplifier. The amplifier as a whole can then be specified by its *average* or *integral noise factor*, defined as the ratio of the (available) output noise power over the total frequency band and the amplified noise power of the driving source over that frequency band.

Because of the assumption that R_g is a thermal noise source at $T = 290$ K, F is a useful quantity for specifying the noise behaviour of the twoport in cases where this actually applies. If not, then the use of F is inappropriate. An example is an amplifier for a television camera. The noise of most camera tubes is shot noise and the noise power is by no means equivalent to the thermal noise that would be supplied by its internal resistance if this were a thermal noise source.

In communication systems a transmission chain usually consists of a cascade of twoports (Figure 8.17). It is then convenient to have a formula for calculating the noise factor of the cascade from the given noise factors of the constituting twoports. Friis has shown that

$$F_{cascade} = F_1 + \frac{F_2 - 1}{A_{pa1}} + \frac{F_3 - 1}{A_{pa1} A_{pa2}} + \frac{F_4 - 1}{A_{pa1} A_{pa2} A_{pa3}} + \ldots \tag{8.48}$$

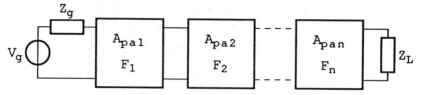

Fig. 8.17 *A cascade of noisy twoports.*

8.12 Optimization of the noise behaviour of a twoport

How can we optimize the noise behaviour of a twoport? First the quantity to be optimized must be selected. In many cases S/N will be the obvious choice. In cases where F is a useful quantity, this quantity can be chosen. Of course both approaches should lead to the same conclusion. The advantage of using F as the quantity to be optimized is that the calculation with this approach often turns out to be somewhat simpler. Having chosen the quantity to be optimized, an expression for this quantity must be found in terms of the parameters of the circuit in question. Subsequently, this expression must be examined for the occurrence of manipulable parameters. When a designable parameter is selected, its value for optimizing the chosen noise quantity must be determined. This strategy can always be applied, assuming that the topology of the circuit is given. The latter will usually be determined by the signal-processing function it has to perform. Fortunately, conflicting topology requirements with regard to signal processing and noise minimization are rare. If such a situation occurs, circuit topology should be reconsidered.

The parameters that can be manipulated for the sake of noise optimization are usually restricted to small-signal parameters of active devices and the turns ratio of a transformer. As can be seen from eq. (8.48), the noise behaviour of a circuit consisting of several stages is dominated by the noise of the first stage, provided the gain of this stage is large enough. For this reason, as a rule, the first stage should be a CE- or a CS-stage and this rarely conflicts with proper signal behaviour.

To exemplify the strategy unfolded above, we consider the optimization of the noise behaviour of the general twoport shown in Figure 8.18. The noise of the twoport is represented, as discussed previously, by a noise voltage source v_n and a noise current source i_n in the input circuit. The circuit is driven by the signal voltage source v_g with internal impedance Z_g, whose resistive part R_g constitutes a thermal noise source v_{gn} at 290 K.

By representing i_n by its Thévenin equivalent, the equivalent circuit of Figure 8.19 is obtained. In this case F is a useful quantity for qualifying the noise behaviour of the twoport. According to the definition of F, we can write

$$F = \frac{\overline{v_n^2} + \overline{i_n^2}|Z_g|^2 + \overline{v_{gn}^2}}{\overline{v_{gn}^2}} = 1 + \frac{R_{nv}}{R_g} + \frac{|Z_g|^2}{R_{ni}R_g} \qquad (8.49)$$

NOISE IN ELECTRONIC CIRCUITS 249

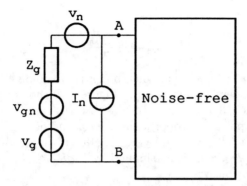

Fig. 8.18 *Equivalent circuit of a noisy twoport.*

where R_{nv} and R_{ni} denote the equivalent noise resistance associated with v_n and i_n. From eq. (8.49) it is clear that F depends on R_g; hence, one method of minimizing F is to manipulate R_g. Since the driving source as such is considered to be given and unalterable, the only way to manipulate R_g is to connect the signal source to the input terminals of the twoport by means of a transformer. If the impedance in the primary circuit of a transformer with turns ratio $1:n$ is Z_g, then the impedance seen at the secondary side is $n^2 Z_g$.

Differentiating F with respect to R_g we find

$$\frac{\partial E}{\partial R_g} = \frac{1}{R_{ni}} - \frac{R_{nv}}{R_g^2} - \frac{X_g^2}{R_{ni} R_g^2}$$

Putting $\partial F/\partial R_g = 0$, we obtain

$$R_g = \sqrt{R_{ni} R_{nv} + X_g^2} \qquad (8.50)$$

If $X_g \to 0$, the optimum value of R_g is

$$R_{g\,opt} = \sqrt{R_{ni} R_{nv}} \qquad (8.51)$$

and

$$F_{min} = 1 + 2\sqrt{R_{nv}/R_{ni}} \qquad (8.52)$$

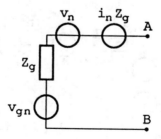

Fig. 8.19 *Thévenin-equivalent for the determination of the noise factor.*

If a transformer is used

$$n_{opt}^2 = \frac{\sqrt{R_{ni}R_{nv}}}{R_g} \tag{8.53}$$

Instead of using F as the qualifying quantity for noise behaviour, we can alternatively use the signal-to-noise ratio. Figure 8.20 shows the equivalent circuit under the assumption that a transformer with turns ratio n is used. Taking into account that signal power is proportional to $n^2 v_g^2$ and noise power is proportional to $\overline{n^2 v_{gn}^2} + \overline{v_n^2} + n^4 R_g^2 \overline{i_n^2}$, we find that SNR is inversely proportional to $R_g + R_{nv}/n^2 + n^2 R_g^2/R_{ni}$. SNR is maximum if $n^2 = \sqrt{R_{ni}R_{nv}}/R_g$, which is identical to the value found previously.

As an example we consider a first stage using a bipolar transistor. In Section 8.10 we found

$$R_{nv} = r_b + \tfrac{1}{2} r_e \text{ and } R_{ni} = 2\beta_0 r_e$$

It must be taken into account that the transformation of the shot noise of the collector current has been performed for the intrinsic transistor. Hence, in the equivalent circuit of Figure 8.14, r_b is still present, albeit as a noiseless resistor. This has to be taken into account in the conversion of the current noise sources i_b and i_{bf} into their Thévenin equivalent voltage sources. It is not necessary to deal with i_b and i_{bf} separately; we can combine them into one (frequency-dependent) current source i_n. For the sake of simplicity we will neglect the source i_{bf}. In many practical situations the frequency band of interest will justify this simplification. Hence, the equivalent voltage source is given by $v_n = i_n(R_g + r_b)$. Applying eq. (8.49) with $X_g = 0$, we find

$$F = 1 + \frac{R_{nv}}{R_g} + \frac{(r_b + R_g)^2}{R_{ni}R_g} \tag{8.54}$$

and

$$R_{g\,opt} = (R_{nv}R_{ni} + r_b^2)^{1/2} \tag{8.55}$$

Substituting

$$R_{nv} = r_b + \tfrac{1}{2} r_e \text{ and } R_{ni} = 2\beta_0 r_e \tag{8.56}$$

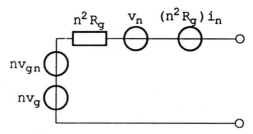

Fig. 8.20 *Equivalent circuit for the case of transformer coupling.*

yields

$$R_{g\,opt} = (\beta_0 r_e^2 + 2\beta_0 r_e r_b + r_b^2)^{1/2} \qquad (8.57)$$

Almost always, $2\beta_0 r_e \gg r_b$ holds. If so, then eq. (8.57) simplifies into

$$R_{g\,opt} = (\beta_0 r_e^2 + 2\beta_0 r_e r_b)^{1/2}$$

EXAMPLE
Let $\beta_0 = 250$, $r_b = 50\,\Omega$ and $I_E = 50\,\mu A$, so that $r_e = 500\,\Omega$. From (8.57) we find $R_{g\,opt} = 8.66\,k\Omega$.

If, for instance, the actual value of R_g is $1\,k\Omega$, then noise matching is achieved by using a transformer with turns ratio $n = \sqrt{8.66} \approx 3$, yielding $F_{min} \approx 1.07$.

In certain cases the use of a transformer is quite feasible, particularly in tuned r.f.-amplifiers, where the transformer action can often be realized by tapping a coil. But in the majority of practical situations a transformer is an unattractive device. Fortunately, an alternative method of noise matching presents itself, since a second manipulable parameter is available, viz. the transistor bias current I_E, and therefore r_e. If eq. (8.56) is substituted into eq. (8.54) and we put $\partial F/\partial r_e = 0$, we obtain

$$r_{e\,opt} = (R_g + r_b)/\sqrt{\beta_0} \qquad (8.58)$$

EXAMPLE
Again let $\beta_0 = 250$, $r_b = 50\,\Omega$ and $R_g = 1\,k\Omega$, then eq. (8.58) yields $r_{e\,opt} = 66.4\,\Omega$, hence $I_E \approx 0.38\,mA$. For F_{min} we find $F_{min} = 1.116$, which is slightly larger than the value found for 'transformer matching'. This is caused by the finite value of r_b. If $r_b \to 0$ (hence $r_b \ll r_e$ and $r_b \ll R_g$), then $R_{g\,opt} = r_e\sqrt{\beta_0}$ and $r_{e\,opt} = R_g/\sqrt{\beta_0}$. And with both matching methods $F_{min} = 1 + 1/\sqrt{\beta_0}$ is obtained. If $r_b \ll R_g$ does not hold, the leading edge of transformer matching increases.

EXAMPLE
Again let $\beta_0 = 250$, $r_b = 50\,\Omega$ and $r_e = 500\,\Omega$, so that $R_{g\,opt} = 8.66\,k\Omega$. If we now consider the case $R_g = 50\,\Omega$, then $n = \sqrt{8660/50} \approx 13$ and $F_{min} = 1.07$. If we achieve noise matching by optimizing r_e, then $r_{e\,opt} = 6.33\,\Omega$, hence $I_E \approx 3.9\,mA$. For F_{min} we find $F_{min} \approx 2.12$, which is considerably larger than the value found previously.

If noise matching is accomplished by optimizing r_e, then $r_b \ll R_g$ should be pursued. In a discrete component implementation it can be difficult to find a transistor with sufficiently low r_b. But in this case use of a transformer may be acceptable if noise optimization is critical. In an IC-implementation, avoiding the use of a transformer is of primary concern. In this case, however, a judicious choice of the transistor geometry can be called in reduce the value of r_b. This approach leads to an input transistor requiring a large silicon area, which must be considered as the cost of avoiding transformer matching.

The above dilemma does not occur if an FET-input stage is employed. If we neglect $1/f$ noise and the noise produced by the leakage current of the gate junction, then, according to eqs. (8.41) and (8.42),

$$R_{nv} = c/g_m$$

and

$$R_{ni} = \frac{g_m}{c\omega^2(C_{gs} + C_{dg})^2}$$

Replacing the current source i_n by its Thévenin equivalent $i_n R_g$ we obtain, where $X_g = 0$,

$$F = 1 + \frac{R_{nv}}{R_g} + \frac{R_g}{R_{ni}} \tag{8.59}$$

$$R_{g\,opt} = (R_{nv}R_{ni})^{1/2} = 1/\omega(C_{gs} + C_{dg}) \tag{8.60}$$

and

$$F_{min} = 1 + \frac{2c}{g_m}\omega(C_{gs} + C_{dg}) \tag{8.61}$$

Theoretically, for low frequencies F_{min} can attain unity value. However, for any finite source impedance, the turns ratio of the input transformer must be very high in order to achieve this optimum. Alternatively, the optimization can be done by using the dependence of R_{nv} and R_{ni} on g_m. Differentiation of eq. (8.59) yields the condition

$$\frac{c}{g_m}\left[\frac{1}{R_g} + R_g\omega^2(C_{gs} + C_{dg})^2\right] = 0, \text{ or } g_m \to \infty$$

which cannot of course be achieved in practice. It is, however, clear that with high impedance sources, the use of an FET as the input element can produce a very low noise figure.

In practice, the impedance of a high-impedance source is nearly always capacitive. An example is a television camera tube. In this case F is not a suitable qualifier for noise behaviour. The camera tube can be modelled as depicted in Figure 8.21 where i_g and i_{gn} denote the signal and the noise produced by the tube,

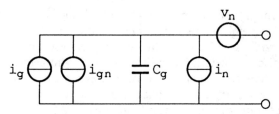

Fig. 8.21 *Noise model for a capacitive signal source.*

while C_g is the output capacitance. Replacing v_n by its Norton equivalent, the total noise power is proportional to $\overline{i_{gn}^2} + \overline{i_n^2} + \overline{v_n^2}(\omega C_g)^2$.

Signal-to-noise ratio is optimum if $\overline{i_n^2} + \overline{v_n^2}(\omega C_g)^2$ is minimum. Using eqs. (8.41) and (8.42), we find

$$P_n = \overline{i_n^2} + \overline{v_n^2}(\omega C_g)^2 = c\omega^2 \left(\frac{C_t^2}{g_m} + \frac{C_g^2}{g_m} \right) \tag{8.62}$$

where $C_t = C_{gs} + C_{gd}$.

For any FET $\omega_T = g_m/C_t$ holds. Substituting $C_t = g_m/\omega_T$ into eq. (8.62), and differentiating with respect to g_m, we obtain

$$\frac{\partial P_n}{\partial g_m} = c\omega^2 \left(\frac{1}{\omega_T^2} - \frac{C_g^2}{g_m^2} \right)$$

and

$$\frac{\partial P_n}{\partial g_m} = 0 \text{ if } g_m^2 = \omega_T^2 C_g^2 \text{ or } C_t = C_g \tag{8.63}$$

Taking into account that ω_T is determined by the fabrication technology of the FET, the conclusion can be drawn that noise matching is accomplished by selecting an FET with C_t equal to the source capacitance.

Until now we have assumed that i_n that v_n are uncorrelated noise sources. In the majority of practical situations, the correlation is low enough to be negligible. If not, then the calculation must be adopted. If there is full correlation, a simple approach can be taken. Any spectral component in i_n can then be written in the form $I_n = Y_c V_n$, where Y_c describes the correlation. If there is partial correlation, i_n can be split up into a component $I_{n(c)}$ that is fully correlated and therefore can be written in the form $I_{n(c)} = Y_c V_n$, and an uncorrelated part $I_{n(nc)}$. Subsequently, the methodology developed previously can be applied. It turns out that, if correlation is present, the optimum value of the source admittance Y_g contains a reactive part: $Y_g = G_d + jS_g$, with $S_g = -S_c$, where S_c is the reactive part of $Y_c = G_c + jS_c$. A case of practical interest is the noise matching of an antenna to a tuned FET-input stage, where the induced gate noise is correlated with the thermal noise of the channel. Obviously, the tuned circuit has to be out of resonance for optimum noise matching. In practice, this possibility of improving noise matching is not often used, but in critical cases it should not be overlooked.

8.13 Noise matching in feedback amplifiers

Some form of feedback is present in the majority of amplifiers in practical use and therefore its consequences with regard to noise behaviour are of concern. In general, feedback as such causes a certain alteration of the noise behaviour, and frequently in an adverse way.

That alterations occur is not at all surprising; the application of feedback introduces additional signal paths and, as a consequence, using the Blakesley transformation leads to changes in the splitting of noise sources. It can be proved that only the application of non-energic feedback paths does not affect SNR. By non-energic components we mean components that neither dissipate nor store energy. The only components in this class are the ideal transformer, the ideal gyrator and the galvanic connection. Practical gyrators contain active elements which themselves constitute sources of additional noise and distortion; therefore gyrators are rarely useful as feedback elements. Capacitances and inductances do not contribute additional noise sources, but resistors do. And, further, local feedback lowers the gain of an amplifier stage, causing the noise of subsequent stages to increase its effect on overall SNR.

In spite of these adverse effects of feedback on the noise behaviour of amplifiers, it is often a powerful tool for optimizing noise matching. As we know, optimum signal transfer requires the input of an amplifier to act as a short circuit ('current input'), an open circuit ('voltage circuit'), or an accurate impedance ('characteristic impedance input'). If the desired input characteristics are realized without the use of feedback, the measures to be taken often conflict with the measures necessary for optimum noise matching. This is particularly obvious in the case of a characteristic impedance input: if the desired impedance level is obtained by connecting a resistor R_{char} to the input terminals, the noise contributed by this resistor equals that of the source impedance. As a consequence, the noise figure increases by 3 dB. Feedback provides the possibility of realizing the desired impedance level without introducing too much additional noise.

EXAMPLE 1

Let the optimization of signal processing require that the input of the amplifier acts as a short circuit ('current input'). Without the use of feedback, this could be accomplished by using a CE-input stage at high bias current. The required bias

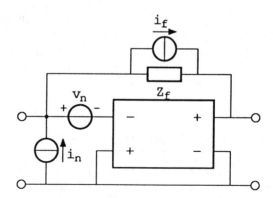

Fig. 8.22 *Transimpedance amplifier with its noise sources.*

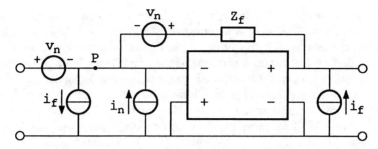

Fig. 8.23 *First transformation of noise sources.*

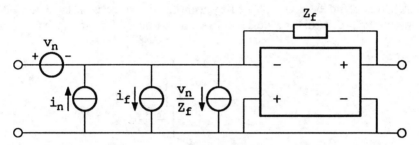

Fig. 8.24 *Equivalent circuit with all noise sources transferred to input.*

current will usually be far from optimum with regard to noise matching. As an alternative a feedback arrangement with parallel connection at the input terminals can be used, for instance a transimpedance amplifier, as shown in Figure 8.22. The resistive part of the feedback impedance Z_f constitutes an additional noise source i_f. Figure 8.23 shows the equivalent circuit after application of the splitting theorem with regard to v_n and i_f. The source v_n in series with Z_f is subsequently transformed into a current source v_n/Z_f, which is, once again, split into current sources at the input and output terminals. The sources at the output of the amplifier can be neglected, because of the gain of the amplifier and its low output impedance. Figure 8.24 shows the resulting equivalent circuit. Replacing the voltage source v_n by a current source v_n/Z_g, where Z_g denotes the internal impedance of the signal source, we obtain the circuit according to Figure 8.25.

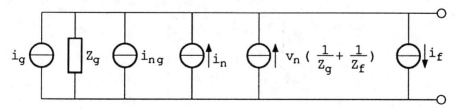

Fig. 8.25 *Signal source and noise current sources representing total noise.*

At this stage we can draw some useful conclusions. First we conclude that, as far as noise is concerned, Z_f can be assumed to be in parallel with Z_g. Obviously, for minimizing the effect of Z_f on noise, $Z_f \gg Z_g$ should be striven for. In order to obtain the desired low input impedance of the amplifier in spite of the large value of Z_f, the loop gain should be large. It can be concluded that for optimum noise behaviour the loop gain should be maximized. Note that the input impedance of the configuration no longer depends on the bias current of the first amplifier stage, which can henceforth be chosen with a view to optimum noise matching.

EXAMPLE 2

If accurate signal processing requires a high-input-impedance amplifier ('voltage input'), this could be accomplished by using a first stage with local feedback, as shown in Figure 8.26. The noise source v_{ne}, associated with R_E, can be moved

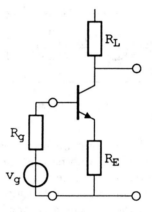

Fig. 8.26 *Single-transistor transconductance stage.*

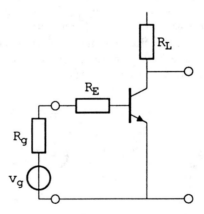

Fig. 8.27 *Circuit that is equivalent to Figure 8.26 with regard to noise.*

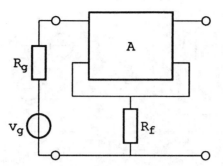

Fig. 8.28 *Application of overall feedback with low-valued feedback resistor.*

through the base–emitter junction, as can easily be seen from the small-signal equivalent circuit, with the aid of the Blakesley transformation. Hence, as far as noise is concerned, the circuit of Figure 8.26 can be replaced by that of Figure 8.27.

Obviously, only if $R_E \ll r_b + \frac{1}{2}r_e$ and $R_E \ll R_g$, is SNR not substantially affected by inserting R_E. However, to obtain the desired high input impedance, R_E should be large. Once again the solution lies in the application of overall feedback (Figure 8.28). If the loop gain is large, R_f can be small, while a large value of Z_i is still obtained.

8.14 Noise behaviour of basic amplifier configurations

The noise properties of CE- and CS-stages have been discussed in the preceding sections. In Section 8.13 we found that a single transistor transconductance stage is much noisier than a CE-stage due to the thermal noise produced by the emitter resistor. By a similar argument a single-stage transimpedance amplifier can be shown to be noisier than the CE-stage. The CB-stage, the emitter follower and the corresponding FET-configurations can be looked upon as CE- or CS-stages with unity feedback. However, in these cases the feedback, being accomplished by galvanic connection, is of the non-energic variety. Therefore, the magnitudes of i_n and v_n are not modified. As a consequence, the value of $R_{g\,opt}$ is the same as for CE- and CS-stages under equal bias conditions. This conclusion can also be arrived at by using the, by now familiar, voltage- and current-source splitting theorems. As an example, Figure 8.29(a) shows an emitter follower with the transistor noise sources v_n and i_n. The current source i_n can be split, as indicated in Figure 8.29(b). The current source at the output can be neglected in view of the high current gain of the transistor. The resulting noise-equivalent circuit is identical to that pertaining to the CE-stage. This fact does not justify the conclusion that, as far as noise behaviour is concerned, the choice of the input-stage configuration does not matter. The CB-stage and the emitter follower exhibit lower gain, and thus using these configurations in the input stage implies that subsequent stages contribute more

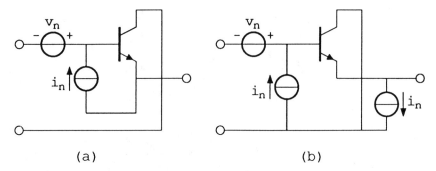

Fig. 8.29 *Emitter follower with noise sources* (a) *and equivalent circuit with transformed noise sources* (b).

noise than in the case of a CE-input stage. It can be concluded that, in principle, the first stage of an amplifier should be CE or CS. Only in cases where arguments related to signal-processing requirements dictate it, can a different configuration be justified.

The congruence of the noise-equivalent circuits of the CE-stage and the emitter follower has a remarkable consequence. Figure 8.30 shows a transistor with equal resistors R in the emitter and the collector leads. The signals obtained at both outputs are, of course, equal except for the phase reversal at the collector side.

However, the amounts of noise showing up at both outputs differ considerably if R is large enough. Using the emitter output, the circuit is employed as an emitter follower; using the collector output it is employed as a series stage. Consequently, the noise properties correspond to those of the circuits depicted in Figure 8.31(a) and (b) respectively. A simple experiment in which both output signals are displayed on an oscilloscope screen confirms the striking difference convincingly.

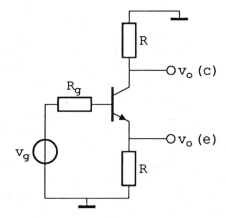

Fig. 8.30 *Amplifier with equal resistors in emitter and collector circuit.*

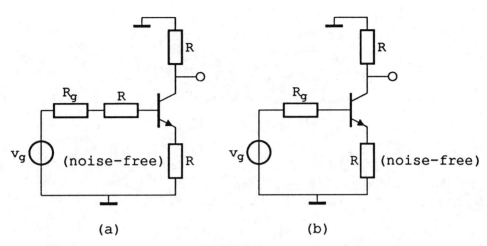

Fig. 8.31(a) *Circuit that is equivalent to CE-stage with local feedback as far as noise behaviour is concerned.* (b) *CE-stage that is equivalent to emitter follower as far as noise behaviour is concerned.*

8.15 Differential pair

Figure 8.32 shows a differential pair with its noise sources. The source i_{on} represents the noise of the current source I_o. The noise sources $4kT/R_L$ represent the noise of the load resistors R_L. They can be transformed into input noise sources $\overline{v_n^2} = 4kT/R_L g_e^2$ and $\overline{i_n^2} = 4kT/R_L \beta_o^2$. If $R_L \gg 2r_e$, these noise sources can be neglected.

If the differential pair is driven single-sided and it is assumed that the basis of $T2$ is grounded, as indicated in Figure 8.32 by a dotted line, then i_n of $T2$ is in parallel with i_{on}, which will usually be much larger than i_n. The voltage source v_n of $T2$ is essentially in series with v_n of $T1$. As a consequence, the equivalent noise resistance R_{nv} of the differential pair is doubled compared with the single stage. The current source i_n of $T1$ can be split into a current source i_n in parallel with i_{on} and a source i_n between the base of $T1$ and the ground point.

If the circuit is symmetrical, the noise source i_{on} is a common-mode source. If the common-mode rejection is high enough it does not therefore contribute to the noise at the differential output. However, due to the asymmetric driving, the circuit exhibits asymmetry. The emitter impedances of $T1$ and $T2$ are $r_e + R_g/\beta_0$ and r_e respectively. As a consequence, an unwanted differential mode current occurs:

$$i_{dn} = \frac{R_g/\beta_0}{2r_e + R_g/\beta_0} i_{on} \qquad (8.64)$$

According to eq. (8.64), the contribution to the output noise of this component depends on R_g. Taking into account the doubling of R_{nv} and the extra noise due

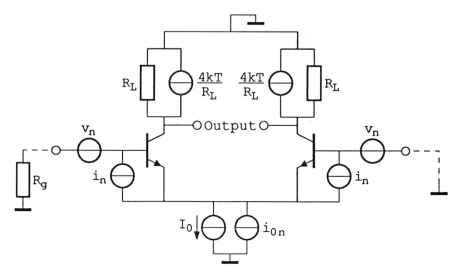

Fig. 8.32 *Noise sources in the single-sided driven differential pair.*

to i_{dn}, we must conclude that the single-sided driven differential pair compares unfavourably with a single CE-stage as far as noise behaviour is concerned.

If the differential pair is driven symmetrically, then i_{on} is a perfect common-mode source. Again, the noise sources v_n add. The current sources i_n can be split into sources in parallel with i_{on} and in sources between the input terminals and the ground point. The sources in parallel with i_{on} are common-mode sources that do not contribute to the noise at the differential output. The sources at the input terminals can be transformed into differential-mode voltage sources $\frac{1}{2}R_g i_n$. Hence, effectively, the equivalent noise resistances pertaining to v_n and i_n both double, so that the optimum value of R_g also doubles, while in principle F_{min} is the same as in the single CE-stage. An alternative way to arrive at this result is shown in Figure 8.33. The starting point is the single CE-stage (Figure 8.33a). The source i_n is transformed into two voltage sources at the input and output terminal of the stage respectively (Figure 8.33b). The validity of the transformation can be checked by using the definition formulas given by eqs. (8.35). Figure 8.33(c) shows the equivalent circuit for the differential pair obtained in this way. Taking into account that series connection of identical twoports implies the modification of two of the a-parameters ($a_{is} = a_i$, $a_{rs} = 2a_r$, $a_{fs} = \frac{1}{2}a_f$, $a_{os} = a_o$), where the subscript s indicates the series connection), and taking into account that the voltage sources are in series with the input and the output, the sources at the output can be retransformed into sources at the input. Figure 8.33(d) gives the expected result, showing that the spectrum of the current noise source is halved with respect to that of the single CE-stage.

For the FET–differential pair the same conclusions hold, albeit that, in the case of asymmetrical driving, the effect described by eq. (8.64) does not occur.

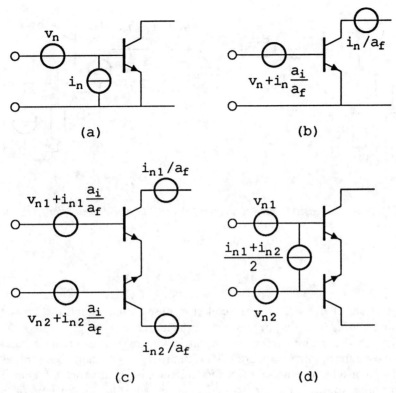

Fig. 8.33 *Derivation of the noise equivalent circuit for the symmetrically driven differential pair: (a) CE-stage; (b) transformation of i_n into voltage sources; (c) differential pair as series connection of CE-stages; (d) resulting equivalent circuit with noise sources at input side.*

8.16 Current sources

Figure 8.34 shows the generalized circuit of a single-BJT current source, together with its noise sources. Obviously, to minimize noise, R_B should be kept minimum. Assuming $R_B = 0$ and $r_b \ll \beta_0 R_E$, straightforward calculation yields, for the collector noise current,

$$\overline{i_{nc}^2} = 4kT \left[\frac{r_b + \tfrac{1}{2}r_e + R_E + R_E^2/2\beta_0 r_e}{(r_e + R_E)^2} \right] \tag{8.65}$$

From this formula the conclusion can be drawn that a large value of R_E is favourable. If $R_E \ll r_e$, then

$$\overline{i_{nc}^2} \approx 4kT \left(\frac{r_b + \tfrac{1}{2}r_e}{r_e^2} \right) \tag{8.66}$$

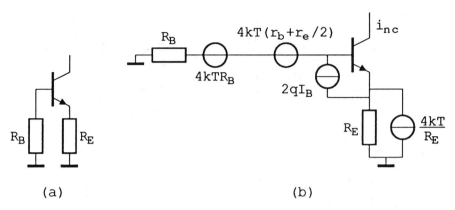

Fig. 8.34 *Single transistor current source* (a) *and its noise equivalent circuit* (b).

And if $R_E \gg r_e$,

$$4kT\left(\frac{1}{R_E} + \frac{1}{2\beta_0 r_e}\right) \qquad (8.67)$$

For large values of $R_E (\gg 2\beta_0 r_e)$, the collector noise current equals the base current noise source $2qI_B$, which is the best we can achieve.

Knowing this result we cannot be optimistic with regard to the noise behaviour of the simple current mirror (Figure 8.35a). In order to be certain about this aspect, let us first consider the noise of a diode-connected transistor. Of course, the dominant noise source is the shot noise associated with the d.c.-current I_D, but in addition the thermal noise of the base resistance r_b must be taken into account. Hence

$$\overline{i_{dn}^2} = 2qI_D + 4kTr_b g_e^2 = 2qI_D(1 + 2r_b/r_e) = 2qI_D' \qquad (8.68)$$

Figure 8.35(b) maps the noise sources of the current mirror. If $T1$ and $T2$ are identical transistors, then $I_D = I_C$ and $2qI_D' \gg 2qI_B$. For the output noise current we

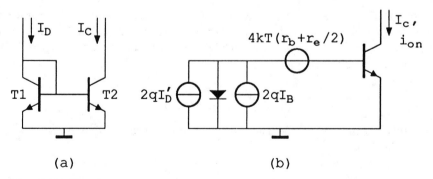

Fig. 8.35 *Simple current mirror* (a) *and its noise sources* (b).

find

$$\overline{i_{on}^2} \approx 2qI_D(1+2r_b/r_e) + 2qI_C(1+2r_b/r_e) \tag{8.69}$$

If $I_C \approx I_D$, eq. (8.69) simplifies into

$$\overline{i_{on}^2} = 4qI_C(1+2r_b/r_e) \tag{8.70}$$

Even when $2r_b/r_e \ll 1$, the output current is afflicted with the shot noise of both transistors.

On the basis of our previous results concerning the noise of a single transistor current source, we may expect that inserting emitter resistors will improve noise behaviour. Figure 8.36 shows the circuit (a) and the relevant noise sources (b). Assuming $\beta_0 \gg 1$, we find

$$\overline{i_{on}^2} = \frac{4kT}{(r_e+R_E)^2}(2r_b+2R_E+r_e) + \frac{2kT}{\beta_0 r_e}\left(\frac{2R_E+r_e}{R_E+2r_e}\right)^2 \tag{8.71}$$

If $R_E \gg r_e + 2r_b$, eq. (8.71) simplifies to

$$\overline{i_{on}^2} = 8kT\left(\frac{1}{R_E}+\frac{1}{\beta_0 r_e}\right) \tag{8.72}$$

If emitter resistors are used, $\beta_0 r_e \gg R_E \gg r_e$ will hold in most cases. In these cases the noise current corresponds to the noise current produced by a resistor $\frac{1}{2}R_E$.

From the foregoing analysis of the noise behaviour of a current mirror it is immediately clear that a current mirror used as an active load contributes much

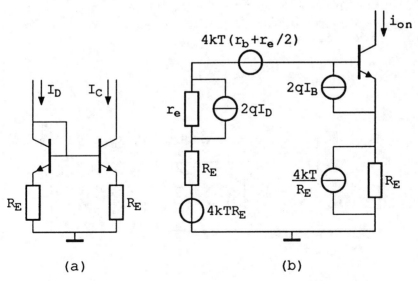

Fig. 8.36 *Current mirror with emitter resistors (a) and its noise sources (b).*

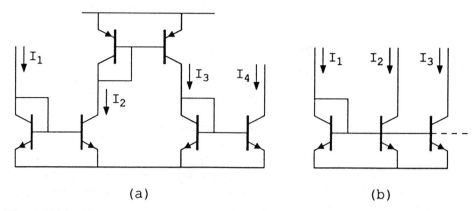

Fig. 8.37(a) *Wrong method of forming biasing currents.* (b) *Correct method of forming biasing currents.*

more noise than a passive load. It is therefore not surprising to find that input stages in low-noise integrated amplifiers usually have passive load resistors.

Perhaps unnecessarily, we note that any noise current that is already present in the input current I_D is fully mirrored by the current mirror and hence adds to the noise current produced by the mirror itself. This makes clear that, if several biasing currents are needed in an IC, these currents should not be derived by repeated mirroring of currents (Figure 8.37a), but by direct mirroring from a primary source (Figure 8.37b).

The behaviour of an MOS-current mirror does not differ essentially from that of the BJT-version. Figure 8.38 shows the circuit and its noise sources. In most cases $R_{ni}/g_m^2 \gg R_{nv}$ will hold, yielding $\overline{i_{on}^2} = 8kTcg_m$. If source resistors are added for reducing the noise current, we find

$$\overline{i_{on}^2} = 8kT \frac{g_m^2}{(1 + g_m R_s)^2} (R_s + R_{nv}) \tag{8.73}$$

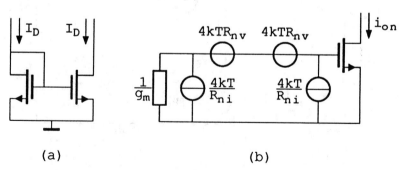

Fig. 8.38 *MOS-current mirror (a) with its noise sources (b).*

With $R_s \gg 1/g_m$, eq. (8.72) reduces to $\overline{i_{on}^2} = 8kT/R_s$, as in the case of the BJT-current mirror.

Further reading

Davenport, W. B. and Root, W. L. (1958) *An Introduction to the Theory of Random Signals and Noise*, McGraw Hill, New York.

Gupta, M. S. (ed.) (1977) *Electrical Noise: Fundamentals and sources*, IEEE Press, New York.

Nordholt, E. H. (1983) *Design of High-performance Negative-feedback Amplifiers*, Elsevier, Amsterdam.

van der Ziel, A. (1970) *Noise (sources, characteristics, measurement)*, Prentice Hall, Englewood Cliffs, New Jersey.

9 Low-pass wideband amplifiers

9.1 The basic problem of wideband amplification

In modern electronic systems signals of large spectral bandwidth frequently have to be dealt with. Though the design of wideband amplifiers does not essentially deviate from that of amplifiers where bandwidth optimization is not a primary objective, the requirement of large bandwidth evokes certain difficulties that call for attention.

It was found in Chapter 5 that the frequency-dependent behaviour of all basic amplifier stages can, to a first approximation, be characterized by one dominant pole, whereas in certain cases some influence is exerted by non-dominant poles and/or zeros. As an example let us reconsider the basic CS-stage used as a voltage amplifier in a cascade of such stages (Figure 9.1). The gain of the stage is given by $G = -g_m R_L$, where R_L denotes a resistive load. The bandwidth in terms of angular frequency is, to a first approximation, given by $B_\omega = 1/\tau$, where $\tau = R_L C_{tot}$, with $C_{tot} = C_{gs} + C_{ds}$. It may be recalled that, in case the influence of C_{gd} cannot be neglected, the expression for the bandwidth has to be modified. Assuming, for the time being, that the effect of C_{gd} has been accounted for by adapting the value of C_{tot} or by an adequate unilateralization method, the product of gain and bandwidth is given by $g_m/C_{tot} \approx \omega_T$, which is fully determined by the properties of the active

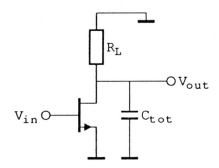

Fig. 9.1 *Basic CS-voltage amplifier stage.*

device. It was found in Chapter 5 that this product of gain and bandwidth is approximately the same for all the basic amplifier stage configurations. It was also concluded that the transit frequency f_T of an active device is a useful figure of merit for specifying its wideband capabilities. Modern monolithic devices can exhibit values of f_T amounting to several GHz, depending on the sophistication of the manufacturing process. However, for economic reasons it is frequently very desirable to use the simplest manufacturing process that is capable of achieving adequate circuit performance through good circuit design.

Given a value of GB, it is obvious that bandwidth can be traded for gain, albeit within certain bounds. Of course, bandwidth cannot increase beyond the value of GB. And without special measures it is not possible to increase gain by simply adding stages in a cascaded configuration, even when the required bandwidth is not close to the limiting value dictated by the GB-product of the devices used. A simple example may illustrate this. If we cascade two identical unilateral stages with a given value of τ, then the voltage transfer function can be written

$$A_v = (g_m R_L)^2 \left(\frac{1}{1 + j\omega\tau}\right)^2 \qquad (9.1)$$

Obviously, bandwidth has been reduced in comparison with the single-stage configuration. From eq. (9.1) we find

$$B_\omega = (2^{1/2} - 1)^{1/2} \cdot \frac{1}{\tau} = 0.64 \, \frac{1}{\tau} \qquad (9.2)$$

Cascading n identical stages we find

$$B_\omega = (2^{1/n} - 1)^{1/2} \cdot \frac{1}{\tau} \qquad (9.3)$$

With increasing values of n, the bandwidth decreases rapidly, for instance where $n = 5$ the narrowing factor amounts to 0.39. Consequently, if we try to compensate for the loss in bandwidth by lowering the stage gain and increasing the number of stages, it will often turn out that no convergent solution is possible.

Obviously, straightforward cascading of basic single-pole stages is not a useful approach to combine high gain and large bandwidth, and we must look for better methods to use the bandwidth capabilities of active devices to their full extent.

9.2 Improving the bandwidth of single stages

The preceding section has indicated that multistage wideband amplifiers should not be built up of several stages with identical or nearly identical poles. If an amplifier has to consist of several stages because of the gain required, the correct approach is to base the design on a preconceived total pole–zero pattern. Methods to specify useful overall p–z patterns will be discussed in the next section. Prior to this we will

268 ANALOGUE ELECTRONIC CIRCUIT DESIGN

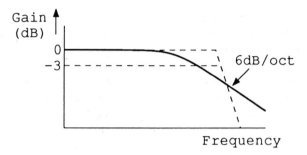

Fig. 9.2 *One-pole frequency response (solid line) and expected frequency response if roll-off is made steeper.*

examine how single-stage amplifiers, which in certain cases can provide adequate gain, can be optimized with regard to their bandwidth capabilities.

The amplitude-frequency characteristic of a one-pole stage shows a 6-dB-per-octave, high-frequency roll-off. Obviously, it is expected that the -3-dB corner frequency will shift upwards if the high-frequency roll-off is made steeper (Figure 9.2). Methods for achieving this are among the oldest techniques for wideband amplification. Figure 9.3 shows a simple configuration based on this idea, called 'inductive shunt-peaking'.

It is immediately clear that the gain for high frequencies has increased since the load impedance, thanks to the added inductance, increases with increasing frequency. Neglecting the (small) effect of r_d on the transfer function we find

$$A_v = -g_m \frac{R_L + sL}{1 + sR_LC_L + s^2LC_L} \tag{9.4}$$

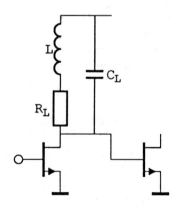

Fig. 9.3 *Improving the GB-product by means of shunt-peaking.*

LOW-PASS WIDEBAND AMPLIFIERS 269

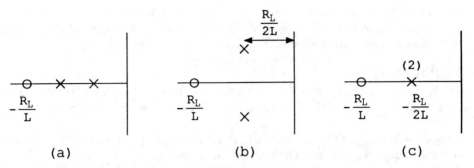

(a) (b) (c)

Fig. 9.4 *Pole–zero patterns of the shunt-peaked amplifier stage:* (a) *two real poles;* (b) *complex poles;* (c) *two coincident poles.*

The p–z pattern comprises two poles and a zero $z = -R_L/L$. The poles are given by

$$p_{1,2} = -\frac{R_L}{2L} \pm \left[\frac{1}{4}\left(\frac{R_L}{L}\right)^2 - \frac{1}{LC_L}\right]^{1/2} \qquad (9.5)$$

If $L < \frac{1}{4}R_L^2 C_L$ the poles are real; if $L > \frac{1}{4}R_L^2 C_L$ they are complex. Where $L = \frac{1}{4}R_L^2 C_L$, the poles are coincident. Figure 9.4 shows the associated p–z patterns.

Adding L to the coupling network does not change the low-frequency gain, whereas the 3-dB bandwidth is increased. If the poles are complex, then the transient response exhibits overshoot and ringing. In most applications overshoot should be kept low. Figure 9.5(a) shows the amplitude-frequency response for various values of the ratio k of the two relevant time constants ($R_L C_L$ and L/R_L). Figure 9.5(b) sketches the corresponding transient responses.

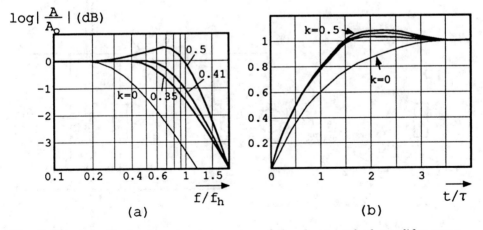

(a) (b)

Fig. 9.5(a) *Amplitude–frequency response of the shunt-peaked amplifier stage for various values of* $k = L/R_L^2 C_L$; A_o *is the low-frequency gain.* (b) *Associated transient responses;* $\tau = 1/2\pi f_h$ *is the time constant* $R_L C_L$.

A commonly recommended value for k is 0.35. For the rise time of a stage without peaking $\tau_r \approx 2.2\, R_L C_L$ holds. Where $k = 0.35$ we find $\tau_r \approx 1.6\, R_L C_L$. The improvement amounts to about 37%, which is quite considerable.

EXAMPLE

Let $\tau_r \approx 51$ ns be required for a video amplifier. With $k = 0.35$ we have $R_L C_L \approx 32$ ns. Let $C_L = 10$ pF, then $R_L = 3.2$ kΩ and $L = kR_L^2 C_L = 36$ µH.

Figure 9.6 shows the case of shunt-peaking in a CE-stage, together with a simplified equivalent circuit. The transfer function is given by

$$\frac{V_{b'e}(T2)}{V_{b'e}(T1)} = \frac{-g_m r_\pi (R_L + sL)}{r_\pi + (1 + sr_\pi C_i)(R_L + sL + r_b)} \tag{9.6}$$

which has the same structure as eq. (9.4).

Further improvement of bandwidth and rise time is possible by adding additional passive elements, thus increasing the order of the coupling network. In the past, when savings in active-device count were at a premium, such methods were elaborated, but at present they are only of historical significance. The simple inductive peaking circuit according to Figure 9.3 can still be a viable option in cases where a single-transistor stage is in place. Stages that are part of an amplifier with overall feedback should in general not be peaked. The presence of several complex poles hampers the design of stable feedback amplifiers considerably. If overall feedback is to be applied, better methods for broadbanding are available. These will be discussed in a subsequent section.

If the use of inductances must be avoided, as an alternative additional poles can be created by the use of capacitive feedback. Figure 9.7 shows a very simple circuit taking this approach, together with the associated p–z pattern. Since the source

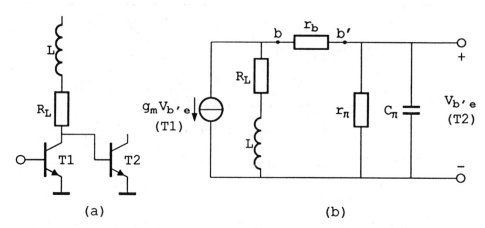

Fig. 9.6(a) *Shunt-peaking in a BJT-amplifier stage.* (b) *Simplified equivalent circuit.*

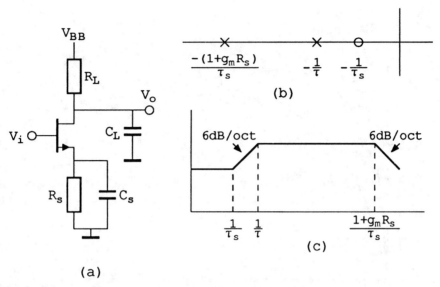

Fig. 9.7(a) *Improving high-frequency response by employing capacitive feedback.* (b) *Pole–zero pattern;* $\tau_s = R_s C_s$, $\tau = R_L C_L$; (c) *Bode plot.*

circuit and the drain circuit do not interact, one pole is exclusively determined by the drain circuit, while the second is determined by the source circuit. Hence, complex poles cannot occur. Moreover, there is a zero located on the right-hand side of the poles. This zero has an undesirable effect on the frequency response, as can be seen in the Bode plot of Figure 9.7(c). In the transient response this zero can provoke the occurrence of considerable overshoot, notwithstanding the absence of complex poles.

EXAMPLE
Calculation of the transient response of the circuit of Figure 9.7
Neglecting the (small) influence of r_d, we find for the voltage transfer of the circuit

$$-A_v = -\frac{V_o}{V_i} = \frac{g_m}{C_L} \frac{(s + 1/\tau_s)}{\{s + (1 + g_m R_s)/\tau_s\}\{s + 1/\tau\}} \tag{9.7}$$

where $\tau_s = R_s C_s$ and $\tau = R_L C_L$.

The transfer is characterized by a zero $z_1 = -1/\tau_s$ and two poles: $p_1 = -(1 + g_m R_s)/\tau_s$ and $p_2 = -1/\tau$.

To find the transient response the right-hand member of eq. (9.7) is multiplied by $1/s$ (Chapter 3, Section 3.3). Translating this function into the time domain yields

$$\frac{g_m}{C_L}\left[\frac{-z_1}{p_1 p_2} + \frac{p_1 - z_1}{p_1(p_1 - p_2)}\varepsilon^{p_1 t} + \frac{z_1 - p_2}{p_2(p_1 - p_2)}\varepsilon^{p_2 t}\right] \tag{9.8}$$

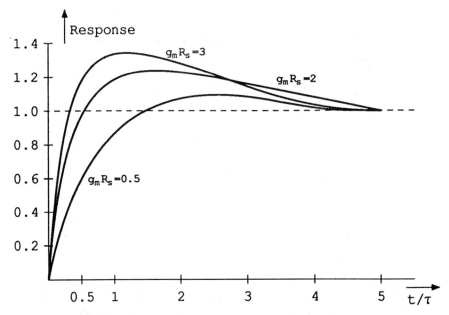

Fig. 9.8 *Transient response for the circuit employing 'source peaking'.*

If $\tau_s = \tau$, the zero is cancelled by the pole p_2. This corresponds to source compensation (Section 5.9). The transient response is described by a single exponential function and, of course, there is no overshoot and no improvement of the rise time. If we now take $\tau_s = 2\tau$ for cashing in an improvement of the rise time, eq. (9.8) takes the form

$$\frac{g_m}{C_L}\left[\frac{\tau}{1+g_mR_s} - \frac{2\tau g_mR_s}{(g_mR_s)^2-1}\varepsilon^{-(1+g_mR_s)t/2\tau} + \frac{\tau}{g_mR_s-1}\varepsilon^{-t/\tau}\right] \quad (9.9)$$

Figure 9.8 shows the transient response for the cases $g_mR_s = 0.5$, $g_mR_s = 2$ and $g_mR_s = 3$. In the latter cases in particular a large overshoot is observed, which will be very undesirable in the majority of applications, the more so because the maximum in the response occurs for rather large values of t. Of course, no 'ringing' occurs in the transient response, since periodic components can only occur when complex poles are present.

If the same method is applied with a BJT as the active device, the effect can be much stronger. As has been discussed in Section 5.9, in this case complex poles can occur, depending on the relative magnitude of the time constants involved. A considerable increase in bandwidth or, alternatively expressed, a decrease in rise time, can be achieved, but again much care has to be exercised in order to avoid excessive overshoot. The reader is advised to convince himself of the surprisingly large effect of capacitive emitter feedback by trying out various values of the emitter capacitor by means of an experimental circuit with an oscilloscope, or by computer simulation.

9.3 Optimum pole–zero patterns for multistage wideband amplifiers

In Section 9.1 it was shown that cascading identical stages has a deleterious effect on the overall bandwidth of an amplifier. The poles and zeros should be distributed such that an optimum overall transfer function is obtained. The first question now is, of course, what criterion for optimum response must be taken. There are several approaches to the solution of this problem.

A first approach is to specify the requirements to be met by the transient response. These requirements should then be related to requirements concerning the p–z pattern. No useful rules have ever been formulated using this approach.

As an alternative, requirements to be met by the frequency response can be formulated and translated into appropriate p–z patterns. This appears to be quite a practical approach. One example of this method is the Chebyshev approximation, according to which a maximum acceptable deviation of a flat frequency response is specified. A response is striven for which touches the maximum tolerance limits as many times as is enabled by the order of the network. In this way an 'equi-ripple' function with maximum bandwidth is obtained, as is sketched in Figure 9.9. This approach assigns priority to bandwidth, whereas transient response is regarded as subordinate. In several areas of application this is an adequate point of view; an example is carrier telephony. The Chebyshev type of response can easily be translated into rules for the associated pole patterns.

If a smooth transient response is of the essence, then the so-called 'maximally flat magnitude' (MFM) type of response is a better proposition. MFM means that the maximum number of derivatives in $\omega = 0$ take zero value. This maximum number depends, of course, on the order of the network. Figure 9.10 sketches this type of frequency response.

Based on network-theoretical considerations, which will be omitted here, a rule for the design of the associated p–z pattern can be derived. Let the complex

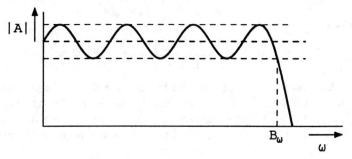

Fig. 9.9 *General shape of 'Chebyshev'-response.*

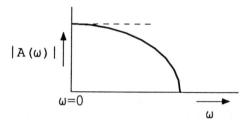

Fig. 9.10 *Sketch of MFM-frequency response.*

transfer function be given by $A(s)$ and let $P(s)$ be defined by $P(s) = A(s) \cdot A(-s)$. The function $P(s)$ is related to the power transfer of the network and can always be written in the form

$$P(s) = C \frac{1 + \alpha_2 s^2 + \alpha_4 s^4 + \ldots \alpha_{2n} s^{2n}}{1 + \beta_2 s^2 + \beta_4 s^4 + \ldots \beta_{2m} s^{2m}} \qquad (9.10)$$

where C is a constant while n and m denote the number of zeros and the number of poles in $A(s)$. For complying with the MFM-requirement it can be shown that

$$\alpha_2 = \beta_2, \ \alpha_4 = \beta_4, \ \ldots \qquad (9.11)$$

should hold.

EXAMPLE

Let a transfer function be characterized by two poles and one zero. A specimen of a relevant circuit is provided by the shunt-peaking circuit considered in the preceding section.

The transfer function can generally be written in the form

$$A(s) = C \frac{s + c}{(s + a)^2 + b^2} \qquad (9.12)$$

hence

$$P(s) = A(s) \cdot A(-s) = C^2 \frac{-s^2 + c^2}{s^4 + 2s^2(b^2 - a^2) + (a^2 + b^2)^2} \qquad (9.13)$$

Applying eq. (9.11) yields

$$\alpha_2 = -\frac{1}{c^2}, \ \beta_2 = 2 \frac{b^2 - a^2}{(a^2 + b^2)^2} \text{ and } c^2 = \frac{(a^2 + b^2)^2}{2(a^2 - b^2)} \qquad (9.14)$$

This is the condition sought for that can be used for the synthesis of the network. If, for instance, the pole positions are given (Figure 9.11), then the position of the zero producing the MFM-response can be found from eq. (9.14).

If a shunt-peaking network is designed according to this rule, the optimum value of the ratio of the two time constants appears to be approximately the same as the one found previously ($k = 0.35$).

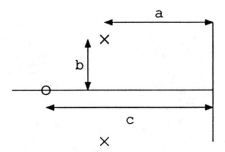

Fig. 9.11 *p–z plot of a transfer function with a complex pole pair and a zero.*

Of particular importance is the case where the transfer function comprises poles only. Such a function is called an 'all-pole function'. Obviously $\alpha_2 = \alpha_4 = \ldots \alpha_{2n} = 0$ holds. To comply with the MFM-requirement $\beta_2 = \beta_4 = \ldots \beta_{2(m-1)} = 0$ should also hold. Hence $P(s)$ can be written in the general form

$$P(s) = \frac{1}{1 \pm s^{2m}} \tag{9.15}$$

The first $2m - 1$ derivatives in $\omega = 0$ are zero, as should be the case. The pole positions follow from $s^{2m} = \pm 1$. If $s^{2m} = +1$, then we can write $s^{2m} = \exp(2jk\pi)$ with $k = 0, \pm 1, \pm 2, \ldots$. Hence

$$s = \exp(jk\pi/m) \tag{9.16}$$

If m is even, two poles are located on the imaginary axis ($k = \pm \tfrac{1}{2}m$). This is not a useful configuration, so only with odd values of m does eq. (9.16) lead to a practicable pole pattern. If m is even, a usable pattern is obtained with $s^{2m} = -1$, or

$$s = \exp\left[\frac{j\pi}{m}(k + \tfrac{1}{2})\right] \tag{9.17}$$

In both cases the poles of $A(s)$ are located and equally spaced on a semicircle. Figure 9.12 shows the pole patterns for $m = 3$ and $m = 4$.

Since the radius of the semicircle corresponds with the bandwidth B_ω of the frequency response, and the poles found from eq. (9.15) are located on a semicircle with unit radius, the bandwidth is normalized at the value $B_\omega = 1$. We can denormalize by replacing ω with ωB_ω or, alternatively, we can forego the normalization by assigning the value B_ω to the radius. All-pole MFM-functions are commonly called Butterworth-functions. Because of their attractive characteristics and their convenient pole pattern they are widely used.

The question can now be posed as to how the desired pole pattern can be implemented in a multistage amplifier. The poles can be subdivided in conjugate complex pairs, while with odd values of m there is in addition a real pole. One method is to realize the complex poles with suitably designed individual stages. If

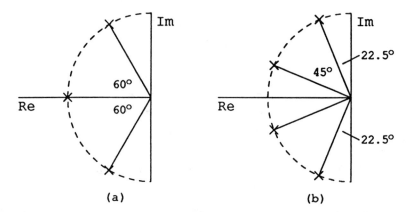

Fig. 9.12 *Pole positions pertaining to MFM all-pole functions:* (a) $m = 3$; (b) $m = 4$.

these stages contribute zeros they can be cancelled by poles in other stages or, alternatively, an attempt can be made to move them into non-dominant locations. With passive networks complex poles can only be realized if they contain both capacitances and inductances. With active networks incorporating feedback, complex poles can be obtained without the use of inductances. Though it is possible to realize the desired MFM-pole pattern by cascading local feedback stages producing complex poles, it is nearly always more appropriate to use overall-feedback configurations incorporating two or more stages. In general, overall feedback is to be preferred above local feedback because of the larger loop gain, with the ensuing advantages of better reduction of distortion, better accuracy, and better noise behaviour (see Section 8.13). On the other hand, feedback loops incorporating more than two dominant poles within the loop are difficult to design because of their inherent proneness to instability. If in a two-stage overall feedback amplifier each of the stages contributes a dominant pole to the loop gain, then by proper design of the feedback loop, the overall response can be made to yield complex poles in the desired positions.

9.4 Complex poles by judicious design of a second-order loop transfer

We consider a feedback arrangement in which the loop transfer possesses two dominant poles. It is assumed that any non-dominant singularities are sufficiently remote to have negligible effect. When the feedback loop is closed, the poles of the closed-loop amplifier are located on the relevant root locus (Section 3.4). If the two open-loop poles are coincident, the root locus is a straight line running parallel to the imaginary axis and intersecting the real axis in the poles (Figure 9.13a). Obviously, possibilities for manipulating the poles into desired positions are virtually absent, since the root locus is rigidly fixed by the open-loop poles.

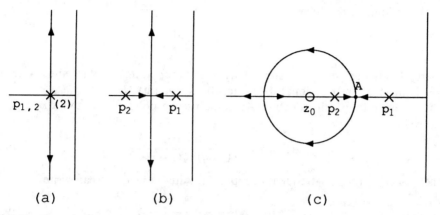

Fig. 9.13 *Root loci in a second-order feedback configuration:* (a) *two coincident poles;* (b) *two separated poles;* (c) *two poles and a zero.*

If the open-loop poles are apart, the root locus comprises the part of the real axis between the poles and, again, a straight line parallel to the imaginary axis (Figure 9.13b). More freedom of manipulation is available now because of the additional degree of freedom provided by the separation distance of the poles. This distance can be manipulated by the judicious use of pole-splitting techniques (see also Section 5.4). However, pole splitting cannot be implemented to any desired extent without loss of inherent bandwidth capabilities. This will be immediately clear for the pole-splitting method in common use, consisting of the deliberate application of capacitive feedback between collector and base (respectively drain and source) of one of the constituting amplifier stages. Adding capacitance inherently increases charge storage and hence adversely affects rise time.

Another method of realizing separate poles is the use of constituent stages with different dominant poles. In this case the design freedom is also limited since large differences can only be achieved at the cost of loss of GB-capabilities in one of the stages.

Though pole splitting can in certain cases provide satisfactory results, more design freedom is often desirable. This can be provided by adding a zero to the open-loop transfer. Since the closed-loop transfer should be represented by an all-pole function, the zero should not show up in the closed-loop transfer. Such a zero present in the open-loop transfer but not showing up in the closed-loop transfer is called a 'phantom zero'. Figure 9.13(c) shows the root locus for this case. The root locus starts, of course, in the open-loop poles, moves along the part of the real axis between the poles, and beyond the meeting point A it moves into the left half-plane along a circle that has the zero as its centre. Where the circle meets the real axis again, the root locus continues along the real axis. One branch ends in infinity, the other in the phantom zero. This can be proved as follows. On the basis of the asymptotic-gain model (Chapter 4), the transfer of a negative-feedback amplifier can be written in the form

$$A_f = -A_{f\infty} \frac{A\beta}{1 - A\beta} \tag{9.18}$$

where $A_{f\infty}$ is the (ideal) transfer A_f when the loop gain $A\beta$ is infinite. Note that $A_{f\infty}$ is inversely proportional to the feedback quantity β.

If the open-loop gain function $A\beta$ contains two poles p_1, p_2 and a zero z_0, we can write

$$A\beta = A_0\beta_0 \frac{1 - s/z_0}{(1 - s/p_1)(1 - s/p_2)} \tag{9.19}$$

where $A_0\beta_0$ is the low-frequency loop gain. Alternatively we can write

$$A\beta = K \frac{s - z_0}{(s - p_1)(s - p_2)} \tag{9.20}$$

where

$$K = -\frac{A_0\beta_0 p_1 p_2}{z_0} \tag{9.21}$$

The quantity $A_0\beta_0 p_1 p_2$ is the product of the low-frequency loop gain and the two dominant poles of the loop. $A_0 p_1 p_2$ is determined by the GB-products of the constituent gain stages. Given the active devices used, this quantity can be considered to be fixed. The product Kz_0 has the dimension $(\text{rad s}^{-1})^2$. For short we will write

$$-A_0\beta_0 p_1 p_2 = \omega_n^2 \tag{9.22}$$

so that $K = \omega_n^2/z_0$.

The root locus sought for is given by $1 - A\beta = 0$, hence

$$\frac{s - z_0}{(s - p_1)(s - p_2)} = \frac{1}{K} \tag{9.23}$$

In order to find the equation for the root locus in the x–y plane we write $s = x + jy$. Substituting this into eq. (9.23) yields

$$\frac{x + jy - z_0}{(x + jy - p_1)(x + jy - p_2)} = \frac{1}{K} \text{ or } (x + jy - p_1)(x + jy - p_2) = K(x + jy - z_0) \tag{9.24}$$

Equating the real and imaginary parts of the right-hand member and the left-hand member of eq. (9.24) and eliminating K from this set yields

$$(x - z_0)^2 + y^2 = p_1 p_2 - z_0(p_1 + p_2) + z_0^2 \tag{9.25}$$

This equation represents a circle with centre z_0 and radius

$$R = [z_0^2 - z_0(p_1 + p_2) + p_1 p_2]^{1/2}$$

The point A where the circle intersects the real axis is located at distance $z_0 - R$ from the origin.

For the purpose of formulating a systematic design method for the type of amplifier under discussion, it is convenient to determine the locus of the closed-loop poles for a fixed value of ω_n^2. Using $K = \omega_n^2/z_0$, we can write eq. (9.23) in the form

$$\frac{s - z_0}{(s - p_1)(s - p_2)} = \frac{z_0}{\omega_n^2}$$

or

$$s^2 - s\left[p_1 + p_2 + \frac{\omega_n^2}{z_0}\right] + \omega_n^2 + p_1 p_2 = 0 \tag{9.26}$$

Again putting $s = x + jy$ and equating the real and imaginary parts of the right-hand and left-hand member, we find for the locus of the closed-loop poles p_{f1} and p_{f2}

$$x^2 + y^2 = \omega_n^2 + p_1 p_2 = \omega_n^2 \left(1 - \frac{1}{A_0 \beta_0}\right) \approx \omega_n^2$$

since normally $A_0 \beta_0 \gg 1$.

The relevant part of this locus is a half-circle with centre O and radius ω_n (Figure 9.14).

Let us now see how the designer of a second-order feedback amplifier can use the available free design parameters.

Since the product $A_0 p_1 p_2$ is essentially determined by the GB-capabilities of the active devices, this product cannot be considered to be a free design parameter. However, within the constraint of the fixed value of this product, a certain freedom exists for choosing p_1 and p_2 (pole splitting). The only really free design parameters are β_0 and z_0. Given $A_0 p_1 p_2$, the magnitude of ω_n, and hence the bandwidth, can be manipulated through β_0. Obviously, the larger β_0 is, the larger the bandwidth of the amplifier. This could be expected, since GB is fixed and A_f is in inverse proportion to β_0. The location of the closed-loop poles on the half-circle can be manipulated through z_0 or by the values of p_1 and p_2 within the constraint of the fixed product $p_1 p_2$.

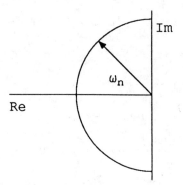

Fig. 9.14 *Locus of the closed-loop poles for a given value of ω_n^2.*

We will now investigate how these parameters should be chosen when the closed-loop poles p_{f1}, p_{f2} have to be in 'Butterworth positions'. In that case:

$$p_{f1}, p_{f2} = \tfrac{1}{2}\sqrt{2}B_\omega(-1 \pm j) \qquad (9.27)$$

Substituting eq. (9.27) into eq. (9.26) and equating the right-hand and the left-hand member yields, assuming again that $A_0\beta_0 \gg 1$,

$$B_\omega = \omega_n \qquad (9.28)$$

and

$$z_0 = \frac{-\omega_n^2}{p_1 + p_2 + \omega_n\sqrt{2}} \qquad (9.29)$$

Since $\omega_n^2 = Kz_0 = -A_0\beta_0 p_1 p_2$, ω_n can be allocated the desired value by assigning the proper value to β_0. The location of the zero then follows from eq. (9.29).

If no phantom zero can be realized, eq. (9.29) provides the condition for correct pole splitting. Putting $z_0 = -\infty$, we find $p_1 + p_2 = -\omega_n\sqrt{2}$. If a phantom zero can be realized, eq. (9.29) states that in principle p_1 and p_2 can assume arbitrary values, provided that $|p_1 + p_2| < \omega_n\sqrt{2}$. In many practical situations ω_n will be much larger than p_1 and p_2. In that case eq. (9.29) simplifies to

$$z_0 = -\tfrac{1}{2}\sqrt{2}\omega_n = -\tfrac{1}{2}\sqrt{2}B_\omega \qquad (9.30)$$

which indicates that the zero tends to determine the bandwidth of the amplifier. The low-frequency gain is determined by the feedback network and is approximately given by $A_{f\infty}$.

The next question is, of course, how a phantom zero can be realized in practice. Since $A_{f\infty}$ is in inverse proportion to β, a pole in $A_{f\infty}$ can be a zero of $A\beta$. An obvious method of realizing a phantom zero is to introduce a zero in the feedback network. As an example, Figure 9.15(a) shows a voltage amplifier with $A_{v\infty} = V_o/V_g = (Z_1 + Z_2)/Z_1$. If Z_1 and Z_2 are resistances, a pole in $A_{v\infty}$ can be obtained by connecting a capacitance in parallel with R_1 (Figure 9.15b). Figure 9.15(c) shows the p–z plot of the voltage transfer of the feedback network for the case where the network is driven from an ideal voltage source and is unloaded. In addition to the desired zero a pole is also created. For the zero to be an effective phantom zero the pole has to be far from the zero. This depends on the ratio of R_1 and R_2. Frequently, $R_1 \gg R_2$ will hold and in this case the pole will belong to the non-dominant group. If the driving impedance and the loading of the network cannot be neglected, these have to be taken into account.

Figure 9.16 shows the case of a current amplifier where a phantom zero has been incorporated into the feedback network. Similar remarks as have been made in relation to the voltage amplifier apply.

In the case of a transadmittance amplifier (Figure 9.17), only one element is available which can be modified for the realization of a phantom zero. If Z is a capacitance, a resistor can be placed in series. If Z is a resistance, an inductance can be placed in series. Again the impedances from which Z is driven and by which

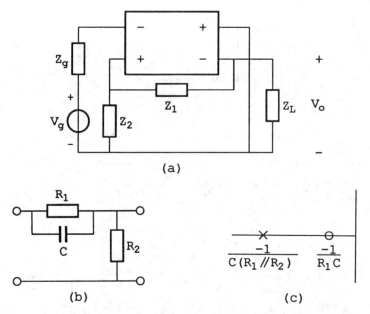

Fig. 9.15(a) *Voltage amplifier with feedback network.* (b) *Realization of phantom zero.* (c) *p–z plot of feedback network.*

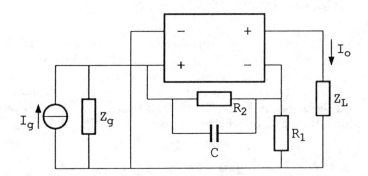

Fig. 9.16 *Current amplifier with resistive feedback network incorporating a phantom zero.*

it is loaded must be taken into account. More design freedom can be created by replacing Z by a π-attentuator. However, to obtain the same value of the transadmittance, larger impedances have to be tolerated in series with the input and the output of the amplifier, which may cause adverse effects on its distortion and noise properties. For the transimpedance amplifier (Figure 9.18), similar considerations apply, albeit that here design freedom can be extended by using a T-attentuator.

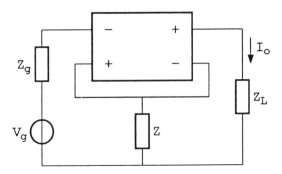

Fig. 9.17 *Transadmittance amplifier.*

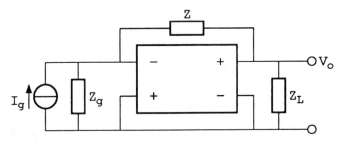

Fig. 9.18 *Transimpedance amplifier.*

Though the realization of a phantom zero by modifying the feedback network is frequently the obvious approach, it is not always feasible or practical. In such cases alternatives must be looked for. Figure 9.19 shows an example where use is made of the properties of the source impedance. Due to the shunt feedback, the input impedance of the amplifier is low. It can easily be seen that, if Z_g is capacitive, a series resistor R is capable of providing a pole at $-1/RC$ in $A_{f\infty}$, which can be an

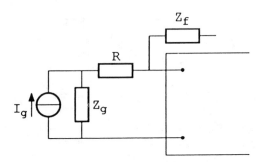

Fig. 9.19 *Creation of a phantom zero using a series resistor in connection with a capacitive source impedance.*

LOW-PASS WIDEBAND AMPLIFIERS   283

effective phantom zero. In other cases use may be made of the properties of the load impedance.

The foregoing discussion was concerned with feedback configurations possessing two clearly dominant poles in the open-loop transfer. In practical situations non-dominant poles will always be present. These should not be totally left out of consideration since one or more of these poles might shift into the dominant group by the feedback arrangement. If so, such a pole could move into a position close to the imaginary axis or even into the right half-plane and hence affect the closed-loop transfer or even give rise to instability.

If there are more than two dominant poles in the loop, the design of the feedback amplifier is essentially done along the same lines as when there are two poles. If there are more than three poles it is advisable to introduce an additional phantom zero, since at least one branch of the root locus is bound to move away somewhere into the left half-plane if the number of poles in $A\beta$ exceeds the number of zeros by more than three. In general it is not useful to have more than two or three dominant poles within the loop. First accomplishing unconditional stability is more difficult because of the large phase shift within the loop. Second, it can be proved that for any Butterworth-type transfer the bandwidth B_ω equals the nth root of the product of loop gain and the product of the n poles. Adding stages will roughly increase this product with a constant factor per additional stage so that overall bandwidth does not significantly increase in general.

If the open-loop transfer possesses three or more dominant poles it can be advantageous to the introduction of a pair of complex phantom zeros. Calculation of the root loci and optimization of the loop parameters 'by hand' is in such cases a cumbersome affair so that resort should be made to computer-aided design. The implementation of complex zeros in the feedback path does not require exotic circuit configurations. Figure 9.20 shows a bridged-T network that can easily be dimensioned to produce a pair of complex zeros. Under the assumption that the network is driven from an ideal voltage source ($R_g = 0$) and is not loaded ($R_L \to \infty$),

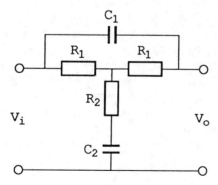

Fig. 9.20 *Bridged-T network for the implementation of complex (phantom) zeros.*

straightforward calculation yields that the condition for complex zeros is

$$4\alpha(1 + R_1/R_2) > 4 + \alpha^2 \qquad (9.31)$$

where $\alpha = \tau_2/\tau_1$ with $\tau_1 = R_1 C_1$ and $\tau_2 = R_2 C_2$. Of course, this second-order network also contributes two poles and this fact aggravates design optimization considerably.

Having now developed the general design strategy for a feedback wideband amplifier, we must turn our attention to the design of the open loop. First, recall that bandwidth is optimized by using stages with maximum GB-product. It must be emphasized that what counts is the GB-product, not the bandwidth of the stage. In a well-designed stage GB is largely determined by the f_T of the active devices. Though it is common practice to strive for large bandwidth of the open-loop amplifier, this approach should not be recommended in view of the ensuing reduction of loop gain. The favoured method for achieving this is the use of low-valued load resistors in the individual stages. There are at least two reasons for not following this practice. First, as has been shown in Chapter 5 (Section 5.4) with BJTs the GB-product is adversely affected by the use of low-value resistances, due to the finite value of r_b. Second, sacrificing low-frequency loop gain implies loss of the beneficial effects of feedback, such as reduction of the influence of noise, interfering signals and distortion. This is not to say that any amount of resistive broadbanding is to be advised against: in certain situations a properly chosen amount of resistive broadbanding in one of the stages constituting the open loop may be applied in order to obtain a desired degree of pole splitting.

What type of stages should be used? If loop gain is to be maximized, CE- and CS-stages are the natural choice. However, as we have seen in Chapter 5, cascading such stages is not without problems. They are not unilateral and they are interactive. By interactivity is meant that the transfer of a stage is not independent of output impedance of the preceding stage and/or the input impedance of its successor. The main drawback of interactivity is that it hampers the designability of the amplifier. Amplifier design is very much facilitated by taking measures for the reduction of non-unilaterality and of interactivity. The obvious method is to add buffer stages that, though hardly or not contributing to the gain of the open loop, provide the desired mutual isolation. Natural candidates are CB(CG)-stages and sometimes emitter (source) followers. Often combinations of amplifying and buffer stages can be treated as one stage. The best-known examples of such combinations are the cascode and the long-tailed pair, which have been investigated extensively in Chapter 5.

The CB-stage and the emitter follower can be considered as unity-gain stages (A_i and A_v respectively) due to full local feedback. Thus their bandwidth, to a first approximation, equals ω_T and consequently the poles they introduce will rarely belong to the dominant group. This is what makes them so useful as buffer stages. The two other local-feedback stages, the series stage and the shunt stage, are less tractable. They rely on external impedances for accomplishing feedback. In general, their GB-product is somewhat smaller than that of the other basic stage configurations. With regard to the shunt stage (Section 5.10), note that, as long as the feedback resistance is fairly large compared with r_b, the reduction in the GB-product is small. In the series stage (Section 5.9) emitter compensation should preferably be applied, but even when this is done the GB-product is somewhat reduced due to the presence of $C_{b'c}$. Nevertheless in certain cases the use of a series stage can be advantageous. A feature of this configuration is its high input impedance. Because of this, it can be the preferred choice for a stage following a CE-stage with deliberate capacitive feedback for the purpose of pole splitting. The capacitance needed for a desired amount of pole splitting can be smaller the larger the load impedance of the stage. In monolithic implementations the total amount of capacitance should be kept low with a view to the necessity of minimizing chip area.

A basic assumption underlying the foregoing discussions is that non-dominant poles and zeros affect the transfer characteristics of the feedback amplifier negligibly. In practical designs it should be checked whether or not this assumption is justified. It should be noted that closing the loop may transfer a pole or a zero from the non-dominant group into the dominant group. The resulting effect can range from a mild peaking in the frequency response to near-instability. If analytical treatment of the problem is impractical, which will often be the case, computer simulation of the closed-loop transfer may provide adequate information.

The systematic design method for wideband negative-feedback amplifiers discussed in the foregoing is based on Nordholt (1983), where it is described in more detail.

9.5 A design example of a wideband feedback amplifier

As an example of the design of an amplifier, the method developed in the preceding section will be applied to a preamplifier for a television camera tube. The requirement is an amplifier with a gain in excess of 1000 and a bandwidth of 5 MHz. As active devices FETs with $f_T = 300$ MHz and BJTs with $f_T = 400$ MHz are available.

A camera tube can be modelled as a current source I_g with a capacitive impedance. The peak-to-peak value of I_g is less than 1 μA. The capacitance C_g is of the order of 10 pF but is not accurately known. In order to prevent it from affecting the information transfer to the amplifier, the best approach is to load the signal source with a virtual short circuit. This leaves two possible configurations, viz. a transimpedance amplifier and a current amplifier (Figure 9.21 a and b), since both configurations use shunt feedback at the amplifier input, and therefore exhibit the desired low-input-impedance character. To minimize the noise contribution of the feedback network, R_1 should be as large as possible. In the case of a transimpedance amplifier the parasitic shunt capacitance of the feedback resistor limits the attainable bandwidth. In the current amplifier (Figure 9.21b), the transfer

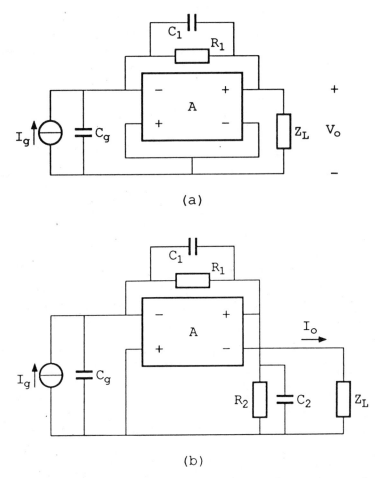

Fig. 9.21 *Two configurations suitable as preamplifiers for a camera tube:* (a) *transimpedance amplifier;* (b) *current amplifier.*

is determined by the ratio of R_1 and R_2. Here, by adding a shunting capacitance to R_2, so that $R_1C_1 = R_2C_2$, it is possible to achieve a transfer independent of frequency. For this reason we choose the current amplifier configuration. Referring to the asymptotic-gain model, we have $A_{i\infty} = I_o/I_g = 1 + R_1/R_2$. The actual gain of the amplifier is given by

$$A_i = A_{i\infty} \frac{-A_L}{1 - A_L} \qquad (9.32)$$

where A_L is the loop gain (see Chapter 4).

With regard to the noise properties of the configuration, R_1 and C_1 can be assumed to be in parallel with C_g (see Section 8.13). In view of the high impedance of the signal source, for optimum noise behaviour an FET is the preferred active device for the input stage of the amplifier block. In Section 8.13 it has been shown that for optimum noise behaviour R_1 should be taken as large as is possible, while $C_{gs} + C_{gd}$ of the FET should equal $C_g + C_1$ (Section 8.12). This can be accomplished by selecting a suitable FET. If necessary, the input capacitance can be enlarged by paralleling two or more FETs. The upper limit for R_1 is determined by the maximum allowable voltage across this resistor, which depends on the supply voltage and on the maximum value of I_g. Let the allowable voltage across R_1 be 1 V, then with $I_g = 1~\mu A$, $R_1 = 1~M\Omega$. It can easily be verified that with any FET featuring an f_T-value of some hundred MHz, the noise contribution of R_1 will be negligible in comparison with the noise of the FET.

As has been shown in Section 8.11, the noise contribution of the second stage is minimal if the first stage has maximum power gain. For this reason we use the CS-configuration. In order to obtain maximum loop gain, the CE-configuration is the obvious choice for the second stage. To achieve near-unilaterality of the input stage and to minimize interactivity between the two stages it is advantageous to cascode the FET. For this purpose a BJT can be used. Since this transistor is used in the CB-configuration the pole it contributes is at approximately ω_T, and hence it belongs safely to the non-dominant group.

The two-stage amplifier obtained in this way is non-inverting. However, to accomplish negative feedback an inverting output is necessary. This can be achieved by configuring the second stage as a different pair. Figure 9.22 shows the conceived amplifier structure schematically, as far as the signal path is concerned.

To determine the loop gain we must assume a suitable controlled source to be independent and then calculate the magnitude of the control signal it evokes. An obvious choice is the controlled current source in the FET. Figure 9.23 shows the equivalent circuit for the calculation of the loop gain. In this diagram Y_1 and Y_2 denote the parallel connection of R_1 and C_1, R_2 and C_2 respectively, while C_t is the total capacitance consisting of $C_{gs} + C_{gd} + C_g \approx 2(C_{gs} + C_{gd})$. Furthermore, β_ω denotes the complex current gain of the CE-connected BJT, hence

$$\beta_\omega = \frac{\beta_0}{1 + j\omega\beta_0/\omega_T} \tag{9.33}$$

From Figure 9.23 we find

$$\frac{I}{I_d} = A_L = -g_m\beta_\omega \frac{Y_1}{Y_1 Y_2 + (Y_1 + Y_2)j\omega C_t} \tag{9.34}$$

Where $Y_2 \approx A_{i\infty} Y_1$, $A_{i\infty} \gg 1$ and $C_t \gg C_1$, eq. (9.34) reduces to

$$A_L = -g_m \frac{\beta_0}{1 + j\omega\beta_0/\omega_T} \cdot \frac{R_1/A_{i\infty}}{1 + j\omega R_1 C_t} \tag{9.35}$$

The low-frequency loop gain A_{Lo} has a value $A_{Lo} = g_m \beta_0 R_1/A_{i\infty}$. The poles of

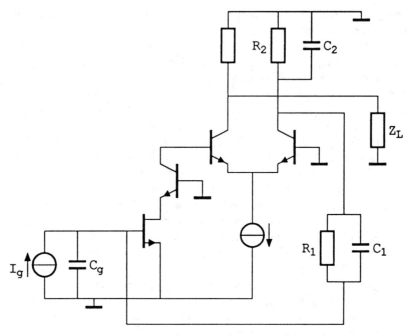

Fig. 9.22 Schematic structure of the camera amplifier.

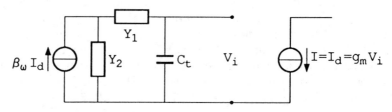

Fig. 9.23 Equivalent circuit for the calculation of the loop gain.

the loop-gain function are

$$p_1 = -\frac{1}{R_1 C_t} \quad \text{and} \quad p_2 = -\frac{\omega_T}{\beta_0}$$

From eq. (9.22) we have

$$B_\omega^2 = A_0 \beta_0 p_1 p_2 = A_{Lo} p_1 p_2 = -\frac{g_m \omega_T}{A_{i\infty} C_t} \tag{9.36}$$

The location of the zero follows from eq. (9.30):

$$z_0 = -\tfrac{1}{2}\sqrt{2} B_\omega$$

Where $B = 5$ MHz or $B_\omega = 10^7 \pi$ and $\omega_{T(BJT)} = 800 \pi\, 10^6$ while $\omega_{T(FET)} = 600 \pi\, 10^6 = g_m/(C_{gs} + C_{gd}) \approx 2g_m/C_t$, eq. (9.36) yields $A_{i\infty} = -2400$, which is comfortably in excess of the required value of 1000. The phantom zero can be realized by inserting a resistor in series with C_2.

If the loop gain would be insufficient for obtaining the required value of $A_{i\infty}$, an extra amplifier stage could be added. If possible, this stage should be designed in such a way that the pole it contributes is sufficiently far away from the dominant poles not to affect the pole positions of the feedback amplifier significantly. If this is not feasible, the loop should be handled as a third-order Butterworth configuration.

The design example given above is a simplified version of a camera amplifier described in Nordholt and de Jong (1983). In the original article a design is described where an additional stage is employed for enlarging the loop gain without adding a pole to the dominant group.

9.6 Use of local-feedback stages

The preceding section dealt with circuit configurations using overall feedback including two or more amplifier stages. Though in general the use of overall feedback is to be preferred above local feedback, or even the absence of feedback, in certain cases local-feedback configurations may be used to good advantage. An obvious reason for choosing this approach can be to avoid proneness to instability. Another reason can be a need for gain control. In a high-loop-gain feedback amplifier the gain is fixed by the feedback network. In principle, gain control is possible by controlling the parameters of this network, but in practice the useability of this method is very much restricted, since any change in loop gain is accompanied by shifts in the locations of the closed-loop poles. We will consider some useful amplifier configurations that utilize local feedback presently.

The by now well-known long-tailed pair and cascode configurations both exhibit a certain degree of local feedback. Both configurations are reasonably well-behaved as far as unilaterality and non-interactivity is concerned. The long-tailed pair is a combination of an emitter follower and a CB-stage, while the cascode is a combination of a CE-stage and a CB-stage (or the corresponding FET-equivalents). A configuration in a similar style which can be useful in certain cases is the combination of a CE-stage and an emitter follower. Figure 9.24 sketches this structure using FETs as active devices.

We will restrict the discussion of the properties of this combination to a qualitative analysis. The basic idea is that the effect of C_{gd} is largely reduced by driving the CS-stages from the low output impedance of the source followers. Moreover, the capacitive load of the drains of the CS-stages is small, thanks to the low input capacitances of the source followers. Whether or not such a configuration is to be preferred above alternative configurations depends on the details of the design problem. In practice the use of cascodes or long-tailed pairs quite often turns

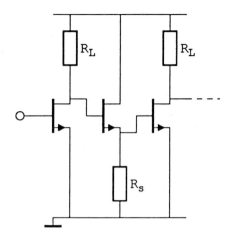

Fig. 9.24 *Alternating CS-stages and source followers.*

out to be the better choice, but the CS–CD combination should not be left unmentioned.

Interesting and often quite practicable configurations are obtained by combining series and shunt stages. The series stage exhibits large input as well as output impedance, whereas the shunt stage has both low input impedance and low output impedance. Alternatively formulated: the series stage behaves as a voltage-controlled current source. Sinking the current provided by a series stage into the input of a shunt stage implies virtual non-interactivity. The same is true for the reverse configuration consisting of a shunt stage followed by a series stage.

Figure 9.25 sketches both combinations. To a first approximation the low-frequency transconductance of the series stage is $1/R_E$, while the low-frequency transimpedance of the shunt stage is R_f. Henceforth, the combination of Figure 9.25(a) acts as a voltage amplifier with $A_v \approx R_f/R_E$ and the combination of Figure 9.25(b) as a current amplifier with $A_i \approx R_f/R_E$. As we have found in Chapter 5, the high-frequency behaviour of both local-feedback stages is somewhat complicated and this is a drawback of these combinations. Further, it should be noted that the individual stages possess only moderate loop gain, so that the beneficial effects of negative feedback are much less prominent than in an overall feedback configuration using the same devices. However, the inherent stability and the excellent designability of these configurations have made them favoured candidates for monolithically integrated wideband amplifying devices. One disadvantage of these configurations is often overlooked. Because of the moderate loop gain in the first stage, its noise behaviour is seriously affected by the noise generated by the feedback resistors. Moreover, frequently the noise of the second stage must also be taken into account due to the moderate gain of the first stage. These combinations should therefore not be used if minimization of the noise figure is at a premium. For the sake of completeness it may be mentioned that all the circuit structures dealt

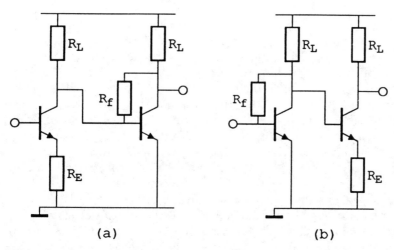

Fig. 9.25(a) *Combination of a series stage and a shunt stage;* $A_v \approx R_f/R_E$.
(b) *Combination of a shunt stage and a series stage;* $A_i \approx R_f/R_E$.

with can be configured both in the single-sided version that has served as the vehicle for their presentation, and in a different version. In monolithic implementations the latter form will frequently be preferred in view of its outstanding features with regard to biasing and common-mode rejection, without resorting to the employment of large capacitors.

9.7 Distributed amplifiers

In the preceding sections several methods have been considered for overcoming the limitations of active devices as building elements for wideband amplifiers. In particular, the bandwidth narrowing due to cascading several active devices with given GB-capabilities could be effectively reduced by judicious positioning of the poles and zeros of the overall transfer function. However, ultimately, the finite GB-capabilities of the active device remain the limiting factor for the gain that can be achieved with a prescribed bandwidth. Clearly, no usable gain can be realized when the required bandwidth approaches, or even surpasses, the GB-value. In particular, in communication systems and in certain instrumentation systems, amplifiers with a bandwidth close to, or even exceeding, the GB-product of the available active devices may be needed. In such cases the common approach of cascading stages for obtaining high gain is of no avail. The way out is to resort to additive amplifiers. In an additive amplifier, active devices are essentially connected in parallel. However, simply connecting active devices in parallel does not increase the GB-product since the transconductances as well as the parasitic capacitances add, so that their ratio, which determines the GB-product, is not improved. A method must be devised for achieving the effective addition of transconductances, without simultaneous addition of the capacitances. An approach that at first sight seems attractive is to split the signal spectrum into several parts. If the spectrum is broken up in n parts, n bandpass amplifiers can be connected in parallel, while their output signals are added (see Chapter 11). Since the bandwidth of each of the amplifiers is B/n, its gain can be n times the gain of an amplifier that handles the complete spectrum. Figure 9.26 depicts the structure obtained in this way. However, the practical implementation of this type of amplifier is hampered by the requirement that the amplitude as well as the phase characteristics of the constituting bandpass amplifiers must match closely. If not, severe deviations of the required flat overall characteristic occur. The required matching is hardly achievable in practice, even in the case where the spectrum is split into no more than two parts.

A much better method is provided by using the principle of distributed amplification. The basic idea comes down to making the capacitances of the active devices part of a lumped-element transmission line by separating the capacitances by means of inductances. Figure 9.27 depicts an implementation with MOSTs. The input capacitances form, together with the inductances, an artificial delay line with low-pass characteristics. The line must be driven from, and loaded by, a resistance equal to the characteristic resistance of the line in order to prevent reflections of the signal. In a similar way the output capacitances are made part of a transmission line. This arrangement has given rise to the alternative name of 'travelling-wave amplifier'.

Transmission line theory teaches that a direct relation exists between the bandwidth and the delay time of the line. Henceforth, if both lines are configured in similar fashion and are designed for the same bandwidth, their delay times are also equal. This is, of course, necessary for proper operation of the configuration. The input signal travels along the input line and when the signal shows up at the input electrode of an active device, it is transferred to its output, whereupon the amplified signal travels along the output line until it reaches the characteristic load of that line.

A detailed discussion of the properties of low-pass transmission lines is beyond the scope of a book on electronics. In order to convey a feeling for the basic design

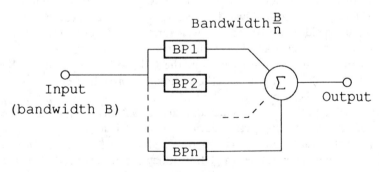

Fig. 9.26 *Principle of the split-band amplifier.*

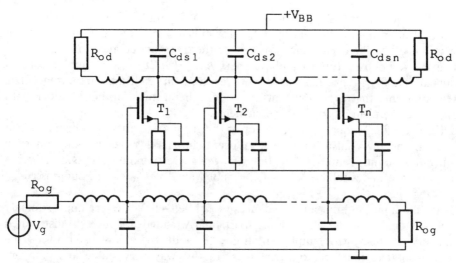

Fig. 9.27 *A distributed amplifier with MOSTs.*

approach, a line consisting of simple so-called constant-k T-sections is used as a vehicle for the discussion. Figure 9.28 shows a T-type section of such a line.

For reflectionless loading the characteristic impedance Z_T is real, but frequency-dependent:

$$Z_T = R_o \sqrt{1 - \omega^2/\omega_c^2} \qquad (9.37)$$

where $R_o = \sqrt{L_k/C_k}$ and $\omega_c = 1/\sqrt{L_k C_k} = 2\pi f_c$, where f_c is the cut-off frequency. In practice Z_T is usually approximated by R_o. Since the input and output capacitances of MOSTs (and of other active devices) are different, for equal ω_c the values of the inductances in the input and output lines are also dissimilar. As a consequence, the characteristic resistance of the drain line R_{od} differs from that of the gate line R_{og}. The effective load resistance is $\frac{1}{2}R_{od}$, since both sides of the line must be characteristically loaded. Hence, the voltage gain A_v of the configuration follows from

$$A_v = \frac{V_o}{V_g} = \frac{1}{4} n g_m R_{od} \qquad (9.38)$$

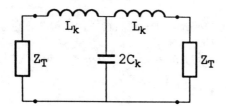

Fig. 9.28 *A simple low-pass T-section.*

where n is the number of active devices operating in parallel. The design procedure follows straightforwardly from the given formulas: if the gain and the bandwidth are specified and the capacitances, as well as the transconductances of the MOSTs are given, L_k follows from C_k and ω_c. Then, $R_{od} = \sqrt{L_{kd}/C_{kd}}$, whereupon n follows from eq. (9.38). In practice, n cannot be given an arbitrary large value. The attenuation and the dispersion that any practical transmission line is ridden with restrict the useful number of active elements.

Dispersion, i.e. frequency dependency of the group delay time, can be minimized by modifying the transmission line. A useful measure is the introduction of a deliberate amount of mutual induction between successive inductances. Further, if necessary, the undesirable effects of the finite value of C_{gd} can be suppressed by cascoding.

Since the input and output impedances of field-effect devices are approximately purely capacitive, these devices are particularly suited for use in distributed amplifiers. In microwave implementations, the inductive couplings between the devices can take the form of strip lines. In this way very stable and compact implementations can be realized. Several successful implementations have been reported providing bandwidths up to 18 GHz with useful gain capabilities.

Bipolar transistors are less suited for use in distributed amplifiers because of their complex input impedance characteristics. Several methods have been devised to overcome the problems evoked by these characteristics. If the number of devices involved in the structure is restricted to two or three, acceptable results can be obtained. However, if possible, application of BJTs in distributed amplifiers is best avoided.

References

Nordholt, E. H. (1983) *Design of High-performance Negative-feedback Amplifiers*, Elsevier Scientific, Amsterdam.

Nordholt, E. H., and de Jong, L. P. (1983) 'The design of extremely low-noise camera-tube preamplifiers', *IEEE Trans. on Instrumentation and Measurement*, **IM 32** (2), 331–6.

Further reading

Cherry, E. M. and Hooper, D. E. (1968) *Amplifying Devices and Low-pass Amplifier Design*, Wiley, New York.

Ghausi, M. S. (1965) *Principles and Design of Linear Active Circuits*, McGraw-Hill, New York.

10 Oscillators

10.1 Introduction

Circuits for the generation of periodic signals are called oscillators. In many electronic systems they perform indispensable functions. In most digital systems information flow is synchronized by a system clock. In communication systems, oscillators are used for generating the carriers needed for the signal transmission by means of modulation, and in time-measuring systems oscillators provide the timing reference.

The most important property of an oscillator is the accurate periodicity of the signal that is generated. In terms of a spectral description this implies that the signal power is concentrated on discrete frequencies that are multiples of the oscillator frequency. In many applications spectral purity of the periodic signal is an essential requirement. Because of the unavoidable presence of noise, some spreading of the oscillator power bands around the intended discrete spectral lines always occurs. Often, the pursuit of spectral purity is an important design criterion. For this reason this aspect deserves ample attention.

10.2 Specification of oscillator properties

As was stated above, the periodicity of the oscillator signal is its most important property. With respect to this property two aspects can be distinguished. They are usually indicated as 'long-term stability' and 'short-term stability'. Long-term stability is related to the steadfastness of the average frequency generated. In many practical oscillators the frequency has a tendency to drift due to temperature variations, the ageing of the constituting components, power-supply variations, and so on. Short-term stability is the term usually employed to qualify the spectral purity, already mentioned in the introduction.

Figure 10.1 sketches the spectral distribution of signal power around the intended spectral line f_0. In general, spectral power decreases rapidly with increasing distance from the intended frequency. In addition to these spectral sidebands an amount of

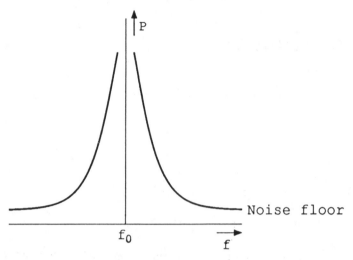

Fig. 10.1 *Sketch of the spectral distribution of oscillator power.*

white noise is present constituting the noise floor. A commonly used measure for specifying the unwanted sideband power is the carrier-to-noise ratio at a given frequency, defined as

$$CNR = 10 \log \frac{Carrier\ power}{Power\ density\ at\ a\ specified\ sideband\ frequency} \text{ (dB Hz)}$$

The sideband power density is the power at the specified sideband frequency per Hz.

A second property to be specified is the waveform of the generated signal. In many applications a harmonic signal is desired, but also square waves, triangular waves or sawtooth-shaped waves can be called for. Aberrations of the desired waveform can be specified as distortion quantities.

The third property to be specified is the amplitude of the generated signal, usually as a peak-to-peak value or, particularly for harmonic signals, as an effective value.

10.3 Harmonic oscillators

The most important class of oscillators is the class of harmonic oscillators that generate approximately sinusoidal signals. Such an oscillator must contain a network comprising at least two energy reservoirs that can exchange energy. An obvious choice is to use an LC-resonator. Figure 10.2(a) shows a parallel LC-resonator. The parallel resistor models the unavoidable losses. In a practical resonator the losses are almost entirely due to the series resistance of the coil (Figure 10.2b). However, for calculation purposes the representation of the losses by a parallel resistance is more convenient. It can easily be shown that for frequencies

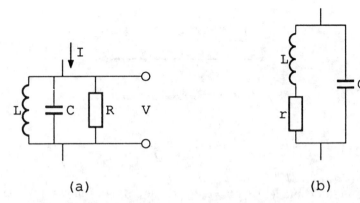

Fig. 10.2 (a) *A parallel LC-resonant circuit.* (b) *Representation of the losses by a series resistance.*

in the neighbourhood of the resonance frequency the series resistance r can be transformed into the parallel resistance $R = Q^2 r$, where Q denotes the quality factor of the resonant circuit. The quality factor is an important parameter of the resonator. It is generally defined by

$$Q = 2\pi \frac{\text{Stored energy}}{\text{Loss per cycle}}$$

provided $Q \gg 1$. This definition applies to resonant systems of any kind. For instance, the quality factor of a bouncing ball that loses 10% of height per bounce is 20π. Applying this definition to the resonator of Figure 10.2(a) we find

$$\text{Loss per period } T = \int_0^T \frac{v^2}{R} = \int_0^T \frac{\hat{v}^2 \sin^2 \omega t}{R} = \tfrac{1}{2}\hat{v}^2 \frac{T}{R}$$

The stored energy in the resonator is $\tfrac{1}{2} C \hat{v}^2$, hence

$$Q = 2\pi \frac{CR}{T} = CR\omega_0 = R\sqrt{\frac{C}{L}} \qquad (10.1)$$

where $\omega_0 = 1/\sqrt{LC}$ is the resonant (angular) frequency. The relation between voltage and current is

$$\frac{V}{I} = Z_p = \frac{R}{1 + jQv} \qquad (10.2)$$

where $v = \omega/\omega_0 - \omega_0/\omega \approx 2(\omega - \omega_0)/\omega_0 \approx 2\Delta f/f_0$; v is called the *detuning*.

According to eq. (10.2) the parallel LC-circuit is a *voltage resonator* in that the quantity exhibiting resonance is the voltage across the circuit.

Figure 10.3 shows the series LC-resonating circuit. Here, current is the resonating

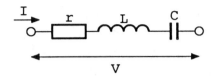

Fig. 10.3 *Series resonant circuit.*

quantity:

$$\frac{I}{V} = Y = \frac{g}{1 + jQv} \qquad (10.3)$$

where $Q = \omega_0 L g$, $\omega_0 = 1/\sqrt{LC}$ and $g = 1/r$.

From eqs. (10.2) and (10.3) $B = \omega_0/Q$ (Figure 10.4) follows for the 3 dB-bandwidth of both circuits.

The pole–zero pattern of the transfer is easily derived from eq. (10.2) or (10.3). Using eq. (10.2) and replacing $j\omega$ by the complex variable s, we find

$$Z = \frac{R}{1 + jQv} = \frac{1}{C} \frac{s}{s^2 + s(\omega_0/Q) + \omega_0^2} \qquad (10.4)$$

so that

$$z = 0 \text{ and } p_{1,2} = -\frac{\omega_0}{2Q} \pm j\omega_0 \sqrt{1 - \frac{1}{4Q^2}} \qquad (10.5)$$

Figure 10.5 shows the pole–zero plot; $\alpha = \omega_0/2Q$ and $\omega_1 = \sqrt{\omega_0^2 - \alpha^2}$. Upon excitation of the circuit, we find, putting $I(s) = I_0/s$,

$$V = \frac{I_0}{C} \frac{1}{(s - p_1)(s - p_2)} \qquad (10.6)$$

Transformation of eq. (10.6) to the time domain yields

$$v(t) = \frac{I_0}{\omega_0 C} \varepsilon^{-\alpha t} \sin \omega_1 t \qquad (10.7)$$

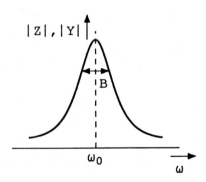

Fig. 10.4 *Shape of the transfer of a resonant LC-circuit.*

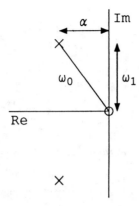

Fig. 10.5 *Pole–zero plot of LC resonant circuit.*

This is a damped harmonic oscillation. As can be seen from eq. (10.7), the damping is inversely proportional to Q. If there were no losses ($Q \to \infty$), the oscillation would last infinitely. A lossless passive system is a physical impossibility. To obtain undamped oscillations, the losses should be compensated for. This requires the use of an active circuit.

At this stage an important conclusion can be drawn. The harmonic signal generated by the circuit will unavoidably be contaminated by noise. The only source of noise present in the resonator is the thermal noise of the loss resistance. The noise is filtered by the resonator, implying that noise spectral power will predominantly be present around the resonant frequency, which is tantamount to deficient short-term stability, as defined in Section 10.2. The obvious conclusion is that high resonator Q is an essential prerequisite for superior short-term stability. Moreover, the larger the losses to be compensated for, the more the undamping circuitry must contribute power. And the active devices that are needed involve additional noise sources.

Addressing now the question of how the loss compensation can be effected, we start with a qualitative view. The lost energy must be supplied at each cycle. This could be done in the way indicated in Figure 10.6. The periodicity of the resonator signal is sensed by an amplifier and the periodic output power is fed back to the resonator. Figure 10.7 gives a formal representation. The output signal of the resonator is amplified by the amplifier A and fed back to the resonator. If the resonator is a voltage resonator, the input signal should be a current, hence the amplifier must be a voltage-to-current amplifier (transconductance amplifier). If x denotes the output signal of the resonator, $x = A \cdot H(s)x$ holds, hence for $x \neq 0$ $A \cdot H(s) - 1 = 0$ should hold.

Where $H(s) = H_0/(1 + jQv)$, we find $-AH_0\omega_0 s + \omega_0 s + Q\omega_0^2 = 0$ or

$$s^2 + \frac{1 - AH_0}{Q} \omega_0 s + \omega_0^2 = 0 \qquad (10.8)$$

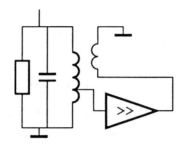

Fig. 10.6 *A general scheme for loss compensation.*

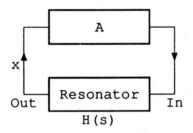

Fig. 10.7 *Formal representation of loss compensation.*

Note that H_0 is the transfer at $v = 0$, hence at $\omega = \omega_0$. The roots of this equation are

$$p_{1,2} = -\alpha \pm j\omega_1$$

where $\alpha = (\omega_0/2Q)(1 - AH_0)$ and $\omega_1 = \sqrt{\omega_0^2 - \alpha^2}$. If $(1/4Q^2)(1 - AH_0)^2 < 1$, the roots are complex. Upon excitation of the circuit, free oscillations are generated according to

$$x = 2K\varepsilon^{-\alpha t} \cos(\omega_1 t - \varphi)$$

where K is a constant accounting for the magnitude of the excitation. If $\alpha > 0$, the oscillations are damped. In this case no full compensation of the circuit losses is achieved. If $\alpha < 0$, oscillations with increasing amplitude are generated. If $\alpha = 0$, undamped oscillations are generated. The condition for $\alpha = 0$ is $AH_0 = 1$. In this case the roots of eq. (10.8) are purely imaginary and the frequency of oscillation is ω_0, the resonant frequency of the resonator. If A is complex, the condition $|AH| = 1$ is fulfilled if

$$|AH(s)|_{s=j\omega_n} = 1 \qquad (10.9a)$$

and

$$\arg[AH(s)]_{s=j\omega_n} = 0 + k \cdot 360° \, (k = 0, 1, \ldots) \qquad (10.9b)$$

In this case, the frequency of oscillation ω_n deviates from the resonant frequency ω_0, since the phase shift arg A has to be compensated by a phase shift arg $H = -\arg A$. The conditions expressed by eqs. (10.9a) and (10.9b) are known under the name *Barkhausen criterion for oscillation*.

Of course it is not possible to design for exact loss compensation, since any practical circuit exhibits unpredictable parameter variations. Therefore, the design should aim at some overcompensation, implying $AH_0 > 1$. If this condition is fulfilled no external excitation is needed for starting the oscillatory action. Any small disturbance, including random noise, suffices to start oscillations with increasing amplitude. Of course this increase cannot go on unlimited. The active circuit inevitably possesses a restricted range of linear operation that will stop further increase of the generated signal. An obvious limiting mechanism is overdrive of active components occurring as a consequence of finite biasing voltages and currents.

In an alternative method of modelling, the loss compensation is represented by a negative resistance. Figure 10.8(a) depicts this representation for a parallel resonator, Figure 10.8(b) for a series resonator. If $R_p = R_n$, $r_s = r_n$ respectively, the resonator is fully undamped enabling it to maintain oscillations for indefinite time.

The negative resistance is effected by means of a feedback amplifier with net positive feedback. To ascertain that continuous oscillations occur, some overcompensation is needed, implying $R_n < R_p$ or $r_n > r_s$. Since R_n or r_n are implemented as internal resistances of feedback amplifiers which necessarily exhibit a finite linear operating range, amplitude limiting occurs. Due to the limiting action a final state is attained in which $R_p = R_n$, $r_s = r_n$ respectively.

Where $R_p = R_n$, the impedance function of the parallel circuit is

$$Z = \frac{R_p}{jQv} = \frac{1}{j\omega C + (1/j\omega L)} \tag{10.10}$$

implying $Z \to \infty$ for $v = 0$ or $\omega = \omega_0$. Similarly for the series circuit

$$Y = \frac{g_s}{jQv} = \frac{1}{j\omega L + (1/j\omega C)} \tag{10.11}$$

where $g_s = 1/r_s$.

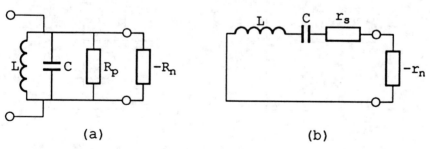

Fig. 10.8 *Representation of loss compensation by a negative resistance:* (a) *parallel circuit;* (b) *series circuit.*

10.4 Practical implementation of harmonic oscillators

At this stage we can address the question of how to implement electronic oscillators. As has become clear in the preceding section, an oscillator consists of a resonant circuit and an amplifier for the compensation of the resonator losses. The requirements to be met by the resonant circuit are obvious. It should exhibit the desired resonant frequency and its Q factor should be as high as is possible, since the losses are responsible for the resonator noise and hence for the short-term stability as expressed by the CNR.

Given the resonant circuit, only the oscillator frequency is fixed. The amplitude is determined by the properties of the amplifier, more particularly by its large-signal behaviour. Many oscillators in common use have not been designed on the basis of an accurate specification of the amplitude of the signal to be generated. This applies to all configurations where no accurate model of the relevant non-linearities is available. As a consequence neither is the waveform well defined since, of course, the type of non-linearity involved has consequences for the waveform of the signal. In practice, even rather hard limiting does not give rise to a strongly distorted output signal thanks to the filtering features of a high-Q resonant circuit. However, strong overdrive of the amplifier, leading to hard clipping, also has consequences for the short-term stability. Signal clipping implies the occurrence of square waves or nearly square waves in the amplifier. Due to the large amount of harmonic frequencies involved, noise bands in the neighbourhood of these harmonic frequencies are mixed down (also called folded back) into the frequency band of the desired ground-wave frequency. For this reason strong overdrive should be avoided and this calls for well-deliberated design of the non-linear effect involved.

In applications where moderate frequency stability can be tolerated, it is natural to favour simple implementations. Obviously, at least one active element is necessary. Figure 10.9 shows two configurations that can be viewed as direct

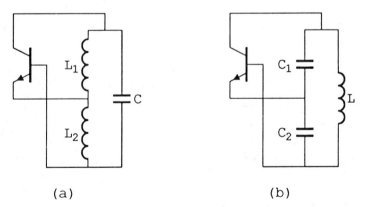

Fig. 10.9 *Two one-transistor oscillator configurations:* (a) *inductive tapping;* (b) *capacitive tapping (so-called Colpitts circuit).*

translations of the principle indicated in Figure 10.6. Since the parallel circuit exhibits voltage resonance, the amplifier should convert voltage into current. In both configurations of Figure 10.9 the desired transconductance character is accomplished. Referring to Figure 10.7, the requirement $AH_0 > 1$ should be met. In the circuit of Figure 10.9(a) the transconductance is $g_m + 1/R = A$ and $H_0 = RL_2/(L_1 + L_2)$, where R represents the total parallel loss resistance, i.e the resonator loss including the additional damping due to the finite input and output resistance of the transistor. Hence, the condition for oscillation is $g_m R \geqslant L_1/L_2$. Similarly for Figure 10.9(b), where the condition for oscillation is $g_m R \geqslant C_2/C_1$.

Inductive coupling can also be accomplished by means of a transformer, as is depicted in Figure 10.10. Transformer coupling can be a useful variant if tuning of the oscillator is done by a hand-driven variable capacitor, since the circuit can easily be biased with one terminal of the capacitor grounded. The requirement $AH_0 = 1$ implies that the loop gain should be positive, i.e the total phase shift should be $k2\pi$ with k integer. Since the CE-configuration exhibits phase reversal, the sign of M should be negative.

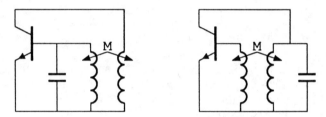

Fig. 10.10 *Two versions of one-transistor oscillators using transformer coupling.*

EXAMPLE

Figure 10.11 shows a simple one-transistor LC-oscillator. Amplitude stabilization is achieved by inserting an RC-parallel combination in the gate circuit of the JFET.

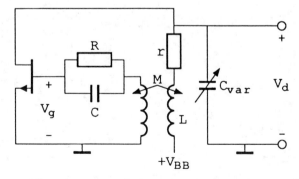

Fig. 10.11 *A simple variable LC-oscillator.*

The sinusoidal gate signal drives the gate junction periodically into forward biasing. Due to the d.c.-component of the gate current a d.c.-voltage builds up across the gate capacitor. The biasing voltage so generated reduces the effective transconductance of the FET until, in the steady state, unity loop gain is attained. This simple method is often practised when simplicity of circuitry is pursued and no high demands on frequency stability have to be met. Design the circuit such that with a variable capacitor of 75–300 pF the frequency range of 1 MHz–2 MHz is covered.

SOLUTION
Since $4\pi^2 f^2 L C_{var} = 1$, L should be 84 μH. If we denote the parallel connection of r_d and $Q^2 r$ by R_p, we have $V_g = (M/L)V_d$. At the resonant frequency, $V_d = -g_m R_p V_g$. Hence, the requirement for unity loop gain is $-g_m R_p M/L = 1$ or $-M = L/g_m R_p$. The negative sign of M is obtained by using an inverting transformer. Let r be about 20 Ω, then the minimum value of Q (at 1 MHz) is 26. Hence $Q^2 r = 13.5$ kΩ. If $r_d = 50$ kΩ, then $R_{p(min)} \approx 10.6$ kΩ. If we strive for a steady state value of $g_m \approx 2$ mA V^{-1}, $|M| > 4$ μH must hold. To ensure that oscillations will be maintained, M is taken larger, for instance $|M| \approx 8$ μH. The time constant of the RC-gate circuit should be several times the oscillation period $1/f$, say $20/f$. For instance, $R = 25$ kΩ and $C = 800$ pF will be adequate.

If a very low frequency is to be generated, the use of a coil can be unattractive. In such cases a resonating circuit using two capacitive energy reservoirs can be a viable option. Figure 10.12(a) shows a configuration in this class. If we choose $R_1 C_1 = R_2 C_2 = \tau$, the voltage transfer is

$$\frac{V_o}{V_1} = \frac{R_2/(2R_1 + R_2)}{1 + jvQ} \qquad (10.12)$$

where $Q = 1/(2 + R_2/R_1)$ and $\omega_0 = 1/\tau$.

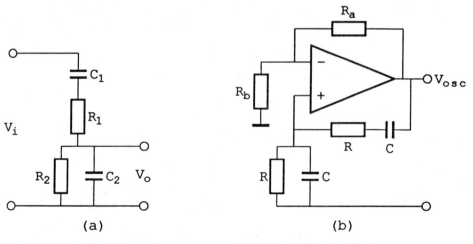

Fig. 10.12(a) *An inductorless resonator.* (b) *Wien-bridge oscillator.*

If also $R_1 = R_2 = R$ eq. (10.12) simplifies to

$$\frac{V_o}{V_i} = \frac{1/3}{1 + jvQ} \text{ with } Q = \frac{1}{3}$$

The obvious disadvantage of this inductorless variant is its low Q-value with the ensuing consequences for short-term frequency stability (low CNR).

Since the transfer is of the voltage-to-voltage type, the undamping amplifier must be a voltage amplifier. Figure 10.12(b) shows a simple implementation using an operational amplifier. This configuration is known as the *Wien-bridge oscillator*. For $AH_0 = 1$ to be valid, the voltage gain of the amplifier should be >3, hence $R_a/R_b > 2$.

EXAMPLE
Let the frequency to be generated be 1 Hz. Using an LC-resonator, when $C = 1000\ \mu\text{F}$, L would amount to 25.3 H, a quite impractical value. Using the Wien-bridge configuration, the desired frequency can, for instance, be obtained with $C = 1\ \mu\text{F}$ and $R = 160\ \text{k}\Omega$.

Another inductorless oscillator in widespread use is the so-called 'phase-shift oscillator'. The resonator consists of three RC-networks (Figure 10.13). If $R_1 C_1 = R_2 C_2 = R_3 C_3 = \tau$ and $R_3 \gg R_2 \gg R_1$, the RC-sections produce equal phase shift. For oscillation the total phase shift should be $180°$, hence each RC-section must contribute $60°$ phase shift. This occurs at $\omega_0 = 1/\tau\sqrt{3}$, while the attenuation is $\frac{1}{2}$ per section. Hence, for the voltage amplification $g_m R$ of the amplifier, $g_m R > 8$ must hold.

In all the configurations dealt with in the foregoing, the biasing of the active devices was not considered. Generally, the design of the biasing circuitry does not introduce serious problems. For practical reasons one terminal of the power supply is grounded. The same applies to the amplifying device. In oscillators that must be tunable by hand it is desirable to also have one of the resonator terminals grounded. And, of course, a path for the biasing current has to be provided. If the resonator

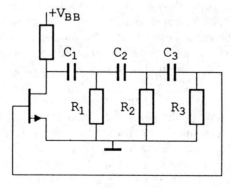

Fig. 10.13 *A phase-shift oscillator.*

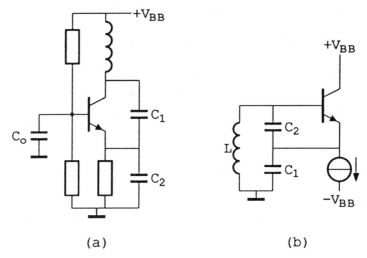

Fig. 10.14 *Two versions of a Colpitts-oscillator:* (a) *base of transistor grounded; C_o is a large bypass capacitor;* (b) *collector grounded, dual-polarity power supply.*

contains an inductor an attempt can be made to configure the circuit in such a way that the inductor is part of the d.c.-path. If it is unavoidable that a capacitor is in series with a current-carrying electrode of an active device, the capacitor can be bridged by a resistor. However, it should be noted that adding resistors to the resonator circuit implies a lowering of Q. As an example, Figure 10.14 shows two versions of the one-transistor Colpitts oscillator of Figure 10.9(b).

10.5 Amplitude stabilization

In the simple oscillator configurations discussed in the preceding section the amplitude of the generated signal is determined by the non-linear large-signal behaviour of the amplifier. Some kind of signal compression or clipping action is bound to occur. However, the amount of compression necessary for maintaining the loop gain at unity value depends on nearly all circuit parameters. As a consequence, judicious design for a specified signal amplitude cannot be accomplished.

A straightforward method of fixing the amplitude level is to use a controllable amplifier. The amplitude of the generated signal is measured and converted into a d.c.-signal which is subsequently used to control the gain of the amplifier. Figure 10.15 depicts this approach schematically. The rectified signal is compared with a reference voltage which represents the desired amplitude level. The difference signal controls the gain of the amplifier. In principle this is the best method for

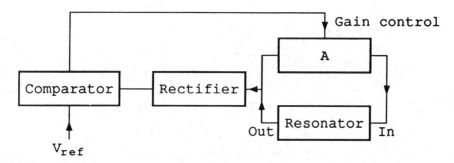

Fig. 10.15 *Schematic representation of the application of automatic gain control (AGC) in an oscillator.*

ascertaining essentially linear behaviour, though it is not without disadvantages. First, circuit complexity is increased considerably. The added circuitry usually involves more components than the oscillator as such. And second, the addition of a control loop introduces a new problem. The rectifier must contain a low-pass filter to suppress the a.c.-signal. Filtering implies delay and this causes the control to lag behind. As a consequence, a stepwise change of the amplitude can give rise to an oscillatory response of the amplitude control. This is called 'bouncing'. A poorly designed control loop can even exhibit instability. For oscillators that must be fast tunable, the bouncing effect is particularly bothersome and often intolerable.

Fortunately, a second method for controlled amplitude stabilization is available which employs simple circuitry and does not involve an additional loop (Boon, 1989). The underlying idea is to deliberately design non-linear behaviour of the amplifier by incorporating the non-linear element in a feedback loop. The method can best be explained by using the model of Figure 10.8(a), where the undamping effect of the amplifier is represented by a negative resistance. Figure 10.16(a) recalls this model, while in Figure 10.16(b) the voltage-to-current transfer of the negative conductance G_n is sketched. Within the range of current values from $-I_L$ to $+I_L$,

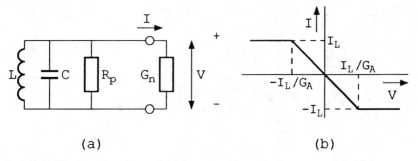

Fig. 10.16(a) *Undamping by a limiting negative conductance.* (b) *Voltage-to-current transfer of the limiting conductance.*

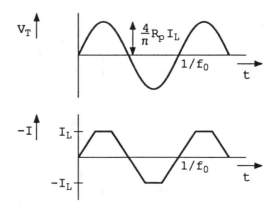

Fig. 10.17 *Voltage and current in the oscillator according to the model of Figure 10.16.*

the element behaves as real small-signal conductance $-G_A$. This conductance should be large enough to ascertain the start up of oscillations, hence $|-1/G_A| < R_p$. The amplitude of the oscillations ceases to increase when the limiting level I_L is reached.

Figure 10.17 depicts the waveforms of the current in the negative conductance and of the resonator voltage. By virtue of the selectivity of the resonator, this voltage is approximately sinusoidal. For the sake of completeness, Figure 10.18 depicts the dual configuration equipped with a series resonator.

Returning to the basic configuration of Figure 10.16(a), the question must be put forward as to how I_L and G_A should be chosen. As has already been stated, G_A has to exceed $1/R_p$ in order to start up oscillations. However, it is advisable to choose G_A not much larger than is needed. The larger G_A, the larger the fraction of the

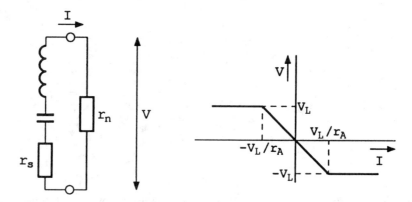

Fig. 10.18 *Undamping of series resonator.*

OSCILLATORS 309

period wherein the current waveform is clipped. Strong clipping incurs not only increasing distortion of the generated waveform, but also lowering of CNR due to aliasing of noise around harmonics. In most cases an excess loop gain by a factor of two provides a good balance between the advantages and disadvantages that are inherent in the use of a limiting amplifier. The total noise present in the oscillator signal can be equally divided into amplitude noise and phase noise. With $G_A \approx 2G_{A\,min}$, the amplitude noise is effectively suppressed, whereas the extra phase noise due to aliasing is still moderate. It can be proved that in this particular case noise by aliasing has the same power as the primary noise around the carrier, so that CNR decreases by only 3 dB with respect to the case of the 'linear oscillator' (Boon, 1989).

How large should I_L be? Obviously, the larger the amplitude of the generated signal, the larger is CNR, since noise is in principle independent of the signal amplitude. Practical constraints are dictated by power supply and dissipation restrictions.

For the formal calculation of the oscillator noise the model of Figure 10.16(a) is extended with the relevant noise sources (Figure 10.19). The current source i_{nR} represents the thermal noise associated with the resonator losses (R_p), while v_n and i_n represent the noise contributed by the amplifier (see Section 8.10).

As has already been found previously (eq. (10.10)), the undamped impedance is given by

$$Z' = \frac{R_p}{2jQ\,\Delta\omega/\omega_0} = \frac{R_p}{2jQ\,\Delta f/f_0} \tag{10.13}$$

For the calculation of the CNR we must find the equivalent noise voltage v_{in} at the input of the amplifier. This can be accomplished by using the voltage-source shifting technique introduced in Section 8.9. Figure 10.20 shows the resulting equivalent circuit.

The voltage source v_n in series with $-1/G_a$ can be transformed into a current source $v_n|G_a|$, whereupon the equivalent circuit of Figure 10.21 is obtained. If $|G_a| > 1/R_p$, the current source $v_n|G_a|$ is only active during a fraction, $1/|G_a|R_p$, of the time. Moreover, it has to be taken into account that the total noise power is equally divided between amplitude noise and phase noise, of which only the latter is of interest for the CNR. The noise sources i_n and v_n can be represented by their

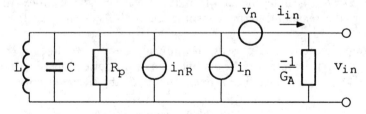

Fig. 10.19 *Equivalent circuit including noise sources.*

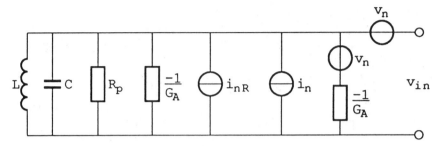

Fig. 10.20 *Transformed equivalent circuit.*

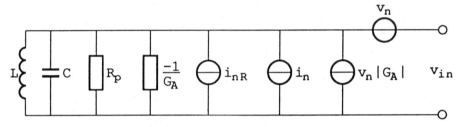

Fig. 10.21 *Further transformation of the equivalent circuit.*

equivalent noise resistances R_{ni} and R_{nv}. Their noise spectra can then be written in the form

$$\overline{i_n^2} = 4kT/R_{ni} \text{ and } \overline{v_n^2} = 4kTR_{nv} \qquad (10.14)$$

Using the Norton–Thévenin transformation and working along the lines put forward in Chapter 8, we obtain for the spectrum of the noise voltage v_{in}

$$\overline{v_{in}^2} = \tfrac{1}{2}\left(\frac{4kT}{R_p} + \frac{4kT}{R_{ni}} + 4kTR_{nv}G_A^2 \frac{1}{|G_A|R_p}\right) R_p^2 \left(\frac{f_0}{2Q\,\Delta f}\right)^2 + \tfrac{1}{2} \cdot 4kTR_{nv}$$

$$= \tfrac{1}{2}\left(4kTR_p + 4kT\frac{R_p^2}{R_{ni}} + 4kTR_{nv}|G_A|R_p\right)\left(\frac{f_0}{2Q\,\Delta f}\right)^2 + 2kTR_{nv} \qquad (10.15)$$

The term $2kTR_{nv}$ is independent of frequency, hence it constitutes the (white) noise floor; the remaining terms constitute the noise sidebands of the oscillator signal.

If no amplifier noise were present, the only noise source would be $\overline{i_{nR}^2} = 4kT/R_p$. Just as we did in the context of the qualification of the noise properties of amplifiers (Section 8.11), we can introduce a noise factor for the sideband noise:

$$F = \frac{\text{Total sideband noise power}}{\text{Sideband noise power due to resonator losses}}$$

Applying this definition, we can write

$$F = \frac{4kT(R_p + R_p^2/R_{ni} + R_{nv}|G_A|R_p)}{4kT/R_p} = 1 + \frac{R_p}{R_{ni}} + R_{nv}|G_A| \quad (10.16)$$

The amplitude V_{max} of the oscillator signal follows from (Figure 10.17):

$$I_L R_p < V_{max} < \frac{4}{\pi} I_L R_p \quad (10.17)$$

and the carrier-to-noise ratio at a given sideband frequency follows from

$$CNR = \frac{1}{\sqrt{2}} \frac{V_{max}}{\sqrt{v_{in}^2}} \quad (10.18)$$

Eq. (10.16) reveals that the noise factor, which is a figure of merit for the short-term frequency stability of the oscillator, depends on R_p, R_{ni} and R_{nv} in essentially the same way as in the case of an amplifier (Section 8.12). Hence the methods employed for optimizing the noise behaviour of amplifiers can be put to good use in optimizing frequency stability of oscillators of the type under discussion. Optimization can be achieved either by adapting R_p by means of tapping the resonator or by adapting R_{in} and R_{iv} by means of an appropriate value of the bias current of the first stage of the feedback amplifier.

10.6 Realization of the negative resistance for undamping the resonator

In the preceding section it was found that the non-linear resistance which undamps the resonator should have an accurate value. As has already been stated, $|G_A| \approx 2/R_p$, constituting an excess gain by a factor of two, is a good compromise.

As we found in the discussion of feedback amplifiers, an accurate input impedance can only be obtained by providing two feedback loops. Of course, for obtaining a *negative* input resistance, one of the loops should constitute positive feedback. And one of the loops should exhibit the desired limiting characteristic for the stabilization of the amplitude of the oscillator signal. With these basic considerations in mind it is an easy task to configure suitable feedback amplifiers. Figure 10.22 shows an example. The non-linear element is located in the positive feedback loop of the parallel–parallel type. The negative feedback loop is of the series–parallel type. The current through the non-linear elements cannot exceed I_o, so that the amplitude of the resonator voltage is limited to a value of $\frac{1}{2}I_o R_p$. For instance, with $I_o = 0.1$ mA and $R_p = 20$ kΩ, the peak voltage amounts to 1 V. Assuming $R_p = 20$ kΩ, the negative input resistance of the feedback amplifier should be -10 kΩ. Where $R_3 = 20$ kΩ, this can be achieved by letting $R_1 = 1$ kΩ and $R_2 = 4$ kΩ. For the proof of this statement reference is made to Figure 10.23. Modelling the amplifier block as an ideal nullor, we have

$$v_i = 2i_i R_3 + v_o \quad (10.19)$$

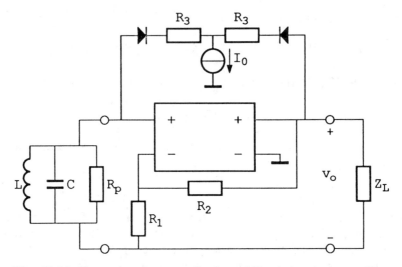

Fig. 10.22 *Example of an amplitude-stabilized, low-noise oscillator.*

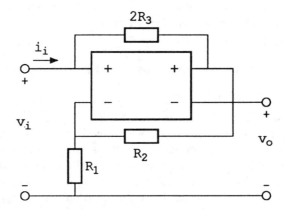

Fig. 10.23 *Schematic for the calculation of the negative input resistance.*

and

$$v_i = \frac{R_1}{R_1 + R_2} v_o \tag{10.20}$$

Eliminating v_o from eqs. (10.19) and (10.20) yields

$$R_i = \frac{v_i}{i_i} = -\frac{2R_3 R_1}{R_2} \tag{10.21}$$

In the example in Figure 10.22, the limiting characteristic has been implemented

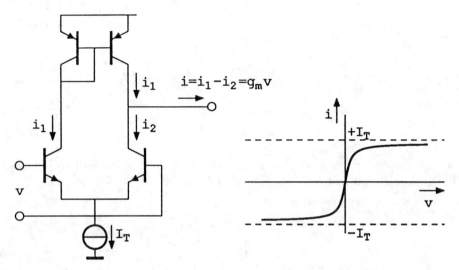

Fig. 10.24 *Differential pair as a symmetrically limiting amplifier.*

by using diodes. Alternatively, transistors can be used as non-linear transactances. Figure 10.24 shows a differential pair operating as a symmetrical current limiting transactance, together with its $v \to i$ transfer.

Figure 10.25 shows the basic oscillator configuration resulting from using this transactance. For small values of v, the transconductance of the differential pair is

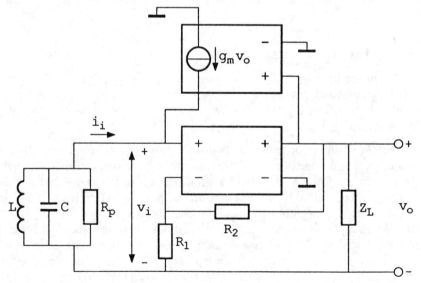

Fig. 10.25 *Basic structure of an oscillator stabilized with a non-linear transactance.*

g_m. When the amplifier block is modelled as a nullor, the input conductance G_i of the amplifier can easily be found. From Figure 10.25:

$$i_i = -g_m v_o \text{ and } v_i = v_o \frac{R_1}{R_1 + R_2}$$

so that

$$G_i = \frac{i_i}{v_i} = -g_m \frac{R_1 + R_2}{R_1} \qquad (10.22)$$

A very simple configuration results when the nullor is realized with one transistor (T1), while $R_2 = 0$ and $R_1 \to \infty$. Figure 10.26 shows this configuration.

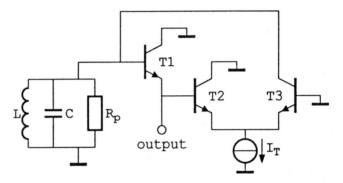

Fig. 10.26 *A simple implementation of the basic structure of Figure 10.25.*

EXAMPLE
An oscillator is required complying with the following specifications:

- low circuit complexity;
- low noise-sideband power;
- stabilized amplitude of about 0.5 V;
- frequency 100 MHz.

The resonator Q should be chosen as high as is possible with a simple coil. In order to minimize the influence of parasitic capacitances the resonator capacitance should not be chosen too low. With a standard coil where $L = 50$ nH we find $C = 51$ pF. With the selected coil, $Q = 96$, so that $R_p = Q/\omega_0 C = 3$ kΩ. Hence a maximum resonator current $\hat{\imath}_r$ of about 200 μA is required. In the circuit of Figure 10.26, $\hat{\imath}_r = \frac{1}{2} I_T$, so that $I_T = 400$ μA. The negative input conductance of the amplifier should be $\geqslant 0.67$ mA V^{-1}. Since, with reference to Figure 10.25, $R_2 = 0$ and $R_1 \to \infty$, the input conductance is approximately $-g_m$. In the implementation of Figure 10.26, $g_m = \frac{1}{4} I_T / V_T$ with $V_T = kT/q = 26$ mV. With $I_T = 400$ μA, $g_m = 3.8$ mA V^{-1}. In order to reduce g_m, emitter resistances R_E in T2 and T3 are inserted. With $R_E = 500$ Ω, the effective value of g_m is reduced to 0.79 mA V^{-1}.

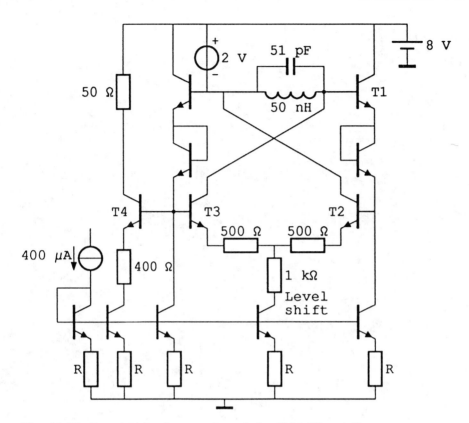

Fig. 10.27 *Practical implementation of the 100-MHz oscillator.*

Figure 10.27 shows an implementation complying with the calculated data. *T*4 is a buffer transistor for separating the load from the oscillator circuit.

Measurements on a monolithic implementation of this oscillator yielded a CNR of 114 dB Hz at a sideband frequency of 9 kHz. As can be calculated from eq. (10.15) where $R_{ni} \to \infty$ and $R_{nv} = 0$, the CNR without the amplifier contributing noise would be 119 dB Hz, so it can be concluded that the degradation of the resonator-determined CNR due to the noise contributed by the active circuit amounts to not more than 5 dB.

In the preceding discussions concerning the design of amplitude-stabilized, low-noise oscillators, circuits with a parallel tuned circuit have been used. It goes without saying that oscillators using a series resonant circuit can be designed along similar lines. In this case the limiting element should exhibit transresistance character, as indicated in Figure 10.28. Such a transresistance can be realized with a differential pair by adding resistors at the input side and the output side (Figure 10.29).

Figure 10.30 shows the schematic of an oscillator equipped with a series

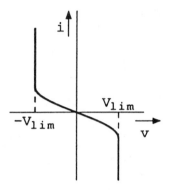

Fig. 10.28 *Characteristic of limiting transresistance.*

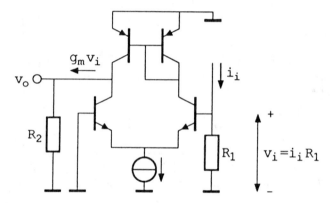

Fig. 10.29 *A possible implementation of a limiting transresistance.*

resonator. As can be found straightforwardly by using the nullor concept, the negative input resistance of the amplifier is given by

$$R_n = -\frac{g_m R_1 R_3}{1 - g_m R_3} \qquad (10.23)$$

where g_m is the transconductance of the differential pair (Figure 10.29).

In determining the equivalent noise sources associated with the feedback amplifiers, the noise contribution of the resistors in the feedback networks should be taken into account. This aspect of feedback–amplifier design has already been dealt with in Section 8.13.

A further remark is to be made about the frequency characteristics of the amplifier. In the foregoing discussion the amplifier and the limiter have been assumed to be memoryless. In terms of sine wave response this implies that no phase shift whatsoever occurs in these elements. In practice, some phase shift is

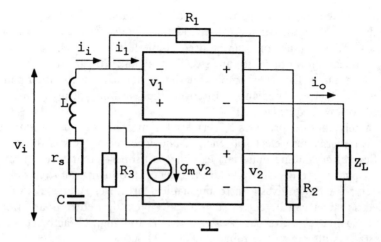

Fig. 10.30 *An amplitude-stabilized oscillator using a series resonator.*

unavoidable because of the finite bandwidth of any amplifier. Since at the frequency of oscillation the total phase shift in the entire oscillator loop should have zero value (or, more generally, $2k\pi$), the phase shift in the amplifier is compensated for by a phase shift in the resonator, which implies that the oscillator frequency deviates slightly from the resonant frequency. Though this effect does not fundamentally detract from oscillator quality, it hampers its designability. For this reason, amplifier bandwidth should preferably exceed the oscillator frequency amply.

10.7 Crystal oscillators

An oscillator is essentially a timing reference. In many applications high demands must be met in regard to the accuracy of this reference. We have seen that, as far as resonator-type oscillators are concerned, short-term stability requires a high value of the resonator Q-factor along with a judicious design of the undamping circuit. Long-term frequency stability depends on the constancy of the elements that constitute the resonator. The Q-factor is associated with losses. Purely electric resonators exhibit these in the form of resistive losses and dielectric losses. Nearly always, the resistive losses dominate by far. Good LC-resonators in the frequency range up to some hundreds of MHz can possess Q-values of the order of 100–200. Higher values are difficult to obtain. The weak element is the inductance due to its resistive losses. Moreover, inductances are temperature-dependent and they are also prone to ageing effects. For these reasons long-term as well as short-term frequency stability of oscillators with LC-resonators is limited and often falls short of the requirements dictated by the specific application.

Mechanical resonators can exhibit much better Q-factors and constancy of resonant frequency. A tuning fork, or even a piano string resonating at some hundreds of Hz, sounds for several seconds, implying that their Q-factors range in the thousands. A tuning fork can be used in an electronic oscillator by converting the mechanical vibration into an electrical signal, for instance by means of capacitive coupling. However, employing this principle leads to rather cumbersome and thus unattractive expedients. Moreover, the feasibility of using such devices is restricted to low frequencies.

Fortunately, nature provides a very useful mechanism for the conversion of mechanical energy into electrical energy, and vice versa. This is the piezoelectric effect. Several crystalline materials exhibit the property that mechanical pressure causes displacement of electric charges, hence producing an electrical voltage. Conversely, a voltage gives rise to mechanical pressure. A crystal exhibiting the piezoelectric effect can be cut in such a way that a desired resonant frequency is obtained. Such a resonator can possess a remarkably high Q-factor. Typical values for quartz-crystals are in the range 20,000–200,000.

Figure 10.31 sketches the make-up of a quartz crystal for use as a resonator. A voltage applied to the metal electrodes causes a mechanical shear stress in the crystal, and conversely. Used as a resonator in an electrical circuit the crystal can be modelled as an electrical resonator circuit as shown in Figure 10.32. The capacitance C_p represents the capacitance formed by the electrodes. Table 10.1

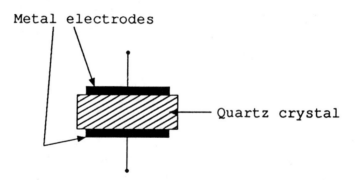

Fig. 10.31 *Sketch of the make-up of a crystal resonator.*

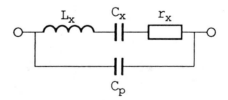

Fig. 10.32 *Equivalent circuit of a crystal resonator.*

Table 10.1 *Parameters for commonly used crystals*

f_0	L_x	C_x	r_x	C_p
100 kHz	52 H	0.049 pF	400 Ω	8 pF
1 MHz	2 H	0.006 pF	24 Ω	3.4 pF
10 MHz	0.01 H	0.026 pF	5 Ω	8.5 pF

gives some values of the parameters of the equivalent circuit for commonly used crystals.

As is evident from the equivalent circuit, the crystal exhibits series resonance as well as parallel resonance. Since $C_x \ll C_p$ always, both resonant frequencies are close to one another. Both types of resonance can be used in oscillator and filter applications. The resonant frequency depends on the thickness of the sheet of crystal. For frequencies in excess of about 20 MHz the sheet would be unpractically thin if the fundamental frequency were used. In high-frequency applications overtone resonant modes are used: in the frequency range 20–60 MHz usually the third overtone, over 60 MHz the fifth overtone. The maximum tolerable voltage is restricted, due to the dissipation in the crystal and to non-linear behaviour at large excitations. Maximum tolerable voltages associated with these limitations are of the order of 0.2–4.0 V.

Because the crystal behaves as a resonant circuit, all oscillator configurations using an LC-resonator can also be implemented as crystal oscillators. Figure 10.33 shows a Colpitts oscillator. If C_1 and C_2 are large compared to C_x, the oscillator frequency is close to the resonant frequency of the crystal.

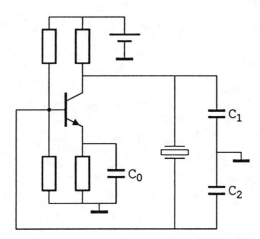

Fig. 10.33 *A Colpitts crystal oscillator.*

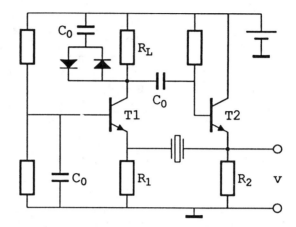

Fig. 10.34 *Crystal oscillator wherein the crystal operates as a series resonant circuit.*

Figure 10.34 shows a configuration where the crystal is used as a series resonator. The crystal impedance is in series with the emitter resistances r_e of $T1$ and $T2$. For an approximate analysis of the operation of the circuit we assume that both $r_{e1} \ll r_x$ and $r_{e2} \ll r_x$, whereas $R_1 \gg r_x$ and $R_2 \gg r_x$. If v denotes the emitter voltage of $T2$, then the current $i_x = v/r_x$ flows into the emitter of $T1$, giving rise to a voltage $i_x R_L$ across R_L. This voltage is transferred to R_2 by the emitter follower $T2$. Hence the loop gain is R_L/r_x. By bridging R_L with the combination of diodes as indicated in

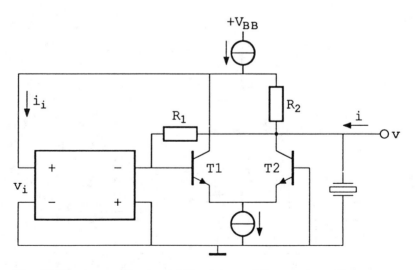

Fig. 10.35 *An amplitude-stabilized 'one-pin' crystal oscillator.*

the diagram, amplitude limiting is achieved. The capacitors C_0 are large enough to be considered as short circuits at the oscillator frequency.

Figure 10.35 shows an example of an oscillator configuration of the general type, discussed in Section 10.6. For the sake of variation, in the example shown the crystal resonator is undamped by the negative *output* resistance of the amplifier. Controlled limiting is accomplished by the differential pair $T1-T2$. If the amplifier is modelled as a nullor ($i_i = 0$, $v_i = 0$), straightforward calculation of the negative resistance v/i seen at the terminals of the crystal is

$$R_n = \frac{v}{i} = \frac{-gR_1R_2}{1 - gR_2} \qquad (10.24)$$

where g is the transconductance of the differential pair, hence $g = \frac{1}{2}g_m$ with g_m the transconductance of $T1$ and $T2$. An attractive feature of this oscillator is that the crystal has one terminal in common with the amplifier ground. This feature is of great advantage in a monolithic implementation, since only one bonding pad is required to connect the crystal to the chip.

10.8 First-order or relaxation oscillators

In the oscillators discussed in the preceding sections, the frequency-determining element was a resonator, which is a device containing at least two energy reservoirs that can exchange energy. The differential equation that describes the state of the system is at least of second order. Consequently, a periodic solution is possible. This requires that at least two poles form a conjugate complex pair. When excited, the resonant circuit reacts with a periodic change of state. The periodicity is fully determined by the location of the poles. If the resonator is properly undamped, continuous oscillation occurs. In the steady state the poles are located on the imaginary axis.

A first-order system, i.e. a system whose characteristic equation possesses only one pole, has no natural frequency. If such a system is to be used as the basis for an oscillator, periodicity has to be accomplished by periodic switching of the mode of operation of the system. Since in a periodic system the state of the system is a known function of time, attaining a certain state can be used to change the mode of operation, thus establishing the oscillatory behaviour. If different modes of operation prevail at different moments, the circuit must obviously contain a memory for retaining information concerning the actual mode of operation. In addition, to mark the moment at which the mode of operation must change, a comparator is needed. Consequently, any first-order oscillator must at least contain a reactance for realizing the pole, a memory and a comparator. Figure 10.36 shows a structure containing all these elements. The reactive element is a capacitor C that is charged by a current source I_1. A second current source I_2 is available for discharging the capacitor. Charging and discharging constitute the two modes of operation. Mode switching is effected by the switch S that is activated by the output

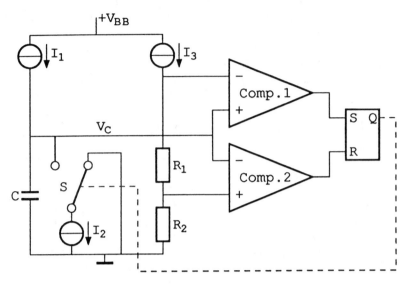

Fig. 10.36 *Basic schematic of a first-order oscillator.*

signal of the R–S latch. Such a latch is a logic circuit whose output state Q is set by its S-input and reset by its R-input. When the capacitor voltage reaches the value $V_B = I_3(R_1 + R_2)$, then comparator 1 activates the setting of the R–S latch, whereupon its output signal V_0 activates the switch S, so that C starts to discharge. If $I_2 > I_1$, the current source providing I_1 need not be switched off. When subsequently the capacitor voltage reaches the lower level $I_3 R_2$, the latch is reset by the second comparator, whereupon S switches back, so that the charging action is resumed. In this implementation, the latch constitutes the memory function, while the capacitor operates as an integrator, implying that the pole is in the origin.

EXAMPLE
In the configuration of Figure 10.36 let the charging and the discharging currents be equal, hence $I_2 = 2I_1$. The slope of the voltage V_C on the capacitor is $V_C/t = I_1/C$. If we denote the upper comparator level V_h, the lower level V_l and $V_h - V_l = \Delta V = I_3 R_1$, then the time needed to let V_C rise from V_l to V_h is given by $\frac{1}{2}T = C \Delta V / I_1 = C I_3 R_1 / I_1$, hence

$$f = \frac{1}{T} = \frac{I_1}{I_3} \frac{1}{2R_1 C}$$

With a view to the integratability of the circuit, let the maximum allowable values of R_1 and C be $R_1 = 50$ kΩ and $C = 100$ pF, so that $2R_1 C = 10^{-5}$ s, and $f = 10^5 (I_1/I_3)$Hz. If, for instance, I_3/I_1 should not exceed the value of 100, then the lowest frequency that can be generated is 1000 Hz. If a lower frequency should be generated, R_1 could be replaced by a Zener diode or a series connection of forward biased diodes.

Several alternative embodiments of the circuit are possible. One of these is to combine the comparators and the latch in a regenerative circuit. An example of a practical realization of this modification is presented in Chapter 12, Section 12.10.

First-order oscillators can easily be designed to be electronically tunable. For example, when the comparator levels in the circuit of Figure 10.36 are left constant, the frequency of oscillation is proportional to the current that charges and discharges the integrating capacitor. This type of oscillator is therefore particularly suited for use as a frequency modulator. Two examples of implementations of first-order oscillators used as frequency modulators will be discussed in detail in Chapter 12, Section 12.10.

In comparison with second-order oscillators, first-order oscillators, also called 'relaxation oscillators', exhibit certain advantages and disadvantages. Advantages are as follows:

- There is only one reactive element. In practice a capacitor is always used. If the capacitor value is not too high, the circuit is fully integratable.
- The frequency of oscillation can be varied simply by changing the charging and discharging currents or by changing the comparator levels. The relation between the charging current and the frequency of oscillation is linear, provided $f_0 \ll f_T$, so that linear frequency modulation can easily be achieved. The amplitude can also be easily fixed by the setting of the comparator levels.
- Since there is only one energy reservoir, variations of the frequency can be accomplished instantaneously.
- The waveforms of the generated signals are readily predictable.

The most important disadvantage of first-order oscillators is their inferior frequency stability. Long-term stability is determined by the accuracy of the current sources, the capacitor value and the comparator levels. With appropriate design effort, satisfactory long-term stability can be obtained for many applications. Short-term stability is inherently rather poor because the circuit contains many noisy elements, notably current sources and transistors. The presence of noise causes uncertainty in the moment at which the comparator responds to the increasing or decreasing capacitor voltage. Within the short time interval round the comparator level, the system is extremely sensitive to disturbances, including noise. Because of the time-dependent behaviour of the circuit, a full analysis of the short-term frequency stability is a complicated matter. It has been shown that configurations can be devised wherein the influence of noise on the switching moments is considerably reduced. Short-term stability comparable to that of resonator-type oscillators can be accomplished, albeit at the cost of additional circuit complexity (Verhoeven, 1990). This is a specialist topic that is beyond the scope of this work.

10.9 Two-integrator relaxation oscillator

In regenerative oscillators the memory function is obtained from a regenerative circuit. Alternatively, a capacitor can be used to perform the memory function. One

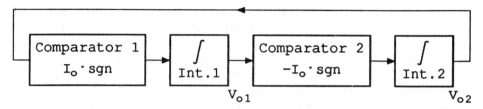

Fig. 10.37 *Basic structure of the two-integrator relaxation oscillator.*

approach is to include the capacitor in a sample-and-hold circuit (see Section 13.5). Another possibility is to use a second integrator. This idea has given birth to an interesting oscillator configuration yielding two signals that are exactly in quadrature, which is quite useful in many applications. Figure 10.37 shows the basic structure of a symmetrical two-integrator oscillator.

The block $I_o \times$ sgn denotes a constant current source that can take either a positive of a negative polarity. The integrator is a capacitor C; its output is a voltage $\pm I_o t/C$. The sign of I_o is determined by the sign of the output of the second integrator. The formation of the 'sign-signal' is accomplished by a comparator. These comparators mutually isolate the integrators.

To understand the operation of the oscillator we assume that at $t = 0$ the output signal V_{o1} is K, while $V_{o2} = 0$. Additionally, we assume that V_{o1} is moving towards zero value. The following equations then apply:

$$V_{o1}(t) = \text{sgn } V_{o2}(t) I_o t/C + K \qquad (10.25a)$$

and

$$V_{o2}(t) = -\text{sgn } V_{o1}(t) I_o t/C \qquad (10.25b)$$

where sgn denotes 'sign of' and I_o is assumed to be of positive polarity. Let T_A be the time needed for V_{o1} to reach zero value. Then $T_A = KC/I_o$. At $t = T_a$, $V_{o2} = -K$. At this moment the sign of V_{o1} changes and V_{o2} starts to move towards zero value. Let $V_{o2} = 0$ at $t = T_A + T_B$. Then $T_B = KC/I_o$ and V_{o1} at $t = T_A + T_B$ is $-K$. Presently, V_{o2} changes sign and hence V_{o1} starts moving towards zero value. The time T_C needed to reach this situation is again KC/I_o. Finally, in a further time interval T_D, V_{o2} reaches zero value, while V_{o1} resumes its starting value K. Obviously, the period of the oscillation is $T = T_A + T_B + T_C + T_D = 4KC/I_o$. The signals V_{o1} and V_{o2} have a triangular shape and are exactly in quadrature. Their amplitude is K. The two integrators can be viewed as memories preserving the starting value, which is exchanged periodically. Since the structure is symmetrical it does not make sense to specify which of the integrators accomplishes the memory function at a given time. Figure 10.38 depicts the output signals as a function of time.

For convenience, we have assumed in our calculation that $V_{o1}(t = 0) = K$ and $V_{o2}(t = 0) = 0$. If, more generally, we assume $V_{o1}(t = 0) = K_1$ and $V_{o2}(t = 0) = K_2$, then a similar reasoning yields $T = 4C(K_1 + K_2)/I_o$, while the amplitude is $K_1 + K_2$.

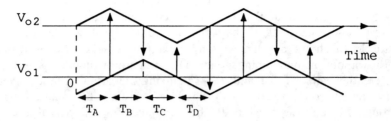

Fig. 10.38 *The relation between the two triangular signals in the two-integrator oscillator.*

When the comparators have no offsets, then at any time the sum of the moduli of the momentary values of V_{o1} and V_{o2} is constant. This is a useful property since it opens the possibility of determination of the amplitude of the output signals at any time.

An ideal integrator has its pole in the origin. Practical integrators exhibit losses, implying a pole on the real axis. If the pole is given by $p = -1/\tau$, it can be shown that for V_o:

$$V_o(t) = \frac{I_o}{C} \tau (1 - \varepsilon^{-t/\tau}) + K\varepsilon^{-t/\tau} \qquad (10.26)$$

indicating that the amplitude of the oscillation decreases exponentially. In order to maintain the oscillations, the integrators have to be undamped. Since the exact value of τ is not known and is, moreover, prone to incidental variations, the undamping must be slightly in excess of the damping. In that case, the amplitude of the oscillations will increase exponentially. The state of affairs is comparable to that in the resonator oscillator, in which the resonator damping is somewhat overcompensated in order to ensure continuity of the oscillatory behaviour. In the oscillator under discussion the same approach can be taken to stabilize the amplitude. The amplitude is measured and compared to a reference value. When the reference value is reached, the undamping is stabilized (Figure 10.39).

As we have seen, a continuous measure of the amplitude of the triangular signals can be derived by summing the absolute values of both signals. In the ideal case of perfect triangular output signals no ac-ripple whatsoever would be present at the

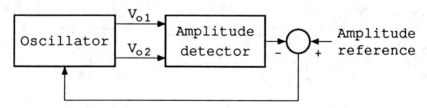

Fig. 10.39 *Amplitude control in a two-integrator oscillator.*

output of a 'sum of moduli' circuit. In practice, aberrations of ideal behaviour occur so that some filtering of the 'sum of moduli' signal may be desirable.

Figure 10.40 shows a practical implementation of the basic two-integrator oscillator without the undamping circuit. The comparator function is performed by the differential pairs $T1-T2$ and $T3-T4$. Fixing the common-mode level at the output terminals could be done with the aid of resistors. By using the transistors $T5$ through $T12$ and the resistors R, the impedance for differential mode signals remains high. Figure 10.41 shows a possible circuit configuration for the amplitude detector. Two differential pairs ($T1-T2$ and $T3-T4$) with emitter resistors act as transconductance amplifiers, converting the input signals V_{o1} and V_{o2} into currents. The currents are fed to four current switches $S1$ through $S4$. The switches are controlled by the voltages V_{o1} and V_{o2}, so that they operate as rectifiers for the integrator signals. The output currents of the switches are proportional to the moduli of the integrator output voltages. The currents are summed through a current mirror. The output signal I_{amp} is a measure for the amplitude of the oscillator signals. It can therefore be used to control the undamping circuit. An example of such a circuit is given in Figure 10.42. In the lower part of the circuit a control current $I_C = I_{ref} - I_{amp}$ is formed. Between the input terminals *in* 1 and

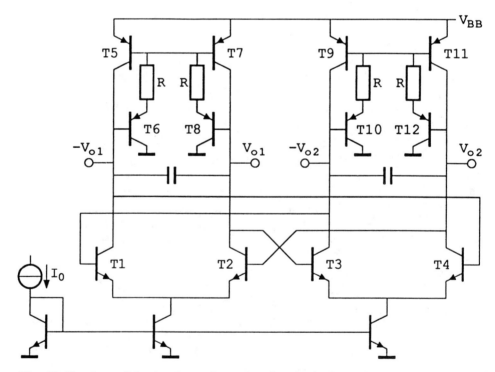

Fig. 10.40 *A possible circuit configuration for the basic two-integrator oscillator without amplitude stabilization.*

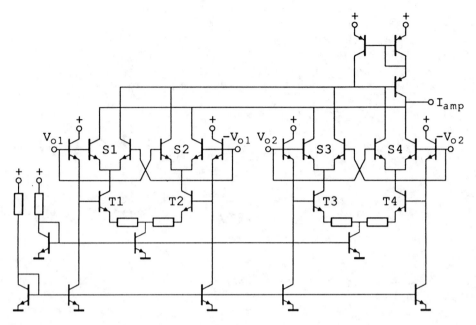

Fig. 10.41 *An amplitude detector of the sum-of-moduli type.*

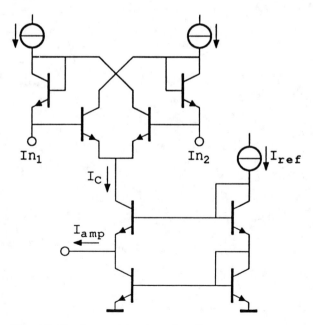

Fig. 10.42 *An undamping circuit.*

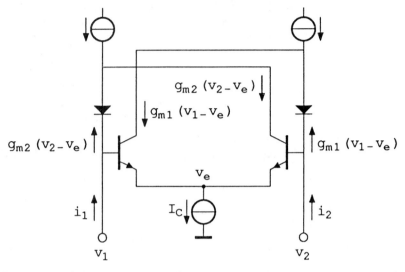

Fig. 10.43 *Simplified schematic for finding the negative input resistance of the undamping circuit.*

in 2 of the upper part a negative input resistance is established that is controlled by I_C. The negative input resistance is due to the positive feedback loop in the upper part of the circuit. To find the relation between I_C and the input resistance, reference is made to the simplified schematic of Figure 10.43. To describe the small-signal behaviour the transconductances of $T1$ and $T2$ are denoted by g_{m1} and g_{m2}, while $r_{e1} = 1/g_{m1}$ and $r_{e2} = 1/g_{m2}$. These quantities depend on I_C, since $g_{m1} = I_{T1}/V_T$ and $g_{m2} = I_{T2}/V_T$ with $I_{T1} + I_{T2} = I_C$. The level shift diodes do not affect significantly small-signal behaviour. When base currents are neglected, the following equations apply:

$$i_1 = g_{m2}(v_2 - v_e) \tag{10.27a}$$

$$i_2 = g_{m1}(v_1 - v_e) \tag{10.27b}$$

$$i_1 = -i_2 \tag{10.27c}$$

Eliminating v_e from these equations yields

$$R_i = \frac{v_1 - v_2}{i_1} = -\left(\frac{1}{g_{m1}} + \frac{1}{g_{m2}}\right) = -(r_{e1} + r_{e2}) \tag{10.28}$$

The larger I_C is, the smaller r_{e1} and r_{e2} and, consequently, the stronger the undamping.

Figure 10.44 shows the oscillator of Figure 10.40 with the undamping circuits

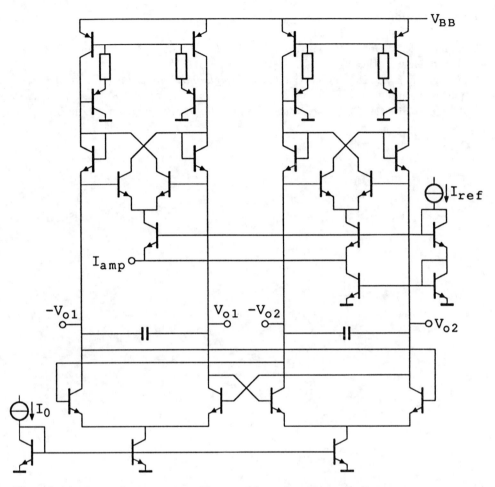

Fig. 10.44 *A two-integrator oscillator with undamping circuits.*

according to Figure 10.42 added to each of the integrators. Instantaneous tuning of the oscillator can be achieved by means of variation of I_o.

References

Boon, C. A. M. (1989) Design of High-performance Negative-feedback Oscillators, PhD thesis, Delft University of Technology.

Verhoeven, C.J. M. (1990) First order Oscillators, PhD thesis, Delft University of Technology.

Further reading

Blackman, R. B. and Tukey, J. W. (1958) *The Measurement of Power Spectra*, Dover, New York.

Frerking, M. E. (1978) *Crystal Oscillator Design and Temperature Compensation*, Van Nostrand.

Robbins, W. P. (1982) *Phase Noise in Signal Sources*, Peregrinus, London.

11 Bandpass amplifiers and active filters

11.1 Introduction

Amplifiers having a specified gain within a certain frequency band, where signals outside this band are attenuated or suppressed, are called 'bandpass amplifiers'. The attenuation requirements for out-of-band signals depend on the type of system that the amplifier is part of. Bandpass amplifiers are indispensable in all systems that use frequency multiplexing for transferring signals from different sources. If high demands have to be met by the bandpass function, tunability is difficult to accomplish. In such cases it is customary to convert the frequency band of the signal that must be selectively amplified to a fixed frequency range, so that a bandpass amplifier with a fixed passband can be used. Examples are intermediate-frequency amplifiers in communication systems (see Chapter 12). For instance, in AM-radio, channel widths of 9 or 10 kHz are standard. A commonly used intermediate frequency is 455 kHz. In FM-radio, channel widths of about 200 kHz are common. An often-used intermediate frequency is 10.7 MHz. In television receivers the intermediate-frequency amplifiers must meet stringent requirements concerning the suppression of certain frequencies, such as the carrier frequencies of adjacent vision and sound channels.

11.2 Methods for achieving bandpass characteristics

For several decades selectivity in amplifiers has been almost exclusively realized by the use of passive filter configurations, i.e. combinations of capacitances and inductances. To obtain steep roll-off of the frequency response the transfer function of the filter network must be of high order. The bandpass character requires the poles to be located in the neighbourhood of the central frequency of the passband on the imaginary axis. Each pole contributes 6 dB octave^{-1} to the slope of the frequency response far from the passband.

Network theory has developed formal methods for the design filters with prescribed characteristics. Design data based on these methods have been laid down in tables, graphs and computer programs (see, for instance, Williams and Taylor, 1988). The availability of all these resources has greatly facilitated filter design. However, in spite of their good designability, the use of these electromagnetic filters is rapidly declining. They cannot be monolithically integrated and in most cases they need adjustments, since it is not possible to produce passive elements of high accuracy. If, as is often the case, the filter function goes along with the need for amplification within the passband, the adjustment problem can be alleviated by spreading the poles over a number of interstage coupling networks. In that case the amplifier stages work as buffers between the various filter sections, which facilitates their adjustment considerably. This has been standard practice in the design of intermediate-frequency amplifiers for many years. However, this approach is certainly not integration-friendly, since all these interstage networks have to be externally attached to the amplifier chip, thus requiring many extra bonding pads. In view of the longstanding tradition of this approach, which can still provide useful solutions in cases where full integratability is not a primordial requirement, the basics of this method will be presented briefly in a subsequent section.

Of primary concern in a bandpass network are two quantities, the midband frequency f_0 and the bandwidth B, usually taken as the 3-dB bandwidth. For a single LC-resonator these quantities are linked by its Q-factor. As has been discussed in the context of resonator-based oscillators (Section 10.3), the Q-factor is associated with the losses of the resonator. The relationship between f_0, B and Q is $Q = f_0/B$. The reciprocal value B/f_0 is called the 'relative bandwidth'. Since with simple LC-resonators Q-values in excess of 100–200 are difficult to obtain, small relative bandwidths cannot be accomplished with simple LC-resonators. Much larger Q-values can be obtained from mechanical resonators such as quartz crystals. Crystal resonators have been in use for the achievement of narrowband filters from the early years of professional radio communication. In consumer applications the cost of quartz crystal filters has hampered their application. However, nowadays low-cost ceramic resonators, providing adequate performance, are available and widely used. The mechanical forces induced by the piezoelectric effect give rise to bulk vibrations, i.e. the entire mass of the crystal is involved in the vibrations.

More recently devices based on the properties of surface acoustic waves (SAW-devices) have come into use. Their operation is based on the interference effects of mechanical surface waves. These waves are excited by the piezoelectric effect. As opposed to the bulk waves excited in crystal and ceramic resonators, in SAW-devices the waves are confined to a thin layer at the surface of the device. The piezoelectric materials most used are quartz and lithium niobate ($LiNbO_3$). Figure 11.1 depicts schematically the construction of a piezoelectric SAW-filter. A metal layer (commonly Al) is deposited on the piezoelectric substrate, whereupon a mask-defined pattern is formed by selective etching. The pattern consists of two pairs of finger-like interspersed electrodes. These so-called interdigital transducers serve as piezoelectric input and output transducers.

Upon application of electrical signals to the input transducer, mechanical surface

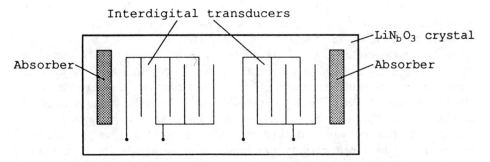

Fig. 11.1 *Schematic drawing of the construction of a SAW-filter.*

waves are generated, which in turn give rise to electrical signals in the output transducer. The transducers can be said to operate as transmitting and receiving antennas for surface acoustic waves. As with electromagnetic antenna systems, the transfer properties are very dependent on the structure of the antenna. By appropriate shaping of the finger-like structures, prescribed bandpass-type transfer properties can be obtained. The midband (centre) frequency, the amplitude response and the phase response are determined by the number of transducer fingers, their lengths and mutual spacing. The fingers can have either the same or varying lengths. In the latter case the transducer is said to be of the 'apodized' type. Since the acoustic waves will not be fully absorbed by the receiving transducer, they travel beyond it and would finally reflect at the edge of the device, thereby producing a backward directed wave that could disturb the proper operation of the device. In order to prevent such reflections, absorbers are placed on the crystal surface between the transducers and the edges (see Figure 11.1).

The design of SAW bandpass filters with a prescribed amplitude- and group-delay response is a specialist subject. It ends with the detailed definition of the transducer electrode patterns. It goes without saying that the initial design costs for SAW-filters are considerable. Once the masks are available, production costs are low. Therefore, this type of filter is particularly attractive for application in mass products. A good example is the intermediate amplifier in a television receiver. For this application many standardized SAW-filters are available nowadays.

It would be of considerable advantage if the SAW-filter structure could be incorporated in a monolithic circuit pattern. However, silicon is not a piezoelectric material. If SAW-filters are to be integrated along with the electronic circuitry, the silicon must be provided with a layer of piezoelectric material at the location of the SAW-filter. Of course such a material has to be technologically compatible with silicon technology. A promising material in this class is zinc oxide (ZnO), but monolithically integrated SAW-filters have not as yet passed the experimental status.

No further attention will be paid to SAW-filters in this chapter. As far as the specific design is concerned, the topic is of a specialist nature and as far as standard types can be used, the application is straightforward. Manufacturers of standard SAW-filters provide ample application information. Though SAW-filters are a welcome contribution to the arsenal of modern filter techniques, their use is restricted. In the present state of the art they lack integrability, their dimensions can become unattractively large if narrow bandwidth must be realized, and manufacturing them in small quantities is not cost effective. It is no surprise that alternative solutions to the filter problem have been looked for, and the development of filter techniques that can be monolithically implemented is being pursued. The methods can be divided into those wherein the signals are processed in their fully intact form of time-continuous analogue signals, those wherein the signals are first converted into time-discrete signals, and those wherein the signals are converted into a purely digital format. Each of these approaches has given birth to a sizeable field of knowledge, far too extended to be incorporated into a general textbook. In subsequent sections, only some basic notions regarding these methods will be presented. For more detailed discussion reference must be made to the relevant specialist works.

11.3 Bandpass filtering by resonators in interstage couplings

In a bandpass amplifier consisting of several stages, the required selectivity can be obtained by using tuned circuits as interstage couplings. These tuned circuits can consist of single or mutually coupled LC-resonators. Together they realize the desired bandpass characteristics. In the design of such amplifiers advantage can be taken of the fundamental relationship between low-pass and bandpass networks. According to network theory a bandpass network with a prescribed centre frequency and a prescribed bandwidth can be derived from a low-pass network with the same bandwidth. The transformation rule can best be formulated in terms of a relationship between the poles and zeros of both networks. It reads: a pole p_l of the low-pass function yields two poles p_{b1} and p_{b2} of the bandpass function, together with a zero in the origin. For the poles,

$$p_{b1}, p_{b2} = \tfrac{1}{2}p_l \pm \sqrt{(\tfrac{1}{2}p_l)^2 - \omega_0^2} \tag{11.1}$$

where ω_0 is the geometric centre frequency of the bandpass function. The geometric mean is defined by $\omega_0 = \sqrt{\omega_l \omega_h}$, where ω_l and ω_h are the lower and the upper -3-dB frequencies of the bandpass transfer. The bandwidth $f_h - f_l$ corresponds with the bandwidth of the model low-pass transfer. Similarly, for the zeros the rule reads: a zero z_l yields two zeros z_{b1} and z_{b2}, together with a pole in the origin. For the zeros,

$$z_{b1}, z_{b2} = \tfrac{1}{2}z_l \pm \sqrt{(\tfrac{1}{2}z_l)^2 - \omega_0^2} \tag{11.2}$$

From eqs. (11.1) and (11.2) it is obvious that the poles and zeros found by using the transformation rule can be either real or complex. A real pole p_l gives rise to

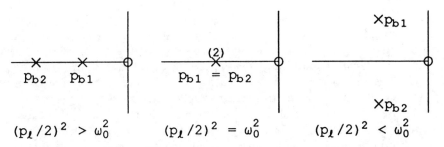

Fig. 11.2 *Low-pass-to-bandpass transformation of a single pole p_l on the real axis.*

either two real poles or a complex pair, depending on whether $(\frac{1}{2}p_l)^2 - \omega_0^2 > 0$ or <0 (Figure 11.2). A complex pair p_{l1}, p_{l2} gives rise to two complex pairs in the bandpass system.

EXAMPLE
Figure 11.3(a) shows schematically a simple low-pass CS-stage. The transfer is characterized by one single pole $p_l = -1/\tau$ with $\tau = RC$. The bandpass equivalent possesses two poles given by

$$p_{b1}, p_{b2} = -\frac{1}{2\tau} \pm \sqrt{\left(\frac{1}{2\tau}\right)^2 - \omega_0^2} \qquad (11.3)$$

A simple network that is capable of these singularities is shown in Figure 11.3(d).

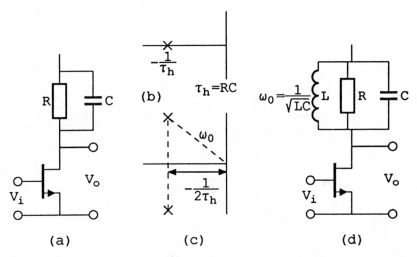

Fig. 11.3 *Transformation of a one-pole low-pass stage into a bandpass stage: (a) low-pass stage; (b) pole pattern of low-pass stage; (c) pole–zero pattern of bandpass stage; (d) bandpass stage.*

The inductance must be chosen such that $\omega_0 = 1/\sqrt{LC}$. Obviously, this circuit is obtained by placing an inductance in parallel with the available capacitance, thereby forming a tuned circuit at ω_0. It can be shown that, without having to resort to pole–zero transformation, a low-pass network can always be transformed into an equivalent bandpass network by placing an inductance in parallel with any capacitance in the network and by placing a capacitor in series with any inductance. In both cases resonance (parallel or series) must occur at ω_0. The practical use of this simple rule is restricted, since it is not always possible to actually perform the tuning. For instance, a diffusion capacitance in an equivalent circuit of a transistor cannot be tuned since it is not externally accessible. The transformation rule for poles and zeros includes the 'tuning rule', but it is more universal. For instance, in our example the use of an LC-resonant circuit is not the only way to obtain the desired p–z pattern. The poles could also be realized with a second-order feedback amplifier and the zero by means of a coupling capacitor.

The equality of the bandwidths of the circuit according to Figures 11.3(a) and 11.3(d) can easily be proven. For the low-pass stage $B_\omega = 1/\tau$, while for the bandpass stage $B_\omega = \omega_0/Q$. With $Q = \omega_0 RC$, $B_\omega = 1/\tau$. If the GB-product should be maximized, C should be kept minimal, just as in the case of the low-pass stage.

In Chapter 9 it was found that an all-pole, maximally flat transfer function possesses attractive response quality in a low-pass multistage amplifier. Applying the low-pass to bandpass transformation provides a simple, but highly effective,

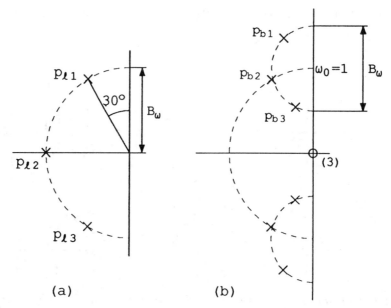

Fig. 11.4 *Low-pass-to-bandpass transformation of all-pole MFM-transfer: (a) low-pass pole pattern; (b) bandpass p–z pattern.*

design concept for multistage bandpass amplifiers. As an example we consider a third-order Butterworth function (Figure 11.4). The poles are given by

$$p_{12} = -B_\omega; \quad p_{11,3} = -\tfrac{1}{2}B_\omega \pm j\tfrac{1}{2}B_\omega\sqrt{3}$$

Transformation of these poles, according to eq. (11.1) yields three pairs of complex poles, as indicated in Figure 11.4(b). The poles are no longer exactly located on semicircles. However, if $B_\omega \ll \omega_0$, i.e. for small relative bandwidth, the deviation is slight.

The three pairs of complex poles and the zeros in the origin can be implemented by three resonant circuits used as interstage networks. Their resonant frequencies and the required dampings are found directly from the calculated pole locations. This method of configuring bandpass amplifiers is called *staggered tuning*, because the bandpass characteristics of the individual stages are different. Together they provide the desired overall passband. This method was in widespread use in the early days of wideband signal processing, for instance in radar engineering.

11.4 Use of coupled resonant circuits

A single resonant circuit contributes no more than two poles. Often, the order of the bandpass characteristics of the interstage network must be larger than can be accomplished with a single resonant circuit. In such cases more complicated interstage networks may be used. Frequently, the use of coupled resonant circuits proves to be propitious. Figure 11.5 shows two resonant circuits coupled through mutual inductance. The coupling can be characterized by the coupling factor $k = M/(L_1L_2)^{1/2}$. Alternatively the inductive coupling can be accomplished as shown in Figure 11.6. The equivalence of both configurations finds its base in the equivalence of the circuits shown in Figure 11.7.

Figure 11.8 shows an analogous configuration, albeit that capacitive coupling is used here. The coupling factor is determined by the value of C_m; the relation between k and C_m is given by

$$k = \left[\frac{C_1 C_2}{(C_1 + C_m)(C_2 + C_m)}\right]^{1/2} \tag{11.4}$$

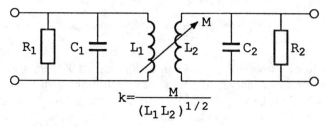

Fig. 11.5 *Inductive coupling through mutual induction.*

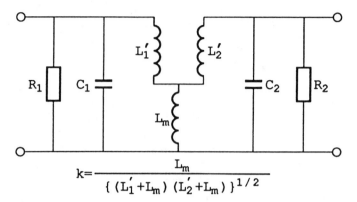

Fig. 11.6 *Alternative circuit with inductive coupling.*

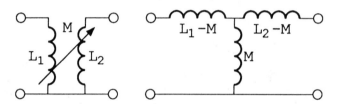

Fig. 11.7 *Equivalence of mutually coupled inductances and T-configuration of inductances.*

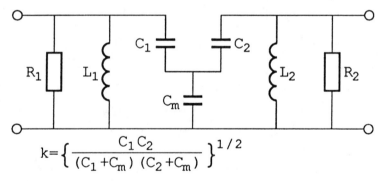

Fig. 11.8 *Capacitive coupling of resonant circuit.*

The configurations shown in Figures 11.6–11.8 possess a Y-structured arrangement of inductances or capacitances. Alternative configurations with equivalent transfer properties are obtained by applying Y–Δ transformation to these core elements. Figure 11.9 summarizes the relevant transformation formulas; Figures 11.10 and 11.11 show the alternative configurations obtained in this way.

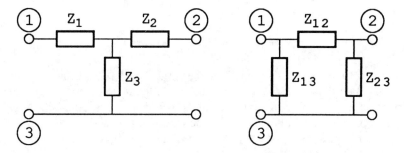

Fig. 11.9 *Y–Δ transformation. Both circuits are equivalent when* $Z_{12} = (Z_1Z_2 + Z_1Z_3 + Z_2Z_3)/(Z_3, Z_{13} = Z_1Z_2 + Z_1Z_3 + Z_2Z_3)/Z_2$, $Z_{23} = Z_1Z_2 + Z_1Z_3 + Z_2Z_3/Z_1$.

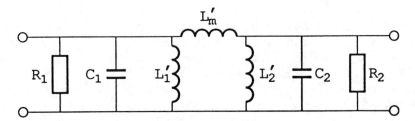

Fig. 11.10 *Alternative form of inductive coupling.*

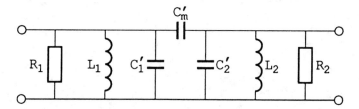

Fig. 11.11 *Alternative form of capacitive coupling.*

An advantage of the last-mentioned configurations is that parasitic capacitances at the terminals can be incorporated into the network, since they are in parallel with capacitances that form part of the regular structure.

Within the frequency range of the passband, the characteristics of the configuration with inductive and with capacitive coupling with the same coupling factor are almost equal, provided the relative bandwidth B/f_0 is small. For frequencies far outside the passband, the transfer is different as a consequence of the different number of zeros in the transfer. An exact calculation of the transfer function of a pair of coupled resonant circuits is somewhat cumbersome. In the case of inductive coupling, the general form of the transfer from a current source

$g_m V_g$ to the secondary voltage V_o is

$$f(s) = A \frac{s}{s^4 + b_3 s^3 + b_2 s^2 + b_1 s + b_0} \quad (11.5)$$

implying that there is one zero in the origin and four poles. An exact analytical solution of the denominator polynomial is difficult to obtain. Of course, a numerical solution can always be obtained. In most practical situations both resonant circuits have equal resonant frequencies, while usually their quality factors are also approximately equal. Moreover, the coupling factor k is nearly always small. If $Q_1 \approx Q_2 = Q$ and $k^2 \ll 1$, an approximate solution for the pole positions can be given:

$$p_1, p_2, p_1^*, p_2^* = -\frac{\omega_0}{2Q} \pm j\omega_0 \left(1 \pm \frac{k}{2}\right) \quad (11.6)$$

Figure 11.12 sketches the pole–zero plot.

The value of the coupling factor affects the shape of the frequency response considerably. If $kQ = 1$, the coupling is said to be critical. If $k < 1$, the maximum in the amplitude response occurs at $\omega = \omega_0$. If $k > 1$, the coupling is said to be overcritical; maximum amplitude response then occurs for two frequencies slightly above and below the resonant frequency. In practice strong overcritical coupling is rarely acceptable, because of the strong overshoot that it incurs in the transient response.

For $k = k_c = 1/Q$, the bandwidth associated with the pole positions given by eq. (11.6) is $B_\omega = \sqrt{2} (\omega_0/Q)$. Computation of the GB-product of the transfer yields for $k = k_c$ the approximate value

$$GB \approx \frac{g_m}{2\pi \sqrt{2 C_1 C_2}} \quad (11.7)$$

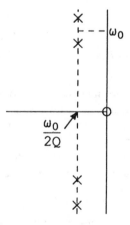

Fig. 11.12 *Pole–zero plot of inductively coupled resonant circuits.*

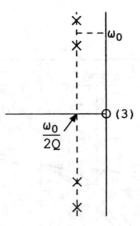

Fig. 11.13 *Pole-zero plot of capacitively coupled resonant circuits.*

where C_1 and C_2 are the capacitances of the primary and secondary circuits. With $C_1 = C_2 = \frac{1}{2}C$, eq. (11.7) yields $GB = g_m/(\pi C\sqrt{2})$, which is a factor $\sqrt{2}$ larger than in the case of a single resonant circuit. This is understandable in view of the improved selectivity due to the use of coupled resonant circuits.

In the case of capacitive coupling the response outside the passband differs from that pertaining to the inductive case. The poles are approximately located as in the inductive case; the difference in response far from resonance is due to the occurrence of a threefold zero in the origin (Figure 11.13). This threefold zero can be found by direct inspection of the circuit of Figure 11.11. For $\omega = 0$ a threefold blocking occurs due to the presence of the primary and secondary inductances and the coupling capacitor C_m'.

In the design of bandpass amplifiers with coupled resonant circuits, use can be made of normalized graphical response curves, which can be found in many handbooks. Figure 11.14(a and b) sketches the general shape of these curves for the amplitude and phase characteristics. The coupling factor is specified by $n = k/k_c$. With strong overcritical coupling the curves show a pronounced dip at $\omega = \omega_0$ and steep slopes at the edges of the passband. In many applications these characteristics are unattractive, since they indicate heavy overshoot in the transient response.

If special requirements hold for the suppression of interfering signals at certain frequencies, extra resonant circuits can be added. One possibility is to make the coupling factor frequency-dependent. Figure 11.15 shows an example. At the resonance frequency of L_3 and C_3 the coupling is blocked. In terms of the pole–zero plot this is tantamount to the introduction of a zero close to the frequency to be suppressed. Figure 11.16 shows another possibility. Here the frequency to be suppressed corresponds to the series resonance of the added 'wingtraps'. When using such suppression circuits one should be aware of their effect on the filter response in the passband.

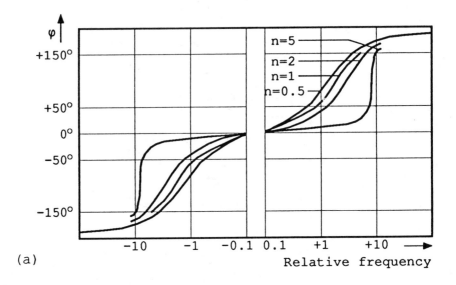

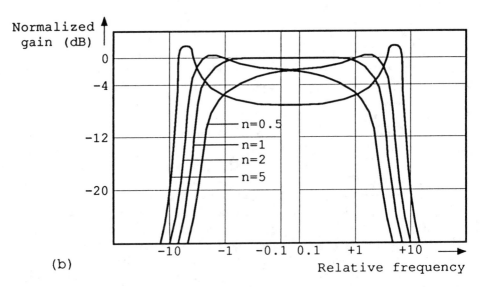

Fig. 11.14 *Shape of the phase* (a) *and amplitude* (b) *response of coupled circuits. The quantity on the horizontal axis is* $(2/k_c) \cdot (\omega - \omega_o)/\omega_o$. *The curves are drawn for* $Q_1 = Q_2 = 20$.

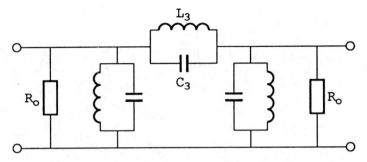

Fig. 11.15 *Frequency-dependent coupling for creating a zero near the passband.*

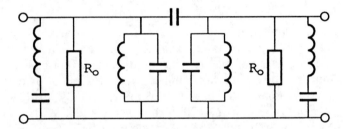

Fig. 11.16 *Use of 'wingtraps' for creating a zero near the passband.*

If a very effective suppression at a certain frequency is required and it is possible to separate the suppression circuit from the rest of the filter elements by adequate buffering, the bridged-T circuit shown in Figure 11.17 may be a useful device. This circuit can provide zero response at the frequency to be suppressed, in spite of the finite series resistance r of the inductance L. Straightforward calculation of the voltage transfer of this circuit yields, for $R_L \to \infty$ and $R = L/2rC$ or $R/r = \frac{1}{4}Q^2$ with $Q = \omega_0 L/r$.

$$A = \frac{V_o}{V_i} = \frac{1-\beta^2}{1-\beta^2 + j\beta\sqrt{a}}, \text{ where } \beta = \frac{\omega}{\omega_0} \text{ and } a = \frac{r}{R} \tag{11.8}$$

The suppression frequency ω_0 is the resonant frequency of the parallel circuit, hence $\omega_0^2 = 2/LC$. The relative bandwidth (Figure 11.7b) is given by

$$B_r = \frac{\omega_2 - \omega_1}{\omega_0} = \sqrt{a} = \frac{2}{Q} \tag{11.9}$$

Eq. (11.9) shows that, although adding the resistor R provides infinite suppression, the 'sharpness' of the suppression depends on the Q-factor of the resonant LC-circuit. If R_L is finite, the zero response remains, but the effective Q-factor is lowered.

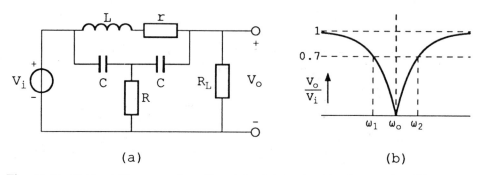

Fig. 11.17 *Bridged-T suppression filter* (a) *with its amplitude response* (b).

EXAMPLE
A suppression filter (notch filter) is required with $f_0 = 4$ MHz and $B < 100$ kHz. According to eq. (11.9), the required bandwidth can be obtained with $Q > 80$. In order to maximize Q it is sensible to make L large, since to a first approximation L increases quadratically with the number of turns, and r increases linearly. However, it must be taken into account that a large coil also exhibits a relatively large parasitic capacitance. Let $C = 20$ F, then $L = 158$ μH. With a well-designed coil a Q-factor of 100 must be practicable. Then $B \approx 80$ kHz. Further, $r = \omega_0 L/Q \approx 40$ Ω and $R = \frac{1}{4}Q^2 r \approx 100$ kΩ. If a substantially lower suppression bandwidth is required, the best approach is to load the suppression filter with a negative resistance R_L. Methods for accomplishing a negative resistance have been dealt with in Chapter 10. Care should be taken that the loading capacitance in parallel with R_L is kept minimal, since the capacitance gives rise to a pole close to the desired zero at ω_0.

11.5 Selective amplifier circuits

Both the CE- and CS-configurations behave approximately as controlled current sources. If they are to be used in frequency-selective amplifiers, the natural approach is to load them with voltage resonators, i.e. parallel resonant circuits. Either single LC-resonators or coupled resonators can be applied. In both cases the simplest mathematical expressions are obtained by using admittances as describing quantities. In this particular case active devices can best be specified by means of y-parameters. The equations for a linear twoport are then expressed as (Figure 11.18a)

$$i_i = y_i v_i + y_r v_o \tag{11.10a}$$

$$i_o = y_f v_i + y_o v_o \tag{11.10b}$$

Any linear twoport can be formally represented by its y-parameters. However, their relation to the parameters rooted in the physical mechanism of operation is frequently rather obscure. For instance, if the input–output relations of a CE-stage

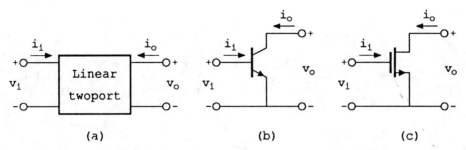

Fig. 11.18(a) *General representation of a linear twoport.* (b) *CE-circuit as a linear twoport.* (c) *CS-circuit as a linear twoport.*

(Figure 11.18b) are described in this way, comparison with the relations derived from the hybrid-π equivalent circuit reveals that the y-parameters are frequency-dependent in a rather complicated way. The y-parameters of the CE-circuit are commonly denoted y_{ie}, y_{re}, y_{fe} and y_{oe}. Because of their frequency dependence they are not often used in calculations for low-pass wideband amplifiers. However, in bandpass amplifiers the relative bandwidth (B/f_0) is small, so that within the passband the frequency dependence of the y-parameters can be made tractable. For the CS-configuration, the y-parameter representation (Figure 11.19) is largely equivalent to the usual physical equivalent circuit. Here $y_{is} = j\omega(C_{gs} + C_{gd})$, $y_{rs} = -j\omega C_{gd}$, $y_f = g_m$, $y_o = 1/r_d + j\omega(C_{ds} + C_{gd})$. Figure 11.19 shows the equivalent circuit representing eqs (11.10).

The loss in an LC-resonant circuit is largely determined by the series resistance r of the inductance. However, for a high-Q resonant circuit, within the passband the loss due to this resistance can be replaced by a parallel loss resistance $R_p = Q^2 r$ (Figure 11.20a), provided $Q \gg 1$.

If a transformer or a tapped inductance (autotransformer) is loaded at the secondary side, the load impedance Z can be transformed to the primary side as $n^2 Z$, where n is the ratio of turns (Figure 11.20b). In the case of capacitive tapping (Figure 11.21a) the transformation is somewhat more complicated. The admittance of the configuration of Figure 11.21(a) can be written in the form

$$Y = \frac{j\omega C_1 (1 + j\omega C_2 R)}{1 + j\omega R (C_1 + C_2)} \tag{11.11}$$

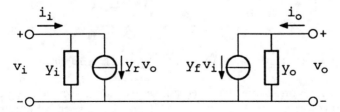

Fig. 11.19 *Equivalent circuit of twoport represented by its y-parameters.*

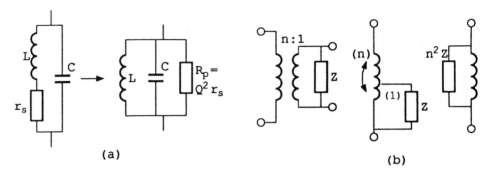

Fig. 11.20 *Transformation of losses in an LC-resonant circuit.* (a) *Representation of series resistance by a parallel resistance.* (b) *Transformation of a secondary impedance to the primary circuit of a transformer or a tapped inductance.*

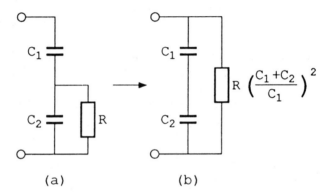

Fig. 11.21 *Capacitive tapping:* (a) *tapped circuit;* (b) *equivalent circuit when* $R \gg 1/\{\omega\sqrt{C_2(C_1+C_2)}\}$.

Splitting Y into its real and imaginary parts yields

$$Y = j\omega \frac{C_1[1 + \omega^2 R^2 C_2(C_1 + C_2)]}{1 + \omega^2 R^2 (C_1 + C_2)^2} + \frac{\omega^2 C_1^2 R}{1 + \omega^2 R^2 (C_1 + C_2)^2} \quad (11.12)$$

If $R^2 \gg 1/\{\omega^2 C_2(C_1 + C_2)\}$, eq. (11.12) simplifies to

$$Y = j\omega \frac{C_1 C_2}{C_1 + C_2} + \frac{1}{R} \frac{C_1^2}{(C_1 + C_2)^2} \quad (11.13)$$

which is equivalent to the inductive case. Figure 11.21(b) shows the transformed impedance.

Using the transformations discussed above, we can conveniently model selective amplifier stages. Figures 11.22 and 11.23 depict basic selective amplifier circuits

BANDPASS AMPLIFIERS AND ACTIVE FILTERS   347

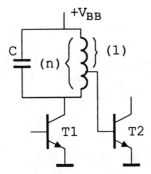

Fig. 11.22 *Basic schematic of selective stage with single resonant circuit.*

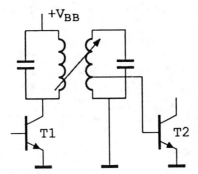

Fig. 11.23 *Basic schematic of selective stage with coupled resonant circuits.*

with a single resonant circuit and coupled resonators respectively. By appropriate tapping, the input impedances of the loading stages are transformed in order to obtain the Q-factors needed in the design. By representing the active devices by their y-parameter-based equivalent circuit, an 'all-parallel' schematic results. Figure 11.24 depicts this model for the case of coupled resonators shown in Figure 11.23.

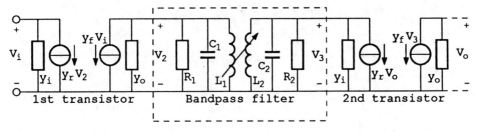

Fig. 11.24 *All-parallel schematic of a bandpass interstage network using coupled resonators.*

EXAMPLE 1

As an illustration of the use of the amplifier model, the transfer of the selective amplifier stage of Figure 11.22 will be investigated. Let the resonant frequency be 500 kHz. Further, $L = 300\ \mu\text{H}$ and $r_s = 10\ \Omega$. The parasitic capacitance C_L of the coil is 10 pF. The tapping is such that $n = 4$. The y-parameters of the transistors in the frequency range of interest are as follows:

$$y_{ie} = g_{ie} + j\omega C_{ie} \text{ with } g_{ie} = 0.25\ \text{mA V}^{-1} \text{ and } C_{ie} = 30\ \text{pF}$$
$$y_{fe} = 80\ \text{mA V}^{-1};\ y_{re} = 0$$
$$y_{oe} = g_{oe} + j\omega C_{oe} \text{ with } g_{oe} = 15\ \mu\text{A V}^{-1} \text{ and } C_{oe} = 3\ \text{pF}$$

With V_1 and V_2 being the signal voltages on the bases of $T1$ and $T2$, the equivalent circuit shown in Figure 11.25 applies. R_p is the parallel representation of the series resistance r_s. Hence $R_p = Q^2 r_s$. Since $Q = \omega_0 L/r_s$, $R_p \approx 89\ \text{k}\Omega$. Transforming $r_{ie} = 1/g_{ie}$ and C_{ie} to the primary side of the tapped coil, we obtain the equivalent circuit of Figure 11.26.

Now, the total capacitance $C_t = C_{oe} + C + C_L + C_{ie}/n^2 = 1/\omega_0^2 L$ with $\omega_0 = 2\pi \times 500 \times 10^3$ and $C_{ie}/n^2 \approx 2\ \text{pF}$. Hence $C_t = 338\ \text{pF}$ and $C = 323\ \text{pF}$. Since $r_{oe} = 67\ \text{k}\Omega$ and $n^2 r_{ie} = 64\ \text{k}\Omega$, the parallel combination R_t of R_p, r_{oe} and $n^2 r_{ie}$ equals 24 kΩ. Hence, the Q-factor of the resonant circuit is

$$Q_t = R_t \sqrt{\frac{C_t}{L}} \approx 25.5$$

The 3-dB bandwidth is $B = f_0/Q_t = 19.6\ \text{kHz}$. Further, $V_2 = -y_{fe} R_t V_1/n$, hence

$$A_v = \left|\frac{V_2}{V_1}\right| = 480.$$

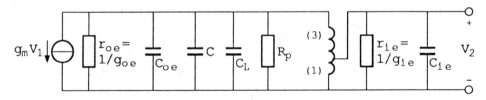

Fig. 11.25 *Equivalent circuit of the stage according to Figure 11.22.*

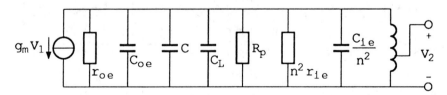

Fig. 11.26 *Transformation of secondary load to the primary side of L.*

Notice that we have assumed $y_{re} = 0$. This is equivalent to the assumption that the effect of $C_{b'c}$ can be neglected. In practice, this is usually not allowed in the case of a simple CE-stage. In a selective amplifier, $C_{b'c}$, though small in a good transistor, jeopardizes the stability of the stage since it constitutes a reactive feedback circuit. Therefore, measures have to be taken to neutralize the undesirable effects of this capacitance. How this is done will be explained in the next section. The net effect of appropriate neutralization comes down to nullifying $C_{b'c}$, which justifies our preliminary assumption.

EXAMPLE 2

The circuit of Figure 11.27 employs two MOS-transistors, the first as CS and the second as source follower. In the selected biasing mode $g_m = 7.5$ mA V^{-1}, $C_{gs} = 6$ pF, $C_{ds} = 1$ pF, $C_{gd} = 0.25$ pF, $r_d = 20$ kΩ. Further, $R_g = 2$ MΩ and $R_s = 1$ kΩ. As in the first example, the effect of C_{gd} on $T1$ is assumed to be negligible. It is required to design the circuit for a resonant frequency of 4.5 MHz with a bandwidth of 900 kHz. The voltage amplification should be maximized. C_c and C_s behave as shorts at the frequencies of interest.

The source follower is included for providing a low output impedance. Its input impedance is capacitive. The input capacitance is $C_i = C_{gd} + C_{gs}/(1 + g_m R_s)$ ≈ 0.9 pF. Figure 11.28 shows the equivalent circuit of the loaded CS-stage, wherein R_p represents the loss of the coil $(= Q^2 r_s)$.

The parallel connection of r_d, R_p, R and R_g can be replaced by the equivalent resistance R_t, while $C_{ds} + C + C_i$ can be replaced by the total capacitance C_t. The Q-factor of the circuit should be $Q_t = 4.5/0.9 = 5$. Further, $Q_t = R_t(C_t/L)^{1/2}$ $= R_t C_t \omega_0$. The amplification is maximized by maximizing R_t, which requires C_t to be as small as is possible, hence C should be minimal. The minimum value of C is determined by the parasitic capacitance of the coil and the wiring capacitance. Its

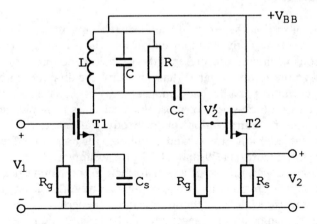

Fig. 11.27 *Selective MOS-stage and source follower.*

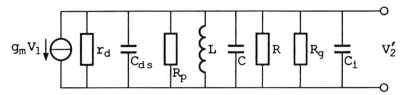

Fig. 11.28 *Equivalent circuit of the CS-stage.*

value can only be estimated. A reasonable assumption is $C = 10$ pF. Then $C_t \approx 12$ pF. Hence $L = 1/\omega_0^2 C_t = 104$ μH. Further, $R_t = Q_t/\omega_0 C_t = 14.7$ kΩ. The value of R_p must be estimated. The Q-factor of the coil is $\omega_0 L/r_s$. For a well-designed coil $Q = 100$ is a reasonable estimate, hence $r_s \approx 29$ Ω and $R_p \approx 290$ kΩ. Hence $r_d \parallel R_p \approx 18.7$ kΩ. In order to obtain the requested bandwidth, the damping resistor R should be chosen such that $R_p \parallel r_d \parallel R \approx R_t = 14.7$ kΩ, hence $R \approx 69$ kΩ. Note that, due to the estimates that had to be made, the calculated values for L and R are approximations. Practical resonant circuits are usually provided with trimming facilities for precise adjustment. With the value $R_t = 14.7$ kΩ, as found above, $|V_2'/V_1| = 110$. The voltage transfer of the source follower is $g_m R_s/(1 + g_m R_s) \approx 0.9$, so that $|V_2/V_1| = 99$.

Note that in this example ω_0 and Q depend heavily on the values of the various parasitic capacitances, and therefore on the constructional details of the practical realization. To ascertain reproducibility and long-term stability, great care in constructional aspects is of the essence. The dependence on parasitic elements and on the internal impedances of active devices can be lowered by adding fixed capacitance, albeit at the cost of the GB-product. In this example such a loss of GB is affordable because of the relatively small absolute bandwidth.

11.6 Neutralization

In the preceding section the effect of the feedback capacitance $C_{b'c}$ or C_{gd} has not been considered. If in both the input circuit and the output circuit of the CE- or CS-stage resonant circuits are present, the capacitive coupling of these resonant circuits through the feedback capacitance can turn the amplifier into an oscillator. Of course, the coupling can be made immaterial by using a cascode or a long-tailed pair instead of a CE(CS)-stage. In most practical cases a simpler method, known under the names of 'neutralization' or 'neutrodynization', is preferred. The basic idea involves the addition of a second coupling with opposite effect, obtained by polarity reversal. Figures 11.29 and 11.30 give examples. In Figure 11.29 phase reversal is obtained with the aid of a transformer; in the symmetrical configuration of Figure 11.30 it is obtained by cross-coupling from both output sides. Note that with bipolar transistors the compensation cannot be perfect, since $C_{b'c}$ is between b' and c and the compensating capacitance C_n between b and c. The compensation can be improved by connecting a resistor in parallel with C_n. In an FET-circuit there is no need for such a measure.

BANDPASS AMPLIFIERS AND ACTIVE FILTERS   351

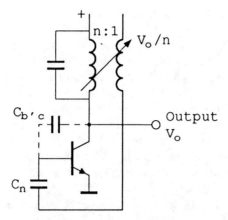

Fig. 11.29 *Neutralization.*

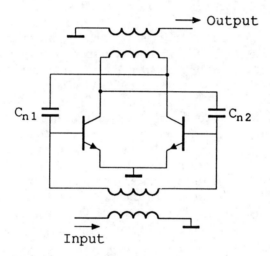

Fig. 11.30 *Alternative method of neutralization.*

11.7 Methods for on-chip filtering

In modern electronics the tendency is towards full on-chip implementation of the required signal-processing functions. As far as filter functions are concerned, this is difficult to accomplish since on-chip realization implies giving up the use of inductances. An amplitude response with steep slopes calls for a transfer function with complex poles. And network theory states that a passive network can only possess complex poles if it contains both capacitive and inductive reactances. If no inductances can be used, resort must be had to active circuits. This introduces at least two fundamental problems:

1. The parameters of active devices are much more prone to spreading, due to incidental variations, to ageing and to temperature effects, than those of well-constructed passive devices. Hence, the sensitivity of the filter function for parameter variations has to be carefully analyzed and measures must be taken to maintain the response within the acceptable tolerance range.
2. Passive reactances are essentially linear and, consequently, free of distortion. In practice, coils with high-permeability cores can exhibit some non-linearity, but judicious design of such coils can keep this effect slight. In addition, passive reactances are essentially free of noise. In contrast, all active elements are neither free from distortion, nor from noise. The implication of these differences is that passive filters can possess a virtually infinite dynamic range, whereas filters using active devices are bound to exhibit a limited dynamic range. In many applications the finite dynamic range of active filters is a very serious drawback that is often hard to overcome.

A third problem that may not pass unnoticed, though it is usually not a great obstacle, is that active devices dissipate energy. This, of course, implies that power has to be supplied.

Any linear transfer can be described by a linear differential equation. A filter function with steep slopes in its frequency response calls for a network of high order, hence it is described by a differential equation of high order. A differential equation can be solved by a number of subsequent integrations. This implies that any linear transfer can be implemented by a judicious arrangement of integrators. This is the way differential equations are solved in an analogue computer. The same method can be employed in the realization of a filter function. In fact, this also takes place in passive filters. An inductance is an integrator that converts voltage into current, whereas a capacitance is an integrator that converts current into voltage. If inductances are to be avoided, the voltage-to-current conversion has to be performed by an active circuit. This is in essence the mode of operation of active filters. The conclusion is that we must search for structures that implement the integration function appropriately, together with the needed voltage-to-current conversion.

Though bipolar circuits as well as MOS-circuits can be devised for executing the necessary operations, MOS-technology provides the best opportunities. This is because in essence MOS-transistors are controlled resistances, while in addition in MOS-technology excellent capacitances can be made. A problem is that MOS-capacitances exhibit spreading from chip to chip. However, within a chip, capacitance ratios can be accurate. With regard to the resistances and/or transconductances involved, much the same applies. And being realized by means of active devices, these constituents exhibit non-linear effects in addition.

In order to cope with the uncertainty of the time constants obtained by means of MOS-capacitances and resistances or transconductances, an elegant method has been developed. Since MOS-resistances or transconductances can be electronically controlled, it is possible to adjust the time constants of which they form part. This opens the way to continuous adjustment of the relevant time constants. At one place in the chip a measuring time constant is implemented. This is continuously compared to an accurate external reference, for instance derived from a crystal controlled clock. Any difference gives rise to an error signal that is used to adjust the measuring device. If all elements in the chip track accurately, this error signal can be used to adjust all time constants. This method is commonly called the 'master–slave method'. The measuring device is the master that controls all the slaves. It can be an oscillator using the measuring time constant as the frequency-determining element. Figure 11.31 shows a possible arrangement. The oscillator containing the measuring time constant is used as a voltage-controlled oscillator (VCO) in a phaselock loop. In this loop a phase detector compares the oscillator signal with an external reference. The error signal controls the VCO and is also used to control all time constants in the filter network.

Alternatively, a simple resonator can be devised out of two time constants. By comparing the resonant frequency with an external accurate reference, an error signal can be obtained that can be used as the control signal for the adjustment of the resistive elements.

The second problem is the non-linearity of the resistive elements. Here, compensation methods may be called to assistance. Figure 11.32 shows an MOS-transistor with an indication of the relevant voltages. The voltage between drain and source is $v_2 - v_1$, the substrate (body) voltage is v_B. If the MOST operates in the triode region, the channel conductance g is given by

$$g = \frac{W}{L} \mu C_{ox}(v_c - v_{tB}) \qquad (11.14)$$

where v_c is the control voltage at the gate and v_{tB} is the threshold voltage occurring with v_B as body potential. To a first approximation, the current is given by the linear relation $I = g(v_2 - v_1)$.

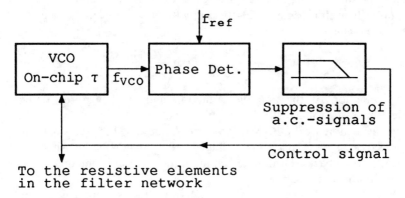

Fig. 11.31 *A possible master–slave arrangement for continuous adjustment of time constants.*

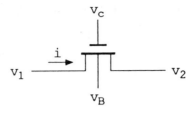

Fig. 11.32 *MOST as a controlled resistor.*

However, a MOST is a non-linear device. In practice, g depends on the voltage across the transistor. By the nature of the characteristics of a MOST, the distortion is mainly of even order. Several methods are available for reducing the distortion. One approach is to add a signal-dependent voltage to the gate control voltage. Another possibility is the use of a balanced configuration. Also, both methods can be combined. Figure 11.33 shows some basic configurations using these principles. In the configurations of Figure 11.33(a and b), compensation is pursued by adding a signal-dependent voltage to the gate control voltage. No full compensation can be accomplished in this way. The balanced configurations of Figure 11.33(c and d) are in principle capable of full compensation of even order distortion. The method shown in Figure 11.33(e) employs a parallel connection of a p-channel and an n-channel MOST. Since true complementarity is impossible, at best partial compensation is achieved. With the double-balanced configuration of Figure 11.33(f) in principle full compensation of even and odd distortion terms can be accomplished. However, this is a cumbersome solution, since no less than four transistors are needed for the realization of one single resistive element.

With regard to the construction of the filters as such, it is advantageous to use fully differential configurations in order to eliminate unwanted effects by interfering signals on the power-supply lines (high PSRR: power-supply rejection ratio). Figure 11.34 shows some possible basic cells for synthesizing filters. Figure 11.34(a) is a simple integrator, Figure 11.34(b) is an integrator for signal v_1 and an amplifier for signal v_2 hence,

$$v_0 = \frac{-g}{C_2} \int_0^t v_1 \, dt - \frac{C_1}{C_2} v_2$$

The combination of the amplifier function and the integrator function is of practical interest in many filter configurations.

The fully symmetrical OPAMP can either be designed in one block, or it can be synthesized as a combination of single-ended OPAMPs, as shown in Figure 11.34(c). A drawback of the latter configuration is the slight difference in signal delay for both output signals.

In the circuits of Figure 11.34 the MOST operates as a controlled conductance, i.e. as voltage-to-current converter. The same can be accomplished with a transconductance; a differential pair can be employed. Figure 11.35 shows

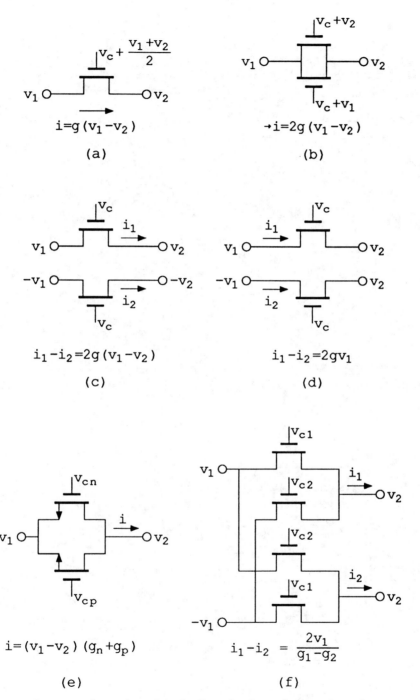

Fig. 11.33 *Some configurations for the linearization of channel conductance.*

356 ANALOGUE ELECTRONIC CIRCUIT DESIGN

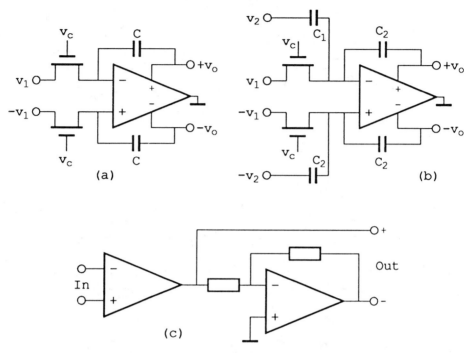

Fig. 11.34 *Examples of filter elements:* (a) *integrator;* (b) *integrating and amplifying element;* (c) *synthesizing a fully balanced OPAMP from two single-ended OPAMPs.*

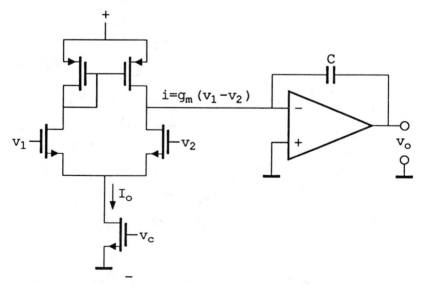

Fig. 11.35 *An integrator using a transconductance as the resistive element.*

the basic integrator obtained in this way. To improve the dynamic range, linearization techniques must be applied here also. The transconductance is controlled by the voltage v_c on the gate of the current source. The time constant realized in this way is C/g_m. An advantage of this circuit is that large time constants can be obtained by choosing I_o low. In the circuit of Figure 11.34(a), a small value of g requires the use of long-channel MOSTs, since g is proportional to W/L and the minimum value of W is dictated by the technology employed. A disadvantage of the transconductance method is that combatting third-order non-linearities is more difficult than with the former method.

The synthesis of low-pass or bandpass filters with prescribed characteristics is performed using the methods network theory developed for active-filter design. As an illustration, Figure 11.36(a) shows a widely used configuration for a two-pole transfer consisting of two integrators. The combination is known as the 'biquad structure'. Each of the blocks represents an integrator with a time constant C/g_m. By cascading such sections higher-order Butterworth responses can, for instance, be synthesized.

The transfer function can be derived by straightforward calculation, yielding

$$\frac{V_o}{V_i} = \frac{g_m^2}{s^2 C_1 C_2 + s g_m C_1 + g_m^2} = \frac{1}{1 + s\tau_1 + s^2 \tau_1 \tau_2} \tag{11.15}$$

The poles are given by

$$p_{1,2} = -\frac{g_m}{2C_2}\left[1 \pm \sqrt{1 - 4\frac{C_2}{C_1}}\right] \tag{11.16}$$

If $4C_2/C_1 > 1$, the poles are complex. The 'resonant frequency' is $\omega_0 = g_m/\sqrt{C_1 C_2}$ and the Q-factor is $Q = \sqrt{C_2/C_1}$ (Figure 11.36b). Hence, the Q-factor is determined by a ratio of capacitances, which can be quite accurate, while ω_0 can be controlled by g_m, which is in turn controlled by the error signal derived from the 'master-circuit'.

The design of integrated active filters, either low-pass or bandpass, is a specialist topic. The material presented in this section has no further pretence than to convey a general idea of the present approaches and the possibilities. A complete discussion could fill a separate textbook. For further study, reference is made to the technical literature (see, for instance, Gregorian and Temes, 1986; Grebene, 1984).

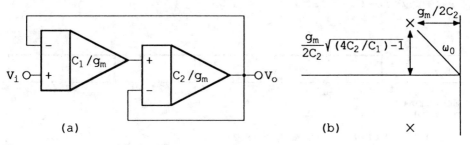

Fig. 11.36 *Biquad structure for the realization of a two-pole transfer.*

11.8 Sampled-data methods

An analogue signal is a continuous function of time. Up to now we have maintained this signal concept strictly. However, for a full description of the signal, knowledge of the signal value at any time is not necessary. Information theory states that a signal that is spectrally confined within a bandwidth B is fully determined by $2B$ equidistant sample values per second. Figure 11.37 illustrates this. In Figure 11.37(a) a signal is depicted that is assumed to be confined to a bandwidth $B < 1/2T$, where T denotes the sampling period. Figure 11.37(b) shows the sampled signal that still contains the full information content of the original signal.

Mathematically, the sampling operation can be described by multiplexing the signal function $S(t)$ with a sequence of narrow sampling pulses representing the sampling moments. If the fundamental frequency of this sequence is $f_0 = 1/T$, then the spectrum of the pulse sequence consists of harmonics of f_0. The multiplication operation in the time domain corresponds to convolution in the frequency domain. The resulting spectrum consists of a two-sided repetition of the spectral content of $S(t)$ around f_0, $2f_0$, ..., etc. Figure 11.38 shows schematically the spectrum of the sequence of samples.

Figure 11.38 clearly shows that, if B exceeds the value $\frac{1}{2}f_0$, overlap of the sidebands around f_0 with the baseband occurs. This is called 'aliasing' or 'folding' of spectral bands. Obviously, the signal $S(t)$ should be free of spectral components beyond $\frac{1}{2}f_0$. This requires filtering of $S(t)$ before the sampling operation takes place. The prefilter is called an 'antialiasing' or 'antifolding' filter. Even if $S(t)$ as such does not contain the undesired spectral components, prefiltering is usually necessary because otherwise noise with spectral components around the harmonics of f_0 would fold back into the baseband.

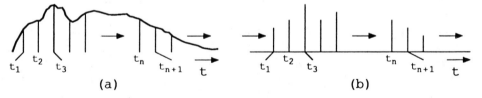

Fig. 11.37(a) *A continuous-time signal S(t).* (b) *The sampled version of the same signal;* $t_n - t_{n-1} = T$.

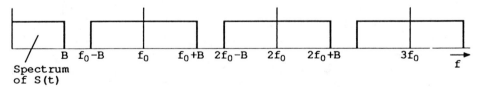

Fig. 11.38 *Spectrum of the sampled signal.*

A further step in discretizing the information stream is discretization of the amplitude levels. In this way a digital signal is created, whereby each sample is represented by a number of bits. Since the starting point for digitization is a discrete-time sampled signal, the requirement for prefiltering applies here, too.

In general at the end of a signal-processing chain, the sampled data signal must be reconverted into a continuous-time signal. This can be simply accomplished by passing the signal through a low-pass filter. This is obvious from Figure 11.38. If the 'smoothing filter' has a bandwidth B, only the original baseband signal is reconstructed.

Converting a signal into its sampled version opens new ways for signal processing, including filtering. The most radical way of working is to convert the signal into full digital format and use a digital filter. In some cases this can be a viable approach, but it should be noted that A/D- and D/A-conversion must be realized and that, dependent on the number of bits per sample, quantization noise has to be taken into account. Digital techniques are beyond the scope of this book and we will give no further attention to them.

The concept of sampled analogue signals has opened a way for integrable realization of transversal and recursive filters. Such filters are built from delay lines. By combining weighted portions of the signal that has undergone various amounts of delay, a filter function can be realized. Figure 11.39 shows the most simple form of a transversal filter that is possible. The signal is delayed over a time span τ, after which it is added to the undelayed signal. A signal $\sin \omega t$ undergoes a phase shift $\varphi = \omega \tau$ in passing the delay line. When it is added to the original signal, cancellation occurs for frequencies where $\varphi = \pi(2k+1)$ with $k = 0, 1, 2, \ldots$, whereas for frequencies where $\varphi = 2k\pi$, both signals are in phase and consequently add up. The amplitude response can be written

$$A(f) = 2\,|\cos \pi f \tau|$$

Figure 11.40 shows this frequency response.

The filter function can be made steeper by adding delay sections. Figure 11.41 shows a filter using two sections with equal delay, together with a sketch of its frequency response. By adding more delay sections and applying appropriate positive and negative weighting factors, further steepening of the filter characteristic can be achieved. Figure 11.42 shows the general form of a transversal filter. The weighting factors can be positive as well as negative. In the latter case the weighting device must include signal inversion.

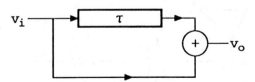

Fig. 11.39 *A simple transversal filter.*

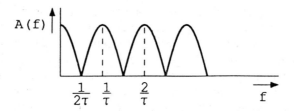

Fig. 11.40 *Frequency response of the configuration of Figure 11.39.*

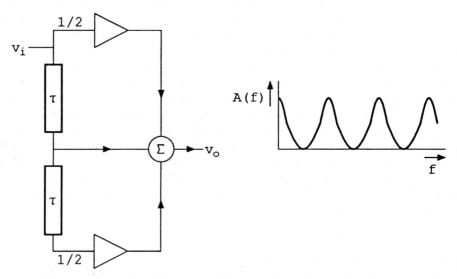

Fig. 11.41 *A filter with delay 2τ and its frequency response.*

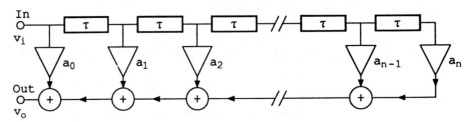

Fig. 11.42 *General form of transversal filter.*

The appropriate mathematical tool for describing the transfer function of a transversal filter is the z-transform. For readers who are acquainted with this transform it may be recalled that the z-transform is related to the Laplace transform by putting $\exp(s\tau) = z$. A delay τ is represented in the z-transform by a factor $1/z$. Hence, the transfer of the configuration of Figure 11.42 can be written

$$H(z) = \frac{V_o(z)}{V_i(z)} = \sum_{m=0}^{n} a_m z^{-m} \tag{11.17}$$

The possibilities for realizing specific transfer functions can be extended by the introduction of feedback. Filter configurations including feedback loops are called recursive filters. Figure 11.43 shows the general form of a recursive filter. The transfer function can be written in the form

$$H(z) = \frac{\sum_{m=0}^{n} a_m z^{-m}}{1 - \sum_{m=1}^{n} b_m z^{-m}} \tag{11.18}$$

In all cases periodic response functions are obtained. If the additional passbands are undesired these must be suppressed by an additional overall filter.

Monolithic technology has created an elegant method for accomplishing the required signal delays. The devices employed for this purpose are called charge-coupled devices (CCDs). Figure 11.44(a) shows schematically the structure of such a device. In a p-type silicon substrate a multigate MOS-structure is formed. The gates are divided in three groups, which are mutually interconnected. Each group is controlled by clock pulses. When a gate is made positive a potential well under that gate is established which is capable of collecting electrons. When the next gate is driven positive, but the positive voltage is removed from the gate under which the electrons were caught, the charge moves to the new potential well. By repeating this action, the charges move from left to right. Since the movement is controlled

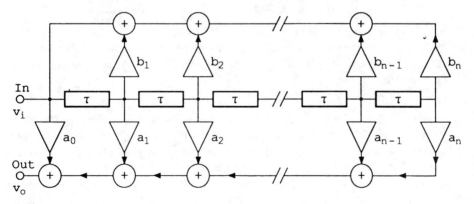

Fig. 11.43 *General form of a recursive filter.*

362 ANALOGUE ELECTRONIC CIRCUIT DESIGN

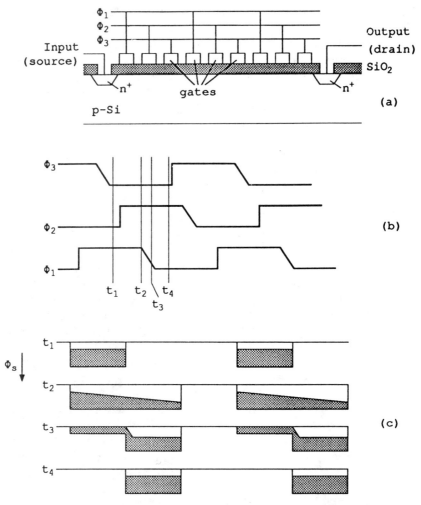

Fig. 11.44(a) *Schematic representation of CCD-structure.* (b) *Shape of clock pulses and mutual time relationship.* (c) *Surface potential ϕ_s under the gates and schematic representation of charge transport. Charge in the potential wells is indicated by hatching.*

by the sequence of clock pulses, the delay per shift is determined by the clock frequency. Figure 11.44(b and c) shows schematically the shape of the clock pulses and the way in which the electron charge is moving. Low-pass as well as bandpass transfer can be configured in such structures.

Variants using two or four clock pulse phases exist, but the general principle of operation is the same. Several methods have been developed for picking up signals from the charges under the gates and for doing so with the appropriate weighting factors. A general treatment of the structure and operation of CCDs is beyond the scope of this book. Rather, the intention is to convey a notion concerning this unique possibility of monolithic technology for realizing filter functions. However, it must be noted that this filtering method does not escape the drawback of all active filter methods in that achieving high dynamic range is difficult. The maximum charge content of a potential well is restricted and several noise-producing mechanisms are active. For a detailed treatment of transversal and recursive filters, reference is made to specialist works (see, for instance, Séquin and Tompsett, 1975).

11.9 Switched-capacitor filters

As has been stated earlier, in active filters a desired transfer function can be synthesized using only RC-time constants and voltage-to-current converters. Though no inductive time constants are needed, the implementation of the frequently required high resistance values can be problematic in monolithic implementations. One approach, which has been discussed in Section 11.7, uses MOST-channel conductances or MOS-transconductances as resistive elements. A great advantage of this method is that no sampling of the signal is necessary; the filter operates directly on the continuous-time signal. Therefore, the full frequency range that can be handled by the MOS-transistors can be effectively used. It is therefore not surprising that this method of making active filters is finding increasing acceptance. However, there are also disadvantages, the most prominent of these being the non-linearity of the devices used as the resistive elements.

In this section an alternative method for the realization of the required resistive elements will be briefly introduced. The essence of the method is that a resistor is replaced by a switched capacitor. The filters implemented in this way are called switched-capacitor filters.

Figure 11.45 shows how the resistive element is simulated. It is assumed that nodes 1 and 2 are low-impedance nodes, so that v_1 and v_2 can be considered to be originating from true voltage sources. When the switch is in position 1, the voltage on the capacitor is v_1; in switch position 2 the capacitor voltage changes to v_2.

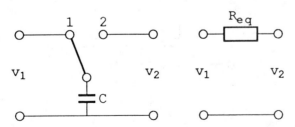

Fig. 11.45 *A switched capacitor and its equivalent resistor.*

Obviously, a transport of charge Q_0 has been involved in this action, amounting to $Q_0 = C\Delta v = C(v_1 - v_2)$. If the clock frequency of the clock signal that drives the switch is f_s, then the charge transport per second is $Q_0 f_s$. If the same charge transport is effected by a continuous current I_{eq}, so that $I_{eq} = Q_0 f_s = C f_s \Delta v$, the equivalent resistance is $R_{eq} = \Delta v / I_{eq} = 1/C f_s$. Of course, this is only true when v_1 and v_2 do not change between subsequent clock pulses, which implies that the clock frequency should be much larger than the signal bandwidth.

If a time constant τ is formed by combining a capacitor C_1 with a resistor that is simulated in the above-mentioned way, then $\tau = R_{eq} C_1 = C_1/C f_s$. Hence, τ is determined by the ratio of two capacitances and the clock frequency. Clearly, both constituting quantities can be accurately realized. Moreover, the chip area needed for the realization of a large time constant can be quite modest. The crucial point in this context is the silicon-area consumption of the smallest capacitance that can be accurately realized. For instance, in an audio processing system let a time constant of 100 μs be required. Since f_s must be much larger than the highest signal frequency f_s, the latter could be chosen to be 500 kHz. Consequently $C_1/C = 50$. For fixing such large capacitance ratios accurately, parasitic capacitances should be carefully taken into account. This is one of the problems facing the designer of switched-capacitor circuits.

A network of switched-capacitor circuits contains only capacitances and switches. The capacitances must preserve their charge free of any leakage during the time interval between subsequent switching actions. It will be clear that MOS-technology is the natural implementation vehicle. It is capable of providing excellent switches and high-quality capacitors. And loading of these capacitors with MOST-inputs does not cause charge leakage.

Figure 11.46 shows a straightforward implementation of a switched capacitor; Φ and $\bar{\Phi}$ denote two non-overlapping phases of the clock cycle. The low impedance nodes, which are conditional for correct operation, can be realized by the virtual grounded inputs of operational amplifiers. As an example, Figure 11.47(a) shows a conventional integrator, while Figure 11.47(b) shows its SC-equivalent. This circuit can be extended to a multiple-input integrator, as shown in Figure 11.48. A further example is the 'lossy integrator', featuring a non-zero pole, shown in Figure 11.49.

The argument, given in the opening of this section, concerning the equivalence

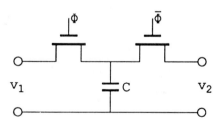

Fig. 11.46 *MOST-implementation of the switched capacitor.*

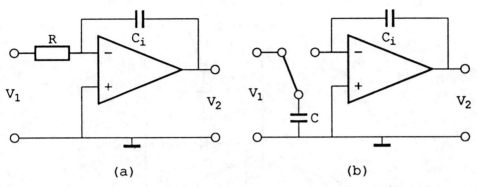

Fig. 11.47 (a) *Conventional integrator.* (b) *SC-equivalent* $\dfrac{V_2}{V_1} = \dfrac{f_s C / C_i}{s}$.

of the switched capacitor and the resistor thereby simulated, is based on a first approximation. A more thorough calculation of the response of an SC-circuit can be performed with the aid of sampling theory. The appropriate mathematical tool is the z-transform. Because of their rather specialist nature, the presentation of such calculations is omitted. Their main value is their capacity to make explicit the errors

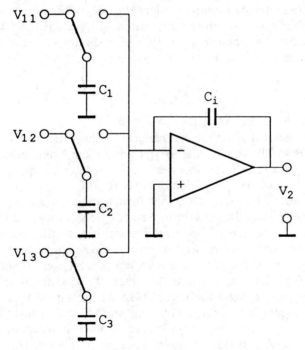

Fig. 11.48 *Summing and scaling integrator.* $V_2 = -(f_s/sC_i)(V_{11}C_1 + V_{12}C_2 + V_{13}C_3)$.

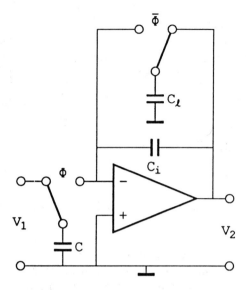

Fig. 11.49 *Lossy integrator* $\dfrac{V_2}{V_1} = -\dfrac{C}{C_l} \dfrac{1}{1 + s\ C_i/f_s C_l}$.

due to the finite switching frequency. Since the integrator is the key element in the synthesis of filter networks, the determination of its exact transfer function is of particular interest. The result is given for the integrator of Figure 11.47(b) in the form of its frequency response ($s = j\omega$) without proof:

$$\frac{V_2}{V_1} = -\left[\frac{C/C_1}{j\omega T_s}\right]\left[\frac{\omega T_s}{2\ \sin(\omega T_s/2)}\right]\left[\exp\left(-\frac{1}{2}\ j\omega T_s\right)\right] \qquad (11.19)$$

where $T_s = 1/f_s$ is the period of the clock signal.

The first factor in the right-hand member corresponds to the approximation found previously. The second factor concerns an amplitude error and the third factor a phase error. The latter error is of particular interest, since in configurations where feedback is present, the additional phase shift can give rise to instability. Anyway it is clear from eq. (11.19) that for accurate operation $f_s \gg f_{max}$ should hold. In general, f_s should exceed the highest signal frequency by a factor of at least 50–100. The consequence of this requirement is that the use of SC-filters is restricted to the domain of rather low frequencies. How large f_s can be chosen depends on the capabilities of the technological process used. In most cases f_s will not exceed a few MHz; in that case the range of signal frequencies that can be handled will be restricted to a few tens of kHz. An advantage of f_s being far above the signal bandwidth is that the antialiasing prefilter can be quite simple.

With regard to the dynamic range that can be accommodated, SC-circuits are certainly not among the leaders. The active circuits are not free from distortion and they introduce noise. Moreover, since switching operations are involved all along in the signal-processing chain, high-frequency noise components are folded back into the baseband of the signal everywhere in the network. This effect cannot of course be eliminated by applying antialiasing filters since it goes along with each separate switching action. Finally, it should be noted that $1/f$ noise, which is quite strong in MOS-devices, also affects noise behaviour significantly.

The synthesis of filter networks with prescribed characteristics from elementary SC-circuits is a specialist topic. A common approach is to start with an appropriate filter configuration as can be obtained from classical filter theory. This filter is then converted into its SC-equivalent. Several methods have been developed for designing SC-circuits with minimum sensitivity for parasitic and stray capacitances. The actual design work can be very much facilitated by the use of the available CAD design packages. For further study, reference is made to the more specialized works (see, for instance, Gregorian and Temes, 1986; Grebene, 1984).

References

Grebene, A. B. (1984) *Bipolar and MOS Analog Integrated Circuit Design*, Wiley, New York.

Gregorian, R. and Temes, G. C. (1986) *Analog MOS Integrated Circuits for Signal Processing*, Wiley, New York.

Séquin, C. H. and Tompsett, M. F. (1975) *Charge Transfer Devices*. Academic Press, New York.

Williams, A. B. and Taylor, F. J. (1988) *Electronic Filter Design Handbook*, McGraw-Hill, New York.

12 Circuits for modulation, demodulation and frequency conversion

12.1 Introduction

An information-carrying signal can be specified as a time-dependent quantity. By virtue of the Fourier transform, it can alternatively be specified by its spectral content. Similarly, the behaviour of signal-processing circuits and transmission channels can be expressed by their response in the time domain and in the frequency domain. Common vehicles for the specification of time-domain behaviour are the transient response and the impulse response; for the frequency-domain behaviour the complex frequency response is the obvious tool. In the preceding chapters ample use has been made of these concepts.

For adequate signal processing and signal transmission a good matching of the properties of the signals and the processing or transmission circuits is essential. If the rise time of a circuit is larger than that of the steepest signal transients, signal quality will deteriorate. Or in spectral terms: if the passband of a transmission channel cannot accommodate the spectrum of a signal, signal quality will suffer.

In other cases circuits and channels are available whose capabilities exceed the requirements posed by a given signal. For instance, coaxial cables, optical fibres, satellite communication facilities and free space, offer transmission bandwidths far beyond the spectral width of any common information-carrying signal.

If mismatches between signals and channels or processing circuits occur, appropriate measures must be taken for improvement. A measure that is quite often used is to shift the spectral content of a signal to a different part of the overall frequency spectrum. This approach is widely employed in communication electronics and, to a lesser degree, in instrumentation electronics. There are several ways in which a signal content can be transferred to a different region of the spectrum. The commonly used term for this process is *modulation*. It is accomplished by choosing a carrier frequency in the selected transmission region while transferring the information content of the signal to be handled to this carrier. One way of modulating the carrier is by varying its amplitude by the modulating signal. Alternatively, its phase or frequency can be modulated. Combinations of these methods can also be devised.

12.2 Amplitude modulation

Amplitude modulation is a modulation technique widely used in communications, particularly in radio and television broadcasting. Let $S(t)$ denote the signal to be transmitted. We will call this signal the modulating signal. Let f_0 be the carrier frequency, then the general expression for the amplitude-modulated signal is

$$M(t) = A[1 + mS(t)]\cos \omega_0 t \quad (12.1)$$

where $\omega_0 = 2\pi f_0$, while A and m are constants. Without loss of generality we can normalize the signal by putting $A = 1$. The constant m is called the modulation depth. It is often given as a percentage, hence $m = 0.3$ corresponds with 30% modulation depth.

To find the spectrum of $M(t)$ we describe $S(t)$ by its spectral content, hence $S(t)$ is viewed as a superposition of its spectral components. To start with, we select one spectral component $S(t) = \cos \mu t$. Then,

$$M(t) = (1 + m \cos \mu t)\cos \omega_0 t = \cos \omega_0 t + \tfrac{1}{2}m \cos(\omega_0 + \mu)t + \tfrac{1}{2}m \cos(\omega_0 - \mu)t \quad (12.2)$$

We conclude that the spectrum of $M(t)$ consists of three components, viz. a component at the carrier frequency ω_0 and two components on both sides of ω_0 at frequencies $\omega_0 + \mu$ and $\omega_0 - \mu$. These components are called the *sidebands* of $M(t)$. Figure 12.1 depicts the waveform of $M(t)$ and Figure 12.2 its spectrum. If $|m| < 1$, then the envelope of $M(t)$ is an exact replica of $S(t)$. If $S(t)$ encompasses a certain spectral width, then all individual spectral components give rise to sideband frequencies, so that both sidebands possess the same bandwidth as $S(t)$ (Figure 12.3).

Using eq. (12.2) we can easily find how the power of $M(t)$ is distributed over the carrier (P_c) and the sidebands (P_l and P_u). If $S(t)$ is a single sine wave, then $P_c : P_l : P_u = 1 : \tfrac{1}{4}m^2 : \tfrac{1}{4}m^2$. The carrier as such does not contain any information

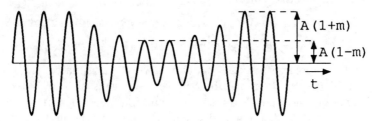

Fig. 12.1 *Amplitude-modulated signal.*

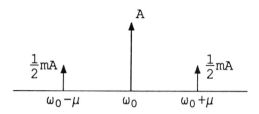

Fig. 12.2 *Frequency spectrum of an AM-signal when $S(t) = \cos \mu t$.*

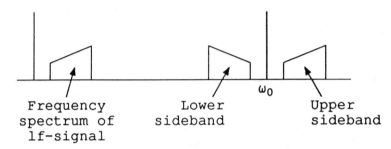

Fig. 12.3 *Frequency spectrum in the case of an arbitrary modulating signal.*

content, but even if $m = 1$ it comprises 66% of the total power of $M(t)$. Moreover, both sidebands contain the same information. Transmission of $M(t)$ in accord with eq. (12.2) is obviously inefficient, with regard to power consumption as well as to spectrum consumption. This method of amplitude modulation is called double-sideband modulation with non-suppressed carrier. In spite of the obvious lack of efficiency of this method, it is widely used in broadcasting because of its aptitude to simple demodulaton implementations. Since, provided $m < 1$, the envelope of $M(t)$ is a copy of the modulating signal, demodulation is possible with a simple rectifier circuit.

A considerable saving of transmitted power is obtained by suppressing the carrier. With $S(t) = \cos \mu t$, then $M(t) = m \cos \mu t \cos \omega_0 t = \frac{1}{2} m \cos(\omega_0 + \mu) t + \frac{1}{2} m \cos(\omega_0 - \mu) t$. However, now the envelope of $M(t)$ no longer corresponds to $S(t)$. Demodulation can yet be accomplished by multiplying $M(t)$ by a locally generated carrier $\cos \omega_0 t$, since

$$M(t)\cos \omega_0 t = \tfrac{1}{2} m \cos(\omega_0 + \mu)t \cos \omega_0 t + \tfrac{1}{2} m \cos(\omega_0 - \mu)t \cos \omega_0 t$$

$$= \tfrac{1}{2} m \cos \mu t + \tfrac{1}{4} m \cos(2\omega_0 + \mu)t + \tfrac{1}{4} m \cos(2\omega_0 - \mu)t \qquad (12.3)$$

The spectral components on $2\omega_0 + \mu$ and $2\omega_0 - \mu$ are situated far beyond the spectrum of $S(t)$ so that they can be removed by means of a low-pass filter.

Even if only one sideband is transmitted, multiplication of $M(t)$ by $\cos \omega_0 t$ can still provide the desired demodulated signal $S(t)$. By the way, note the finding that transmission of one sideband suffices to fully convey the information content of

$S(t)$; this is in agreement with the basic laws of information theory. From the point of view of spectrum utilization single-sideband transmission is the most efficient modulation method.

To accomplish the multiplication with $\cos \omega_0 t$, this carrier signal must be locally generated if it is not transmitted along with the information-carrying sideband signal. If the spectral content of $S(t)$ is to be exactly demodulated, information about the exact carrier frequency has to be transmitted in some way along with the sideband signal. This can be done in several ways. If during small time intervals the transmission of $S(t)$ can be interrupted, these intervals can be used for transmitting a few cycles of a reference signal at the carrier frequency. A time-continuous carrier signal can then be regenerated at the receiving site from the short bursts of carrier information. This is usually done with the aid of a so-called phaselock loop. This method is used in colour television for the regeneration of the subcarrier that conveys the colour information.

Another possibility is to transmit a low-power continuous pilot signal in a different part of the spectrum, for instance at half the carrier frequency. This method is practised in the transmission of stereo information in FM-audio broadcasting.

12.3 Receiver architectures

A receiver for a communication system that employs frequency multiplex must perform two operations: selection of the desired signal from the mix of received signals and demodulation of the selected signal. The selection is performed by a frequency-selective filter. Figure 12.4 shows a receiver architecture that executes the desired operations straightforwardly. The input signal $\Sigma M(t)$ is the superposition of a multiplicity of modulated carriers. The tunable bandpass filter selects the desired signal that is subsequently amplified and demodulated. Since the various signals constituting $\Sigma M(t)$ usually have widely different signal strengths and the demodulator should preferably operate at an approximately constant signal level, the gain of the amplifier must be adapted to the signal level of the selected signal. This is called *automatic gain control* (AGC). This type of receiver is called a *tuned*

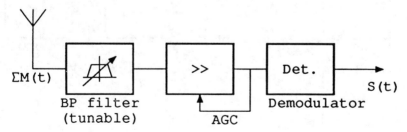

Fig. 12.4 *Architecture of the tuned radio frequency (TRF) receiver.*

radio frequency (TRF) receiver. Although this type of receiver architecture is theoretically entirely sound, its practical implementation is nearly always unworkable. As an example consider a simple medium wave broadcast receiver. The tuning range in such a receiver encompasses the frequency range from about 500 to 2000 kHz. The signals are of the double-sideband variety with unsuppressed carrier. Channel width is 9 or 10 kHz. The main problem is the exceptionally large dynamic range of the input signal. The weakest signal that can be received is determined by the noise floor. Let us assume that a signal level 30 dB above the noise floor is capable of providing acceptable signal quality. Due to the large difference between the field strength of a weak carrier and a very strong one at a given receiver location, the amplitude of the strong carrier can exceed that of the weakest carrier by about 100 dB. The total dynamic range to be handled thus amounts to 130 dB. For reception of the weakest signal, free of interference from a strong signal in an adjacent channel, the channel selectivity should amount to at least 120 dB. A filter providing this selectivity requires a large number of reactive elements that have to be simultaneously tunable. The realization of such filters is virtually impossible.

The required selectivity can be realized much more effectively in a fixed-frequency bandpass filter. The use of such a filter becomes possible by shifting the carrier of a desired channel in frequency so that it coincides with the central frequency of the filter passband. This operation is called 'frequency conversion'. Circuits that are capable of performing this operation are called 'frequency converters' or 'mixers'. The principle employed by these circuits is simple. Let $M(t) = [1 + mS(t)]\cos \omega_0 t$ be the r.f. signal to be demodulated. If we multiply this signal by a locally generated auxiliary harmonic signal $\cos \omega_1 t$, we obtain

$$M(t)\cos \omega_1 t = \tfrac{1}{2}[1 + mS(t)]\cos(\omega_1 + \omega_0)t + \tfrac{1}{2}[1 + mS(t)]\cos(\omega_1 - \omega_0)t \quad (12.4)$$

The modulating signal is transferred to two new carrier frequencies $\omega_0 + \omega_1$ and $\omega_0 - \omega_1$. Figure 12.5 shows the architecture of a receiver using this principle. The mixer circuit M is fed by the full spectrum $\Sigma M(t)$ on the one hand and the local

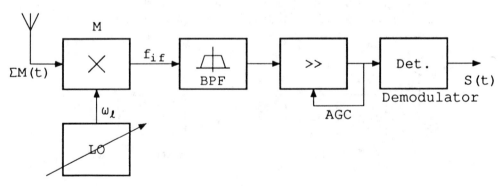

Fig. 12.5 *Principle of the heterodyne receiver.*

oscillator (l.o.) signal $\cos \omega_l t$ on the other hand. Its output signal is bandpass filtered by a filter with channel selectivity tuned at the frequency $\omega_l - \omega_0$. This frequency is called the *intermediate frequency*. Receivers which employ the method of operation just described are often called 'heterodyne receivers'. By tuning the local oscillator, all channels within the mixture $\Sigma M(t)$ can be converted to the intermediate frequency (i.f.). All signals within $\Sigma M(t)$, except the signal in the desired channel, are suppressed by the i.f.-filter. However, there is one exception. A carrier at $\omega_0' = \omega_0 + 2\omega_{if}$ also yields, after conversion with $\omega_l = \omega_0 + \omega_{if}$, the intermediate frequency. The receiver is found to be equally sensitive at two channels $2\omega_{if}$ apart. The unwanted carrier frequency is called the *image frequency* of the wanted one. For proper operation of the receiver the image channel has to be suppressed in front of the mixing operation. By choosing ω_{if} high enough, the separation distance $2\omega_{if}$ can be made sufficiently large to make the suppression of the image channel possible with a relatively simple bandpass filter.

At this stage two situations can be distinguished. As an example we consider again the medium wave receiver. A first possibility is to choose the intermediate frequency below the lowest frequency to be received, for instance 470 kHz, as is often used. The image frequency associated with the uppermost desired carrier of 2000 kHz is then 2940 kHz; the image frequency associated with the lowest carrier of 500 kHz is 1440 kHz. The latter frequency is situated within the medium wave band. This shows that the filter that precedes the mixer must be tunable. Its tuning should track the tuning of the local oscillator at distance f_{if}. Figure 12.6 shows the architecture of the receiver with image suppression according to this strategy.

The realization of such a tracking tuning over a wide band of frequencies is quite possible. However, it is not an attractive approach if the receiver is to be implemented as much as is possible in monolithic fashion. Furthermore, note that the suppression of the image channels in such a 'down-conversion' receiver becomes more difficult the higher the signal frequencies to be received. For example, the image frequency associated with a frequency of 16 MHz is 16.94 MHz. The *relative*

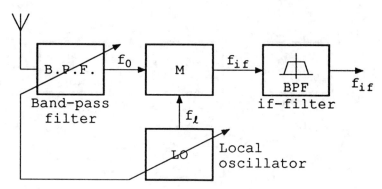

Fig. 12.6 *Architecture of the heterodyne receiver with image suppression.*

distance between both frequencies is much smaller than in the case of the medium wave receiver and consequently achieving adequate image frequency suppression is much more difficult.

An alternative solution is to use 'upconversion'. In this case f_{if} is chosen well in excess of the highest frequency to be received. Taking again the medium wave receiver as an example and choosing $f_{if} = 10$ MHz, the band of image frequencies ranges from 20.5 to 22 MHz. This is far beyond the upper edge of the medium wave band, so that the image band can easily be suppressed by means of a simple low-pass filter that needs no tuning whatsoever. Figure 12.7 shows the architecture of this type of receiver. However, as to be expected, this type of receiver is also afflicted with problems, albeit of a different kind. First, note that if the frequency band of interest is large, the intermediate frequency must be chosen quite high. For instance, if the upper frequency of interest is 30 MHz, for image suppression to be possible with a simple low-pass filter, the intermediate frequency should be of the order of 100 MHz. Achieving channel selectivity at such a high frequency is not a simple matter. A second and more fundamental problem is that the mixer has to accommodate the entire r.f. input spectrum with its enormous dynamic range. The mixer must be capable of handling this range. The consequences of this requirement will be dealt with in the next section. For the receiver operating according to the down-conversion principle, these requirements are considerably lower, since the tracking image-suppression filter partly suppresses strong undesirable components in the r.f.-spectrum.

A final note is appropriate on the use of the frequency conversion principle. The spectral distance between the desired and the image channel is $2f_{if}$. This distance is nullified if f_{if} takes zero value, implying that the r.f. signal is multiplied by a local oscillator frequency equalling the desired radio frequency. The relevant receiver architecture is shown in Figure 12.8. This type of receiver is called a *homodyne* receiver. At first glance this kind of receiver architecture is quite simple. The selectivity is implemented by a low-pass filter with half the channel bandwidth (in the case of a double-sideband modulated r.f. signal). However, in the majority of

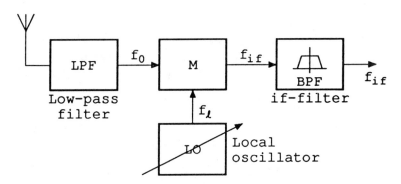

Fig. 12.7 *Architecture of the upconversion receiver.*

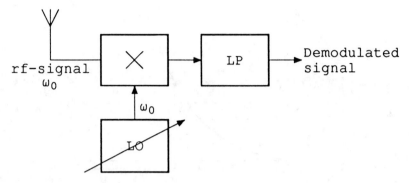

Fig. 12.8 *Homodyne architecture.*

practical situations the local generation of the carrier frequency leads to nearly unsurmountable problems. The locally generated carrier has to be in phase synchronism with the received r.f.-carrier. Since there is no preselection whatsoever, the phaselock loop that must provide this carrier has to cope with the full input spectrum with its enormous dynamic range. Up to now no satisfactory solutions have been found for this problem except in situations where the channels constituting the r.f.-spectrum are far apart.

12.4 Consequences of non-linearity in non-selective parts of the receiving system

The r.f. mixture $\Sigma M(t)$ encompasses a broad spectrum of modulated signals. As long as the desired channel is not yet separated from the mixture, the desired operation of the receiver is jeopardized due to non-linear effects. In order to find out what can happen, we consider a non-linear transfer characteristic, for instance an I–V characteristic (Figure 12.9). Around the biasing point P the characteristic can be expanded into a Taylor series.

$$I = f(V_o) + vf'(V_o) + \tfrac{1}{2}v^2 f''(V_o) + \tfrac{1}{6}v^3 f'''(V_o) + \ldots \quad (12.5)$$

where $v = V - V_o$ denotes the small signal handled by the characteristic. Where $i = I - I_o$

$$i = vf'(V_o) + \tfrac{1}{2}v^2 f''(V_o) + \tfrac{1}{6}v^3 f'''(v_o) + \ldots \quad (12.6)$$

It turns out that the small-signal transfer can be represented by a power series that can be written in the general form

$$i = a_1 v + a_2 v^2 + a_3 v^3 + \ldots \quad (12.7)$$

Assume now that v corresponds to

$$\Sigma M(t) = M_1(t) + M_2(t) + M_3(t) + \ldots$$

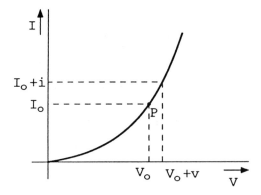

Fig. 12.9 *A non-linear transfer characteristic.*

Substituting this into eq. (12.7) yields

$$i = a_1(M_1 + M_2 + M_3 + \ldots) + a_2(M_1 + M_2 + M_3 + \ldots)^2$$
$$+ a_3(M_1 + M_2 + M_3 + \ldots)^3 + \ldots \quad (12.8)$$

Writing out this expression, we obtain a large number of terms consisting of products of the signals M_n and powers of these signals. These constituents are called *intermodulation products*. The spectral location and the power of each of these undesired intermodulation products is determined by the magnitudes of the individual factors. For instance, an intermodulation product $M_1 M_2$, where M_1 and M_2 represent signals with carrier frequencies ω_1 and ω_2, will produce spectral components around the frequencies $\omega_1 + \omega_2$ and $\omega_1 - \omega_2$, with sidebands related to both modulating signals. Any spectral components located in a spectral region where the receiver is sensitive, hence at least the desired channel and the image channel, give rise to unwanted i.f. signals that cannot be removed in subsequent parts of the receiver.

Since the coefficients a_n in eq. (12.7) usually decrease rapidly for increasing values of n, in practice the hardest problems will almost always be caused by the finite values of a_2 and a_3. In order to get a feeling for what can happen, consider the simple case of only two signals being present: $M_1 = S_1 \cos \omega_1 t$ and $M_2 = S_2 \cos \omega_2 t$. If the non-linear transfer is approximated by

$$i = a_1 v + a_2 v^2 + a_3 v^3$$

then the following terms arise

$$i = a_1(S_1 \cos \omega_1 t + S_2 \cos \omega_2 t) + a_2 S_1^2 \cos^2 \omega_1 t + a_2 S_2^2 \cos^2 \omega_2 t$$
$$+ 2 a_2 S_1 S_2 \cos \omega_1 t \cos \omega_2 t + a_3(S_1^3 \cos^3 \omega_1 t + S_2^3 \cos^3 \omega_2 t \quad (12.9)$$
$$+ 3 S_1^2 S_2 \cos^2 \omega_1 t \cos \omega_2 t + 3 S_1 S_2^2 \cos \omega_1 t \cos^2 \omega_2 t)$$

The products of $\cos \omega_1 t$ and $\cos \omega_2 t$ and their power functions can also be written as goniometric functions of sums and differences of ω_1 and ω_2 and multiples of these frequencies. A particularly irksome term is $3a_3S_1S_2^2 \cos \omega_1 t \cos^2 \omega_2 t$. If this term is decomposed, among others, a component $\tfrac{3}{2} a_3 S_1 S_2^2 \cos \omega_1 t$ shows up. If M_1 is the desired signal, a disturbing signal is apparently generated at the carrier frequency of M_1 but modulated with a distorted version of S_2. If the desired signal M_1 is weak and the undesired signal M_2 is strong, then this interfering term can be much larger than the desired term $a_1 S_1 \cos \omega_1 t$. When tuning the receiver at M_1, the demodulator produces, apart from a weak desired signal S_1, the distorted signal S_2 also. This is called *cross-modulation*, which is a very annoying effect. Our analysis shows that it is caused by odd terms in the expansion of the non-linear characteristic.

EXAMPLE
A signal to be received generates a voltage of 10 μV amplitude in an antenna. A nearby transmitter generates 0.1 V amplitude in the same antenna. The cross-modulation caused by the strong signal should not exceed the -40 dB level. If eq. (12.7) is used for the representation of the non-linearity of the receiver front end, what is the requirement for a_3?

Let A denote the amplitude of the weak signal, then after transfer the desired signal is $a_1 A$. The amplitude of the cross-modulation signal is $10^8 \cdot \tfrac{3}{2} a_3 A^3$. Hence, for the cross-signal to be 40 dB weaker than the desired signal

$$a_3 A^3 < \tfrac{2}{3} \cdot 10^{-10} a_1 A$$

or

$$a_3/a_1 < \tfrac{2}{3} \cdot 10^{-12} (\mu V)^{-2}$$

illustrating the extreme linearity requirements to be met by the non-selective part of the receiver front end.

The question arises as to what can be done to avoid the intermodulation effects, including cross-modulation. An effective method is to realize all the necessary selectivity in front of the first device that can exhibit a non-linear transfer (amplifier or mixer). In practice, this method is unworkable. It is therefore necessary to conquer the non-linearity as much as possible, particularly with regard to the odd terms. By simultaneously applying a moderate amount of filtering it is possible to alleviate the requirement concerning the non-linearity. An amplifier can be linearized by the application of feedback; but for a mixer, effective feedback schemes can hardly be devised because the input and output signals are in different ranges of the frequency spectrum. The design of good mixers is therefore of crucial importance in receiver engineering.

To judge the quality of a receiver the concept of *intermodulation free dynamic range* (IMFDR) is useful. The basic idea is that an intermodulation component that is below the noise threshold can be considered as unharmful since it is masked by the noise. To determine the IMFDR two input signals of equal amplitude are fed to the receiver. The maximum signal that can be received without perceivable

378 ANALOGUE ELECTRONIC CIRCUIT DESIGN

intermodulation has an amplitude that gives rise to an intermodulation signal equalling the noise threshold. The IMFDR is the ratio of this maximum signal and the noise threshold. It is usually expressed in dB.

As an illustration, Figure 12.10 depicts the relevant intermodulation characteristics of a practical receiver. This diagram is equivalent to Figure 4.5, and for convenience it is repeated here. The 0-dB level is arbitrarily assigned to the maximum signal that can be accommodated by the receiver. It is determined by, for instance, the available supply voltage or bias current. The noise threshold is assumed to be at -130 dB. Consequently, the full dynamic range of the receiver is 130 dB. The relation between input and output signal is represented by the upper line in Figure 12.10. The levels of the intermodulation signals due to second-order (IM2) and to third-order (IM3) non-linearities are also indicated as a function of the input signal. The IMFDR with regard to IM2 and IM3 can be found directly from the diagram.

In the example shown in Figure 12.10, IMFDR2 is lower than IMFDR3. This may seem improbable, since in general the coefficients in eq. (12.7) are expected to decrease with increasing order. However, the use of balanced circuits can drastically reduce even order effects, whereas for odd order effects no comparable measures can be devised.

At this stage we can draw some general conclusions with regard to the requirements to be met by circuits for modulation, demodulation or frequency conversion. All these functions come down to multiplying two signals, of which one carries information and consequently encompasses a certain spectral bandwidth. Ideally, the circuits should produce no other output signal than the desired product signal. The amount of non-linearity that can be tolerated depends on the actual system architecture, particularly on the amount of filtering that is applied in front of the multiplier. In modern electronics maximum integration in single-chip fashion is pursued. In the present state of the art, filters that can handle signals with a large

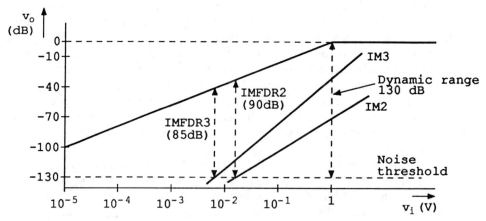

Fig. 12.10 *Dynamic range and intermodulation-free dynamic range.*

dynamic range cannot be monolithically implemented. Architectures depending on the availability of such filters should therefore preferably be avoided, particularly if they have to be tunable and/or need adjustment. In view of this state of affairs the architecture employing up-conversion seems to offer the best perspectives for maximum integratability. However, this architecture also poses the most exacting requirements concerning the suppression of the effects of non-linearity.

12.5 General principles of circuits for modulation and frequency conversion

In the preceding sections it was found that the functions of modulation, demodulation and frequency conversion all come down to the multiplication of two signals. We can write these signals in the form $S_1 = A \cos \omega_1 t$ and $S_2 = B \cos \omega_2 t$. One of these signals encompasses a certain spectral bandwidth. For convenience we will assume that this applies to S_1. In the case of a modulator, S_1 is the modulating signal, in the case of a demodulator or a frequency converter, S_1 is a modulated carrier.

A first distinction can be made between implementations of the *additive* or the *multiplicative* variety. In an additive circuit both signals operate on the same transfer characteristic. To produce a product term this characteristic has to be non-linear. The general principle is indicated in Figure 12.11. The block *NL* represents a non-linear element. If its transfer is given by $i = a_1 v + a_2 v^2 + a_3 v^3 + \ldots$, then

$$i = a_1 (A \cos \omega_1 t + B \cos \omega_2 t) + a_2 (A \cos \omega_1 t + B \cos \omega_2 t)^2 + \ldots \quad (12.10)$$

Writing out eq. (12.10) we obtain, among many other terms, the desired product $2AB \cos \omega_1 t \cos \omega_2 t$. All further terms are essentially undesirable and must be suppressed. This can be done by means of filters. Of course, this is only possible if none of the signal components to be removed is located within the spectral band of the desired signal. The requirements concerning the filtering can be alleviated considerably by using a non-linear element with an approximately quadratic

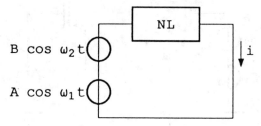

Fig. 12.11 *Principle of additive mixer circuits.*

characteristic. In this case

$$i = a_1 A \cos \omega_1 t + a_1 B \cos \omega_2 t + a_2 A^2 \cos^2 \omega_1 t + a_2 B^2 \cos^2 \omega_2 t \\ + 2 a_2 AB \cos \omega_1 t \cos \omega_2 t \quad (12.11)$$

A further reduction of undesired signal components can be obtained by the application of compensation techniques. Let an identical circuit be driven by signals $-S_1$ and $-S_2$. The output signal of this circuit is

$$i' = - a_1 A \cos \omega_1 t - a_1 B \cos \omega_2 t + a_2 A^2 \cos^2 \omega_1 t + a_2 B^2 \cos^2 \omega_2 t \\ + 2 a_2 AB \cos \omega_1 t \cos \omega_2 t \quad (12.12)$$

Adding i and i' yields

$$i + i' = 2 a_2 (A^2 \cos^2 \omega_1 t + B^2 \cos^2 \omega_2 t + 2 AB \cos \omega_1 t \cos \omega_2 t) \quad (12.13)$$

The undesired terms at ω_1 and ω_2 have been suppressed. If, alternatively, the terms at $2\omega_1$ and $2\omega_2$ are to be suppressed, the second circuit should be driven by $S_1 - S_2$, yielding

$$i'' = a_1 A \cos \omega_1 t - a_1 B \cos \omega_2 t + a_2 (A^2 \cos^2 \omega_1 t + B^2 \cos^2 \omega_2 t - 2 AB \cos \omega_1 t \cos \omega_2 t).$$

Subtracting i'' from i yields

$$i - i'' = 2 a_1 AB \cos \omega_2 t + 4 a_2 AB \cos \omega_1 t \cos \omega_2 t \quad (12.14)$$

Driving the second circuit by $S_2 - S_1$ yields suppression of the term at ω_2 instead of that at ω_1.

Figure 12.12 depicts the principle of a balanced circuit employing MOSTs, which possess approximately quadratic characteristics. Subtraction of the output signals can be accomplished with the aid of a balanced transformer, as indicated in Figure 12.12. Alternatively, a current mirror can be employed.

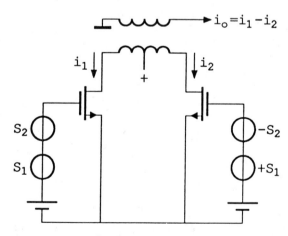

Fig. 12.12 *Principle of a balanced additive mixer using MOSTs.*

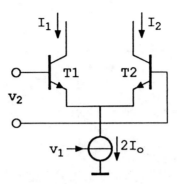

Fig. 12.13 *Principle of the controlled transconductance mixer.*

In designing such circuits it should be taken into account that devices with accurate quadratic characteristics do not exist. Some third-order distortion will always occur.

An essentially better approach is to use a multiplicative circuit. In such a circuit, both signals S_1 and S_2 operate on different transfer characteristics, which can be individually linear. The circuit is configured in such a way that the signal transferred by the first characteristic is subsequently handled by the second characteristic in proportion to the quantity operating on this second characteristic. The most widespread implementation of this principle is multiplication by means of a controllable transconductance. Figure 12.13 indicates the basic structure of an implementation using BJTs. The signal S_1 is applied as a voltage v_1 that modulates the current source with quiescent current $2I_o$ by means of a linear voltage-to-current converter. This current splits up in the currents through $T1$ and $T2$. The signal $S2$ is applied as a voltage between the bases of $T1$ and $T2$. Using the exponential voltage-to-current relation of the BJT we find

$$I_1 = \frac{2I_o}{1 + \exp(-v_2/V_T)} \quad \text{and} \quad I_2 = \frac{2I_o}{1 + \exp(v_2/V_T)}$$

where V_T denotes the thermal voltage kT/q. Subtracting I_2 from I_1, we find for the differential output current

$$i_o = I_1 - I_2 = 2I_o \tanh \frac{v_2}{2V_T} \tag{12.15}$$

Figure 12.14 depicts this transfer.

$$\text{If } v_2 \ll V_T, \text{ then } i_o \approx 2I_o \cdot \frac{v_2}{2V_T} = I_o \frac{v_2}{V_T} \tag{12.16}$$

If the current source is modulated by v_1, so that it can be represented by $2I_o + g_{m1}v_1$, then

$$i_o = (I_o + g_{m1}v_1) \frac{v_2}{V_T} = I_o \frac{v_2}{V_T} + g_{m1} \frac{v_1 v_2}{V_T} \tag{12.17}$$

382   ANALOGUE ELECTRONIC CIRCUIT DESIGN

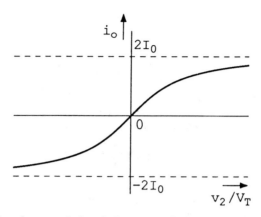

Fig. 12.14 *Transfer characteristic of the controlled transconductance mixer.*

If the term $I_o v_2 / V_t$ is undesired, it can be suppressed by employing a second circuit, driven by $-v_1$. The differential output current of this circuit can be written

$$i'_o = I_o \frac{v_2}{V_T} - g_{m1} \frac{v_1 v_2}{V_T}$$

Subtracting i'_o from i_o yields

$$i_o - i'_o = 2 g_{m1} I_o v_1 v_2 / V_T \qquad (12.18)$$

The multiplicative property of this circuit rests on the linear relation between i_o and I_o, as expressed by eq. (12.15). This relation is in turn a consequence of the precise exponential transfer of the BJT. With MOSTs this feature is absent. When representing the I–V characteristic of a MOST by $I = \frac{1}{2}\beta(V_{GS} - V_t)^2$, we find for the circuit of Figure 12.15

$$I_1 = I_o + \tfrac{1}{2}\beta v_2 \sqrt{\frac{2I_o}{\beta} - \frac{1}{4} v_2^2} \qquad (12.19a)$$

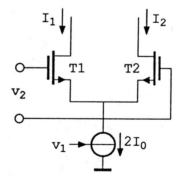

Fig. 12.15 *A controlled transconductance mixer with MOSTs.*

$$I_2 = I_o - \tfrac{1}{2}\beta v_2 \sqrt{\frac{2I_o}{\beta} - \tfrac{1}{4} v_2^2} \qquad (12.19b)$$

and

$$i_o = I_1 - I_2 = \beta v_2 \sqrt{\frac{2I_o}{\beta} - \tfrac{1}{4} v_2^2} \qquad (12.20)$$

Hence, the relation between i_o and I_o is non-linear and consequently, if operated as assumed, the circuit cannot be used as a linear frequency converter.

12.6 Switching modulators

In a modulator or a frequency converter two signals are always involved. One of these is the information-carrying signal; the other is a periodic signal of fixed frequency. What the circuit essentially accomplishes is a spectral shift of the information-carrying signal. The periodic signal acts as a reference for this shifting operation. Up to now we have assumed that the periodic signal is a sine wave. However, this is not compulsory: what matters is just the frequency. Therefore, the periodic signal can alternatively be used to drive an electronic switch. An obvious advantage of this mode of operation is that only two discrete states of the circuit are involved.

Figure 12.16 indicates the principle of a switching modulator. The switch might be implemented by a MOST (Figure 12.17). The operation of the switch can be represented by a switching function $Sw(t)$, as shown in Figure 12.18, where 0 corresponds to the off-state of the switch and 1 to the on-state. If $Sw(t)$ is symmetrical (0.5 duty cycle), then its Fourier expansion does not contain even terms, hence

$$Sw(t) = b_0 + b_1 \cos \omega_0 t + b_3 \cos 3\omega_0 t + \ldots \qquad (12.21)$$

Alternatively, the switch can be of the polarity-switching type (Figure 12.19). In that case the term b_0 in (12.21) is absent (Figure 12.20). The output current of the

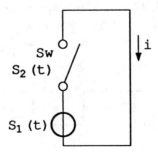

Fig. 12.16 *Principle of the switching modulator.*

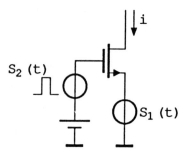

Fig. 12.17 *MOST as a switching modulator.*

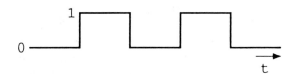

Fig. 12.18 *Switching function Sw as a function of t.*

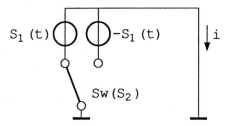

Fig. 12.19 *Polarity-switching modulator.*

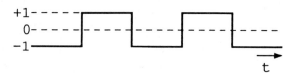

Fig. 12.20 *Sw(t) for polarity-switching modulator.*

modulator is given by

$$i = S(t) \cdot Sw(t) \tag{12.22a}$$

With $S(t) = A \cos \mu t$, eq. (12.22a) yields

$$i = A \cos \mu t [b_0 + b_1 \cos \omega_0 t + b_3 \cos 3\omega_0 t + \ldots] \tag{12.22b}$$

Besides the desired term $M(t) = Ab_1 \cos \mu t \cos \omega_0 t$ higher-order terms $Ab_3 \cos \mu t \cos 3\omega_0 t$, etc., also occur. The operation generates a multiplicity of spectrally shifted versions of the spectrum of $S(t)$. They are at distances of $2\omega_0$ apart and therefore in most cases the suppression of unwanted spectra will not be problematic.

For the sake of completeness note that in the implementation care has to be taken that the switch is driven by the periodic signal $S_2(t)$ only. A simple additive circuit, such as that in Figure 12.17, can thus only function correctly if $S_1(t) \ll S_2(t)$.

Switching modulators and frequency converters are widely used. Frequency converters are nearly always of the switching variety. In most systems the conversion operation takes place in the front section where the signal level is low. This implies that the noise properties of the converter are quite important. It can be shown that the noise behaviour of a switching converter is much better than that of a non-switching one. On the other hand, the switching converter introduces a new problem. Not only the desired carrier and its image frequency are converted to the intermediate frequency, but also a multiplicity of other carriers at mutual distances $n\omega_1$ and the image frequencies corresponding with these carriers. Therefore, using a switching converter enhances the requirements concerning the preselection, i.e the selectivity in front of the frequency conversion.

The type of switching frequency converter most widely used is the variable-transconductance mixer of which the basic structure has already been depicted in Figure 12.13. In the case of a switching mixer, v_2 is applied as a square wave signal. If the switching action is instantaneous, both transistors are conducting alternately, implying that both transistors are never in their active region simultaneously. This observation explains the superior noise behaviour of the circuit. With regard to the noisiness of the circuit, it behaves as a single current-driven CB-stage. The output noise current of the mixer is therefore given by

$$\overline{i_{on}^2} = 2qI_B \Delta f \tag{12.23}$$

Hence, the only noise source that is relevant to the mixer operation is the base current noise source. The output noise can be transformed into an equivalent input noise current by dividing it by the squared conversion gain factor. The conversion gain is the ratio of the i.f.-signal output current and the r.f.-signal input current. In the case of a symmetrical square wave we have

$$i_o = I_1 - I_2 = \frac{4}{\pi} I_o(2I_o + i_{rf}) \left[\cos(\omega_{lo}t) - \frac{1}{3}\cos(3\omega_{lo}t) + \ldots \right] \tag{12.24}$$

If $i_{rf} = \hat{i}_{rf} \cos \omega_{rf} t$, the amplitude of the output signal at $\omega_{lo} - \omega_{rf}$ is $2/\pi\, \hat{i}_{rf}$. Hence, the conversion gain is $2/\pi$ and consequently, the equivalent input noise source is $(\pi^2/2)\, qI_B\, \Delta f$.

In practice the switching action cannot be instantaneous due to the finite bandwidth of the driving circuit. The minimum achievable rise time of the switching signals is determined by the parasitic capacitance of the switching input of the mixer. The relative deviation from the ideal situation is, of course, larger the higher the local oscillator frequency. The finite rise time causes other noise sources to contribute to the output noise to a certain degree. When designing a high-performance mixer one must strive for minimization of the rise time.

EXAMPLE

What is the minimum magnitude of the switching signal capable of switching the current from $T1$ to $T2$ and vice versa? Using eq. (12.15), the question is one of finding the value of v_2, for which $\tanh v_2/2V_T \to 1$. With $v_2 = 5V_T \approx 130$ mV, $\tanh \frac{5}{2} = 0.987$, so that 99% of the current flows in one transistor.

To determine the IMFDR of this circuit, besides the noise floor, the distortion produced by the circuit also has to be known. Assuming ideal switching, the distortion corresponds to the distortion of a current-driven CB-stage. In such a stage distortion mainly occurs due to two effects. First, current transfer from emitter to collector is slightly current-dependent. And, second, the finite output impedance of the driving r.f.-modulated current source, together with the current-dependent finite emitter impedance of the switching transistors, gives rise to current-dependent transfer of the r.f.-modulated current. This effect is larger the smaller the output impedance of the current source. Consequently, at high frequencies, where the output capacitance of the current source lowers its output impedance, this effect can dominate in the total distortion. A quantitative analysis of the IMFDR features of this mixer arrangement is rather complicated and beyond the scope of this book (Nauta, 1986).

Regarding noise, it goes without saying that minimization of the noise floor is of primordial importance, since it determines the sensitivity of the receiver directly. As has been argued, in this respect the switching multiplier is far better than the non-switching variety. The minimum r.f.-signal level that can be caught is determined by the noise floor. Ideally, the noise should be restricted to the noise in the source signal. Practically, the noise added by the receiver should be less than the noise that the source signal is afflicted with.

In this context a further matter of concern deserves to be mentioned, viz. the spectral purity of the local oscillator signal. The signal generated by a practical oscillator exhibits noise sidebands. This sideband noise is transferred to the intermediate frequency along with the converted r.f.-signal. Apart from affecting the sensitivity of a receiver, oscillator sideband noise can also affect its selectivity. This happens when in an adjacent channel an r.f.-signal is present that is much stronger than the signal that the receiver is tuned to. The conversion product of the strong signal and the sideband noise at channel distance from the local oscillator

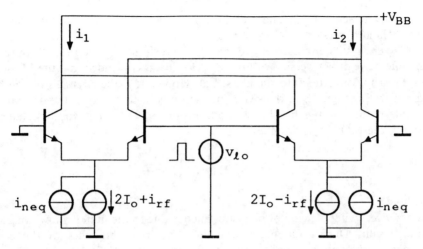

Fig. 12.21 *Double-balanced switching mixer. The (differential) output signal is $i_1 - i_2$.*

frequency gives rise to an interfering i.f.-signal. This effect is commonly denoted 'reciprocal mixing'. Minimizing this effect calls for the use of a high-performance local oscillator.

EXAMPLE
Let a signal in an adjacent channel be 70 dB above the level of the signal to be received. What is the requirement for the carrier-to-noise ratio (CNR) pertaining to the phase noise of the local oscillator, if the interfering i.f.-signal should be at least 40 dB below the wanted signal?

If the CNR at channel distance is denoted by α, the interfering i.f.-signal level is $(70 - \alpha)$ dB with respect to the wanted signal. From $70 - \alpha < -40$ we find $\alpha > 110$ dB.

The circuit according to Figure 12.13 is called a single-balanced mixer. A disadvantage of this configuration is that its output signal contains many undesired signal components. Besides the desired signal at $\omega_{lo} \pm \omega_{rf}$ the local oscillator signal at ω_{lo} and its harmonics, together with modulated signals at $3\omega_{lo} \pm \omega_{rf}$, $5\omega_{lo} \pm \omega_{rf}$, etc. are also present. By using a second single-balanced circuit of the same structure, but driven with the inverted r.f.-signal, and subtracting the output signals of these two single-balanced circuits, the so-called double-balanced mixer is configured. Figure 12.21 shows the circuit obtained in this way. It can easily be seen that the l.o.-signal and its harmonics are no longer present in the output signal. The modulated signals at $3\omega_{lo} \pm \omega_{rf}$, $5\omega_{lo} \pm \omega_{rf}$, etc., are not eliminated and these have to be removed by the i.f.-bandpass filter. The single-sided parts contribute equal amounts of noise as indicated by the equivalent input noise sources i_{neq} in the diagram. Both noise sources are uncorrelated and, consequently, output noise

388 ANALOGUE ELECTRONIC CIRCUIT DESIGN

increases by a factor of $\sqrt{2}$. The useful output signal doubles, so that by using the double-balanced version, S/N is improved.

The mixer circuits discussed above can also be realized with MOSTs. With regard to the sine-wave driven version of Figure 12.15, it was concluded that the r.f.-signal is not linearly converted into the i.f.-signal. However, if the circuit is driven by a square wave l.o.-signal and the switching action is instantaneous, the operation is identical to that of the bipolar version. From eq. (12.19) it follows that, for operating the MOSTs as switches,

$$|v_2| > 2\sqrt{\frac{2I_o}{\beta}} \qquad (12.25)$$

must hold.

In practice, switching cannot be instantaneous due to the finite bandwidth of the driving circuit. During the finite switching time both branches of the circuit contribute to the output signal, so that a certain amount of distortion continues to be present. If the rise time is small compared to the period of the switching signal, the distortion is of minor importance. All in all, it can be concluded that, particularly in high-frequency applications, the MOS-switching mixer is less suitable.

A modulator of the switching type that was very popular in former days, but can still be a useful proposition, is the so-called ring modulator (Figure 12.22). The amplitude of the carrier signal is made so large that it switches the diodes between the states of full conduction and cutoff. The amplitude of the modulating signal is made small compared to the amplitude of the carrier, so that the switching action is nearly fully controlled by the carrier signal. If point A (Figure 12.22) is positive and B is negative, $D1$ and $D2$ are conducting. The modulating signal $S(t)$ is then

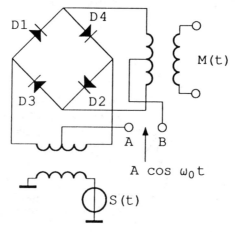

Fig. 12.22 *A ring modulator using four diodes.*

connected to the output transformer through $D1$ and $D2$. When, a half-period later, A is negative and B is positive, $D3$ and $D4$ are conducting, so that $S(t)$ is connected to the output transformer through these diodes. Hence, the current invoked by $S(t)$ in the transformer changes polarity in rhythm with the carrier.

12.7 Single-sideband modulators

The spectrum of the product function $\cos \mu t \cos \omega_0 t$ always includes two sidebands, because $\cos \mu t \cos \omega_0 t = \frac{1}{2}\cos(\omega_0 + \mu)t + \frac{1}{2}\cos(\omega_0 - \mu)t$. In a frequency-conversion circuit the occurrence of two 'sidebands' shows up as sensitivity at two frequencies, the desired one and the image frequency.

For the purpose of spectrum conservation, in many communication systems it is preferred to transmit only one sideband. In certain cases it is possible to suppress the undesired sideband by means of a filter. However, if the spectral distance between both sidebands is small, which is the case when the modulating signal contains low-frequency components, this method is not feasible. Figure 12.23 shows the principle of an alternative method for suppressing one of the sidebands. The configuration contains two balanced modulators that produce the signals $\cos \mu t \cos \omega_0 t$ and $\sin \mu t \sin \omega_0 t$ respectively. Addition of these signals yields

$$\cos \mu t \cos \omega_0 t + \sin \mu t \sin \omega_0 t = \cos(\omega_0 - \mu)t$$

The other sideband can be obtained by subtraction of both signals.

Deriving the quadrature signal $\sin \omega_0 t$ from $\cos \omega_0 t$ is simply accomplished by means of a $90°$ phase shifter. The problem of this type of single-sideband modulator is the derivation of the quadrature signal of the modulating signal $S(t) = \cos \mu t$. This signal occupies a certain spectral band and it is required that all spectral components within this band are shifted by $90°$. The resulting signal is called the Hilbert-transformed signal $H[S(t)]$ of $S(t)$, since the corresponding mathematical operation is called the Hilbert transform. Several methods have been devised for the approximate formation of $H[S(t)]$ from $S(t)$. The available methods yield satisfactory results if the spectral bandwidth of $S(t)$ is relatively

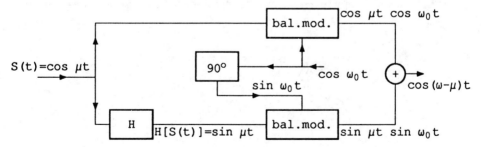

Fig. 12.23 *Block diagram of a single-sideband modulator.*

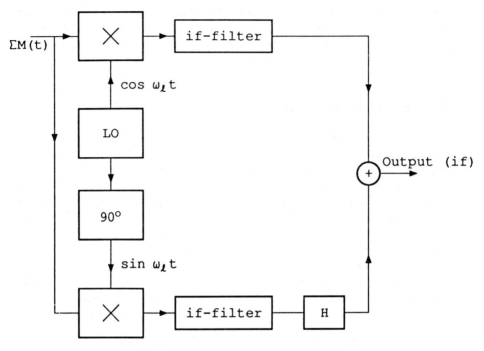

Fig. 12.24 *A frequency converter with inherent image rejection.*

small, as is the case with voice signals. With wideband signals complex networks are needed for achieving reasonable performance.

This method can in principle also be used for suppressing the image band in frequency-conversion circuits. The Hilbert transform has then to be applied to the i.f.-signal. Since this signal has a small relative bandwidth, performing the transform does not involve complex circuitry. Figure 12.24 shows the structure of such a mixing circuit with inherent image rejection. This method rarely finds application since the intermediate frequency can nearly always be chosen such that image rejection can be simply accomplished by proper filtering.

For the sake of completeness it may be mentioned that in some cases a format

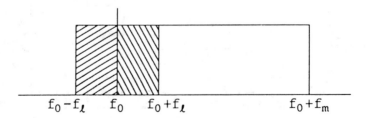

Fig. 12.25 *Spectrum of a vestigial-sideband signal.*

for the modulated signal is preferred that is between double-sideband and single-sideband. The low-frequency part of the spectrum of the modulating signal is transmitted in double-sideband fashion, whereas the high-frequency part is transmitted in single-sideband fashion. Figure 12.25 shows the spectrum of such a signal. This method is called 'vestigial-sideband modulation'. It is generally used in broadcast television transmission.

12.8 Demodulation of AM-signals

In Section 12.2 we have found that the envelope of a double-sideband AM signal with small modulation depth ($m < 1$) is a replica of the modulating signal (Figure 12.1). In this particular case demodulation is possible with a simple rectifier circuit, as shown in Figure 12.26. The capacitor C charges up to the amplitude of the carrier during the positive half of the carrier period. If there were no charge leakage the capacitor voltage would maintain this value. Due to the resistor R in parallel with the capacitor, its voltage slowly decreases. When, during the next period $M(t)$ again approaches its maximum value, the diode starts to conduct again (Figure 12.27). Due to the leakage through R, the capacitor voltage is capable of following the relatively slow amplitude variations of $M(t)$ caused by the modulation. It is clear

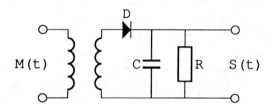

Fig. 12.26 *An envelope demodulator.*

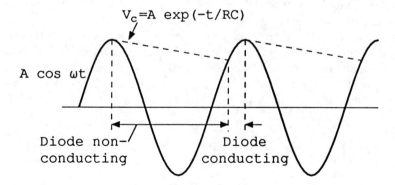

Fig. 12.27 *Operation of the envelope demodulator.*

that the time constant RC should be such that the capacitor voltage can follow the amplitude variations. It is also obvious that following the amplitude variations of $M(t)$ is more difficult the larger the modulation depth m. It can be shown that for proper operation $1/RC > m\mu_{max}$ should hold, where μ_{max} denotes the maximum frequency in the spectrum of the modulating signal $S(t)$. On the other hand, the value of RC should not be taken smaller than is necessary, since then the discharge between two successive maxima of the carrier is too strong, with the result that the output signal contains a large component at the carrier frequency, placing heavier demands on the post-demodulation filtering.

This simple method of demodulation is widely used in broadcasting receivers. It can be applied there since in AM broadcast systems double-sideband modulation with low modulation depth ($m \approx 0.3$) is the rule. In all other cases demodulation must be realized by multiplying $M(t)$ with a locally generated reference carrier $\cos \omega_0 t$. The circuits used for this purpose are conceptually identical to the circuits used for modulation and frequency conversion. In the case of a vestigial-sideband modulation signal, envelope demodulation leads to certain errors. Whether or not these can be tolerated depends on the details of the modulation format and the nature of the actual application. In older television receivers it was the usual demodulation method. With the advent of IC-technology with its ensuing reduction of the costs of electronic circuitry, the use of synchronous demodulation, i.e. demodulation by multiplication, has become current practice.

12.9 Angle modulation

Generally, a carrier signal, whether modulated or not, can be written in the form $A \cos \Psi$, where A is the amplitude and Ψ the angle or argument. By definition, the time derivative of the argument is the (momentary) angular frequency ω, hence

$$\omega = \frac{d\Psi}{dt} \text{ and } \Psi = \int_0^t \omega \, dt \qquad (12.26)$$

In the case of amplitude modulation ω is constant ($\omega = \omega_0$) and A is modulated by the modulating signal $S(t)$.

It is also possible to keep A constant and to modulate Ψ by $S(t)$. If Ψ is a function of time, then according to eq. (12.26), the same applies to ω, indicating that the momentary frequency depends on $S(t)$. The momentary frequency varies around a central or average value, which can be seen as the carrier frequency of the angle-modulated signal. Denoting it by ω_0, we can write

$$\Psi = \omega_0 t + \varphi$$

Since ω_0 is assumed to be constant, the momentary (angular) frequency follows from

$$\omega = \frac{d\Psi}{dt} = \omega_0 + \frac{d\varphi}{dt} \qquad (12.27)$$

Two varieties of angle modulation are in widespread use. In the first one φ is modulated by the signal $S(t)$, hence $\varphi(t) = m_p S(t)$, so that

$$M(t) = \cos[\omega_0 t + m_p S(t)] \qquad (12.28)$$

This type of modulation is called *phase modulation*. This name is somewhat misleading in that it suggests that the frequency is constant. From eq. (12.28) it is seen that this is not the case, since

$$\omega = \frac{d\Psi}{dt} = \omega_0 + m_p \frac{dS(t)}{dt} \qquad (12.29)$$

indicating that the momentary frequency is modulated by the time derivative of $S(t)$. The proportionality factor m_p is called the 'phase modulation index'.

A phase modulator is simply a phase shifter that is controlled by the modulating signal. The phase shifter is fed by the unmodulated carrier. To accomplish the phase shift any component or circuit can be used whose reactance can be controlled by a voltage or a current. Examples will be given later when dealing with frequency modulation.

Demodulation of a phase-modulated signal can be accomplished in several ways. Generally, m_p is chosen such that for any value of $S(t)$, $m_p S(t) \ll 2\pi$. In that case, phase demodulation can be simply done by multiplying $M(t)$ by $\sin \omega_0 t$:

$$M(t) \sin \omega_0 t = \cos(\omega_0 t + m_p S)\sin \omega_0 t = \tfrac{1}{2}\sin(2\omega_0 t + m_p S) - \tfrac{1}{2}\sin m_p S \quad (12.30)$$

If $m_p S \ll 2\pi$ then $\sin m_p S \approx m_p S$. The term at $2\omega_0$ is removed by filtering.

In another type of phase demodulator the phase modulated signal is converted into a square wave by means of a limiter. The same is done with the reference carrier. Figure 12.28 depicts a possible way of operation. Let v_1 be the square wave version of $M(t)$ and v_2 that of the carrier. A digital circuit generates a square wave signal whose width is determined by the time interval between positive transients in v_1 and v_2. The d.c.-component in the signal obtained in this way is proportional to the width of these pulses and hence to the phase difference between v_1 and v_2.

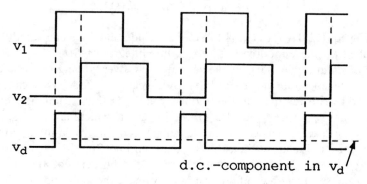

Fig. 12.28 *Principle of the square wave phase demodulator.*

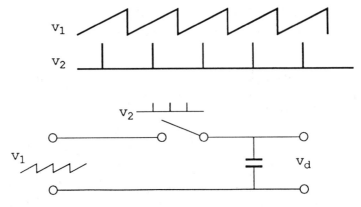

Fig. 12.29 *Phase demodulation by means of a sample-and-hold circuit.*

Figure 12.29 shows the principle of yet another demodulation method. One of the signals is transformed into a sawtooth wave, the other into a switching pulse that drives an electronic switch. The sawtooth signal is fed to a sample-and-hold circuit that is driven by the pulse-type signal. When the switch is closed, the hold capacitor takes the momentary voltage of the sawtooth signal and retains this voltage until the next time the switch closes.

A type of angle modulation that has found very wide application is known as *frequency modulation*. As is suggested by the name, in this case the momentary frequency is modulated by the modulating signal. Hence,

$$\omega = \omega_0 + \Delta\omega \cdot S(t) \qquad (12.31)$$

If $S(t)$ is normalized so that $-1 < S(t) < +1$, then $\Delta\omega$ represents the maximum difference between the carrier frequency ω_0 and the momentary frequency ω. The quantity $\Delta f = \Delta\omega/2\pi$ is called the *frequency deviation* of the FM-signal. Where $\Psi = \int_0^t \omega \, dt$ we can write

$$M(t) = \cos\left[\omega_0 t + \Delta\omega \int_0^t S(t) \, dt\right] \qquad (12.32)$$

Comparing eq. (12.33) with eq. (12.28) we conclude that frequency modulation can be described as phase modulation by the time integral of $S(t)$.

The momentary (angular) frequency can vary over a range with width $2\Delta\omega$. However, the spectral width of $M(t)$ is larger than $2\Delta\omega$. This can be easily appreciated by considering that the frequency varies in the rhythm dictated by $S(t)$. The wider the frequency band of $S(t)$, the larger the widening of the spectrum of $M(t)$. The ratio $\Delta\omega/\mu$, where μ is the (angular) frequency of a modulating signal, is called the 'modulation index'. By representing $S(t)$ by its spectrum and expanding eq. (12.32), the spectrum of $M(t)$ can be derived. It turns out that essentially the spectrum has infinite width. However, the sidebands of higher order rapidly diminish in power. If $\Delta\omega/\mu > 5$, the relevant part of the spectrum of $M(t)$

amounts to about 2.5 $\Delta\omega$. This is called the 'Carson bandwidth', which implies that for distortionless transmission of an audio signal with 15 kHz bandwidth, if $\Delta f = 75$ kHz, a channel bandwidth of about 200 kHz suffices. Obviously, the channel bandwidth needed for FM-transmission is much larger than for AM-transmission. The advantage of FM is that it is less prone to noise and interfering signals. Elaboration of this aspect is not a topic for a book on electronics. A qualitative argument may suffice here. First, it will be clear that all types of disturbances that cause amplitude variations only have no effect whatsoever on the transmission of $S(t)$. With regard to disturbing signals of a purely statistical nature (i.e. noise), it is clear that they will affect the amplitude as well as the location of the zero crossings, and consequently the phase relations. However, if variations of the modulating signal are producing very large phase variations in $M(t)$, then the relative effect of noise-induced phase variations will be small. This is precisely what is achieved by a large-frequency deviation, since the phase variations induced by the modulating signal are proportional to $\Delta\omega$. The disadvantage of a large value of $\Delta\omega$ is, of course, the larger channel bandwidth it requires. This result is entirely in agreement with the findings of information theory, which teaches that for a specified transmission quality the signal-to-noise ratio can be smaller the larger the available bandwidth.

12.10 Circuits for FM-modulation

Having acquired some basic knowledge concerning frequency modulation, we can turn our attention to the electronic circuits for modulation and demodulation. An FM-modulator is an oscillator whose frequency is controlled by the modulating signal. The modulating signal can be applied either as a voltage or as a current. In the former case, the oscillator is labelled a voltage-controlled oscillator (VCO), in the latter a current-controlled oscillator (CCO).

In LC-oscillators the frequency is determined by the resonant frequency of the tuned circuit. By modulating one of the constituting reactive elements by $S(t)$ an FM-modulator is created. A widely used method is the use of a so-called varicap as a controllable reactance. A varicap is a reverse biased pn-diode. The capacitance depends on the magnitude of the bias voltage. In the design of a VCO using a varicap, two things require attention. First, the fact that the relation between voltage and capacitance is non-linear should be taken into account. The exact relation depends on the doping profile of the pn-junction. In an LC-tuned circuit the relation between capacitance and frequency is also non-linear. If a linear modulation characteristic is desired, which is usually the case, these non-linearities have to be carefully balanced. The second point of interest is that the oscillator signal itself is also present at the varicap. This causes distortion of the oscillator signal. The circuit must therefore be arranged such that the amplitude of the oscillator signal on the varicap is small. This can, for instance, be achieved by connecting a fixed capacitance in series with the varicap.

An alternative method uses a circuit that is capable of bringing about a phase shift under the control of the modulating signal. It goes without saying that such a circuit can be used as a phase modulator, but if inserted into an oscillator configuration it can accomplish frequency modulation. A well-known example of such a circuit is the so-called reactance circuit. Figure 12.30(a) shows the principle of this circuit, in which Z_1 and Z_2 constitute a phase shift network. If Z_1 and Z_2 are chosen such that the voltage across Z_2 is shifted over $90°$ with respect to the voltage across the series connection of Z_1 and Z_2, $V_g = \pm jkV_d$ holds, where k represents the amplitude attenuation. The drain current is given by

$$I_d = g_m V_g = \pm jk g_m V_d \tag{12.33}$$

so that

$$Z_d = \frac{V_d}{I_d} = \mp j \frac{1}{kg_m} \tag{12.34}$$

The total impedance present between drain and source is composed as indicated in Figure 12.30(b). The impedance Z_d represents an inductance or a capacitance, depending on the relevant sign. This impedance is controlled by g_m, which is approximately linearly dependent on V_g. If Z_1 and Z_2 are chosen high, their effect on the total impedance, as indicated in Figure 12.30(b), can be neglected. If Z_3 represents a tuned circuit, its resonance frequency can be varied by varying Z_d, and hence by V_g. The frequency of an LC-oscillator in which Z_3 is the tuned circuit can be modulated by using V_g as the modulating signal. The phase-shifting network Z_1–Z_2 can be quite simple. If Z_1 is a resistor and Z_2 a capacitor (Figure 12.30c), holds

$$\frac{V_g}{V_d} = \frac{1}{1 + Rj\omega C}$$

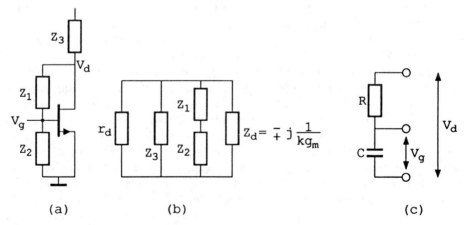

Fig. 12.30 *Reactance circuit:* (a) *configuration;* (b) *impedance between drain and source;* (c) *simple phase-shifting network.*

If $R\omega C \gg 1$, then $V_g/V_d = -j/R\omega C$, so that $Z_d = V_d/I_d = j\omega RC/g_m$. In this case we have an inductive reactance circuit with inductance RC/g_m. If Z_1 is a capacitor and Z_2 a resistor, a capacitive reactance is obtained.

A class of oscillator circuits that is quite suitable for configuration as VCO or CCO is the class of relaxation oscillators. Since these oscillators can be implemented without the use of inductances, they are attractive candidates for monolithic implementation. The operation principle of relaxation oscillators has been dealt with in Chapter 10. For convenience it will be recalled briefly here. A capacitance is charged by a current source. When a predetermined voltage is reached, the charging action stops and a discharging action starts. When a second predetermined voltage is reached the discharging ends, after which the charging action is resumed. In this way a periodic triangular voltage is generated. The amplitude is determined by the difference of both reference voltages and the frequency by the magnitude of the charging and discharging currents, together with the difference between the reference voltages. Figure 12.31 illustrates this operation principle.

The element indicated by ST is a Schmitt-trigger. This is a regenerative circuit that can find itself in two discrete states, dependent on the input voltage. Switching of the state occurs when the voltage exceeds V_h and when the voltage sinks below V_l.

The operation of the relaxation oscillator is as follows. Assume $V_c = V_l$. The capacitor charges according to $V_c = V_l + I_1 t/C$. Let $V_c = V_h$ at $t = \tau_1$. Then $\tau_1 = C\Delta V/I_1$ with $\Delta V = V_h - V_l$ holds. Subsequently C discharges until $V_c = V_l$. The discharging time τ_2 follows from $\tau_2 = C\Delta V/(I_2 - I_1)$. The period time follows from

$$T = \tau_1 + \tau_2 = C\Delta V \left(\frac{1}{I_1} + \frac{1}{I_2 - I_1}\right)$$

If, for instance, $I_2 = 2I_1$, then

$$T = 2C\Delta V/I_1 \text{ and } f = 1/T = I_1/2C\Delta V \tag{12.35}$$

so that the frequency depends linearly on I_1. In passing, note that proportionality between f and I_1 requires that ΔV does not depend on I_1.

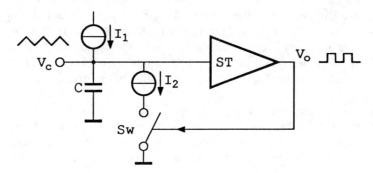

Fig. 12.31 *Principle of the relaxation oscillator.*

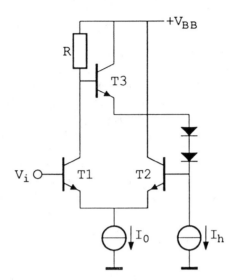

Fig. 12.32 *A simple Schmitt-trigger.*

Figure 12.32 shows a practical implementation of the Schmitt-trigger. Either $T1$ or $T2$ is conducting and then carries the current I_o. If $T2$ is conducting, its base voltage V_{B2} is $V_{BB} - 3V_d$, where V_d denotes the voltage across a forward-biased junction. When V_i reaches this value, $T1$ takes over, hence $V_h = V_{BB} - 3V_d$. In this state $V_{B2} = V_{BB} - I_o R - 3V_d$, if we neglect the effect of base currents. When V_i reaches this value, $T2$ resumes conduction. Hence, $V_l = V_{BB} - I_o R - 3V_d$, so that $\Delta V = I_o R$.

The switch Sw in Figure 12.31 can be conveniently implemented as a long-tailed pair operating as a current switch (Figure 12.33). If $V_4 > V_{BB} - V_{ref}$, then $T4$ carries the current $2I_r$; if $V_4 < V_{BB} - V_{ref}$, then $T5$ is conducting. Choosing V_{ref} such that $V_{BB} - V_{ref}$ is between V_l and V_h, it is assured that $T4$ is conducting when $T2$ is conducting, while $T5$ takes over when $T1$ is in the conducting state. Figure 12.34 shows a simple method for deriving V_{ref}. If $I_h < I_o$, then V_{ref} is with certainty between V_h and V_l. Figure 12.35 shows the complete configuration of the relaxation oscillator according to the model of Figure 12.31. The double pnp-current mirror supplies through $T8$ a charging current I_r and through $T6$ and the npn-current mirror $T9-T10-T11$ the current $2I_r$ for the current switch $T4-T5$. Linear frequency modulation is accomplished by modulation of I_r. The basic circuit of Figure 12.35 can be improved in several respects. For instance, in practice it will often be desirable to compensate for the effects of finite base currents by adding extra transistors.

A type of current-controlled relaxation oscillator that has found widespread application in monolithic and discrete versions is the emitter-coupled multivibrator. In this configuration the switching mechanism and the comparator mechanism are interwoven. The advantage of this approach is the simplicity of the circuit. A

MODULATION, DEMODULATION AND FREQUENCY CONVERSION 399

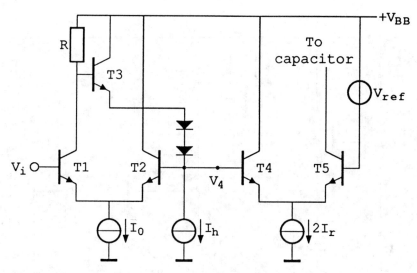

Fig. 12.33 *A Schmitt-trigger with a current switch.*

disadvantage is that the current that is responsible for establishing ΔV and the charging current are linked. If ΔV is proportional to the charging current, according to eq. (12.35) the frequency cannot be modulated by varying this current. A possible way out is the use of non-linear load impedances, providing a voltage drop independent of current. Figure 12.36 depicts an implementation using this idea. The circuit has two metastable states, i.e. states that cannot last for indefinite time. In one state $T1$ is on and $T2$ is off; in the other the reverse holds. Current is supplied by a number of coupled current sources, derived from the primary current I by means of emitter junction scaling. Assume that $T1$ is conducting. The

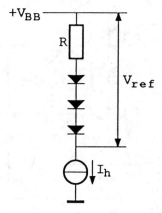

Fig. 12.34 *Formation of V_{ref}.*

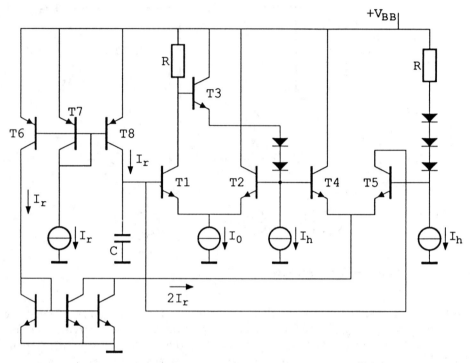

Fig. 12.35 *Example of a relaxation oscillator used as a CCO.*

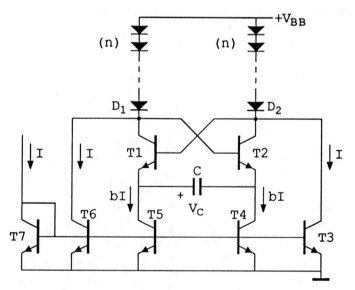

Fig. 12.36 *Emitter-coupled multivibrator as CCO.*

current in $T1$ is then $2bI$, half of which is supplied via the capacitor. The current in the load of $T1$, consisting of a string of n pn-junction diodes $D1$, is then $(2b+1)I$; the current through the string $D2$ is I. Hence, the base voltage of $T1$ is $V_{BB} - nV_T \ln(I/I_s)$, where $V_T = kT/q$. The voltage on the base of $T2$ is $V_{BB} - nV_T \ln[(2b+1)I/I_s]$. The difference between these base voltages is $nV_T \ln(2b+1)$. Due to the charging current bI, the capacitor voltage increases. When the voltage V_C between both emitters equals the voltage between both bases, $T2$ starts to conduct, and by regenerative action a change of state quickly takes place. Due to the reversed current the capacitor starts to discharge, and so on. The capacitor voltage possesses the by now familiar triangular shape and its peak-to-peak value amounts to $V_Z = 2nV_T \ln(2b+1)$. In practice a slightly different value will be found, since we have simplified the calculation by neglecting base currents and other second-order effects. In addition, note that the change of state does not occur exactly upon the attainment of the equality of the emitter–base voltages of both transistors. For the onset of the regenerative action, the loop gain must exceed unity value. In the transient situation both transistors operate in their active region, causing a slight aberration of the triangular waveform.

Figure 12.37 depicts V_C as a function of time. Since the current through C is mI, the time duration of a half-period is $\tau = V_Z C/bI$. Hence, the oscillator frequency is

$$f = \frac{1}{2\tau} = \frac{bI}{4CnV_T \ln(2b+1)} \qquad (12.36)$$

Hence, f is proportional to I so that linear frequency modulation can be accomplished by varying I.

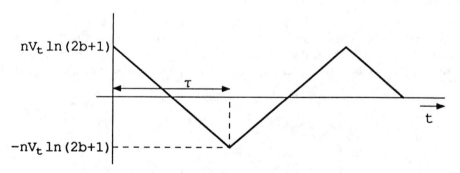

Fig. 12.37 *Capacitor voltage as a function of time.*

EXAMPLE

A CCO is to be designed for the frequency range 800 kHz–1200 kHz. In view of the chip area involved, the value of b should be kept within reasonable limits. An acceptable value is $b = 4$. For the same reason and in view of the demands on the power supply voltage, n should also be kept low. In any event, the minimum value of n is 2 in view of the requirement that loop gain should exceed unity value.

For the sake of keeping dissipation within acceptable limits, a moderate value of the current I must be pursued. Further, C should be well in excess of parasitic capacitances. Taking $n = 3$ and $I = 0.25$ mA, eq. (12.36) yields $f = 1$ MHz with $C = 1.52$ nF. The maximum and minimum values of I amount to 0.3 and 0.2 mA.

Inspection of eq. (12.36) reveals that f is also inversely proportional to V_T, and thereby to T. This feature can be put to good use in a temperature sensor operating as a temperature-to-frequency converter. A frequency is a convenient quantity for transmitting information: it is robust and it can be easily measured. However, if frequency modulation independent of temperature is desired, the temperature dependence must be compensated for. An obvious idea is to make I also linearly dependent on T, implying that I should be PTAT (see Section 6.4). However, the primary current generated by a PTAT-current source cannot be modulated. A controllable PTAT-current can be obtained with the aid of a current amplifier with variable amplification. Figure 12.38 gives an example. The circuit is differentially driven by $I_0 + i$ and $I_0 - i$, where i is proportional to the modulating signal. The base–emitter junctions of transistors $T1$ through $T4$ form a closed loop. Hence, neglecting base currents,

$$\sum_{1-4} V_{BEn} = \sum_{1-4} V_T \ln\left(\frac{I_n}{I_s}\right) \qquad (12.37)$$

Further, $I_2 + I_3 = 2I_T$, where I_T is PTAT. $\qquad (12.38)$

Substituting $I_1 = I_0 + i$ and $I_4 = I_0 - i$, and solving eqs. (12.37) and (12.38) yields

$$I_2 = I_T\left(1 + \frac{i}{I_0}\right), \quad I_3 = I_T\left(1 - \frac{i}{I_0}\right) \qquad (12.39)$$

which shows that I_2 and I_3 have the desired character.

In high-quality FM transmission systems, the carrier frequency must be crystal controlled. A crystal oscillator cannot be frequency-modulated if a substantial frequency deviation must be realized. Several methods have been devised for coping

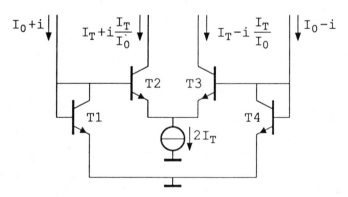

Fig. 12.38 *Formation of a controllable PTAT-current.*

with this difficulty. In one method a low-frequency crystal oscillator signal is phase-modulated with a signal corresponding to $\int S(t)\,dt$. According to eq. (12.32), this is equivalent to frequency modulation by $S(t)$. The modulated signal obtained in this way is fed to a frequency multiplier. The multiplication factor is chosen such that the required frequency deviation is achieved.

EXAMPLE

Let the desired carrier be 95.4 MHz and the frequency deviation $\Delta f = 75$ kHz. A convenient primary crystal-controlled frequency is 200 kHz. By phase modulation with $\int S(t)dt$ a small frequency deviation can be realized, for instance $\Delta f = 9.155$ Hz. To achieve a final frequency deviation of 75 kHz a multiplication factor of $8192 = 2^{13}$ is needed. In order to obtain the desired carrier frequency as well as the desired frequency deviation, the multiplication is done in two steps. Between these the carrier frequency is converted to a lower frequency. A convenient arrangement might be, first, multiplication by 2^7, yielding $f_0 = 25.6$ MHz and $\Delta f = 1.172$ kHz. This signal is mixed with a (crystal controlled) sine wave at 24.11 MHz, yielding an intermediate frequency of 1.49 MHz with $\Delta f = 1.172$ kHz. Frequency multiplication by 2^6 then yields $f_0 = 95.4$ MHz and $\Delta f = 75$ kHz.

12.11 Demodulation of FM-signals

In FM-transmission the amplitude of the modulated signal does not convey any useful information. Amplitude disturbances can therefore be effectively eliminated by limiting the modulated signal $M(t)$ before demodulation. Figure 12.39 shows a very simple limiter circuit. The diodes behave as shorts for voltages exceeding their threshold voltage, so that the output signal is confined between V_t and $-V_t$. A disadvantage is that the input signal must exceed the threshold voltage, which is rather large. A more sensitive signal limiter can be realized by means of an overdriven differential pair (Figure 12.40). As found previously,

$$I_1 - I_2 = 2I_o \tanh \frac{v_i}{2V_T}$$

If $|v_i| > 6V_T$, $|I_1 - I_2| \approx 2I_o$. The sensitivity can be increased by cascading several differential stages. In the case of weak input signals, the first stages operate as amplifiers. The input signal of the last stage should exceed $6V_T \approx 150$ mV.

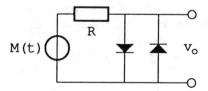
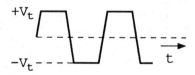

Fig. 12.39 *A simple limiter circuit.*

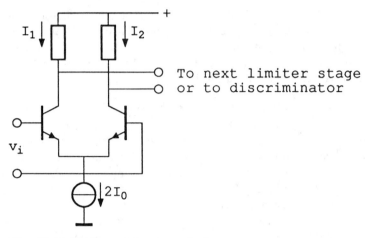

Fig. 12.40 *Differential pair as limiter.*

The circuits used for the demodulation as such can be grouped into four classes as follows:

(a) discriminator circuits with diodes;
(b) φ-detectors;
(c) counting demodulators;
(d) phaselock-loop demodulators.

Discriminator circuits with diodes
In this type of demodulator the frequency variations are first converted into phase variations, which are subsequently converted into amplitude variations. The latter are demodulated with simple rectifier circuits. At first sight this seems to be a cumbersome approach. However, it turns out that this method can yield quite good results. In the past this type of demodulator has found widespread application. It is not often used now since it is incompatible with monolithic technology. Figure 12.41 shows an example of this type of demodulator. Both resonant circuits are tuned at the carrier frequency. They are weakly inductively coupled. If two resonant circuits are weakly coupled, the secondary voltage is 90° phase-shifted with respect to the primary voltage. The coupling capacitor C_c is chosen so large that it represents a short circuit at the resonant frequency. The voltages V_{s1} and V_{s2} are mutually in antiphase, while both are in quadrature with respect to V_p. Both diodes are part of a rectifying circuit, as indicated in Figure 12.42. They act as peak detectors for the signals $V_p + V_{s1}$ and $V_p + V_{s2}$ respectively.

When the momentary frequency equals the carrier frequency, the output signal $|V_a| - |V_b|$ has zero value (Figure 12.43a). When the momentary frequency deviates from the resonant frequency of the tuned circuits, V_s is no longer in quadrature with V_p (Figure 12.43b and c). The deviation from the quadrature

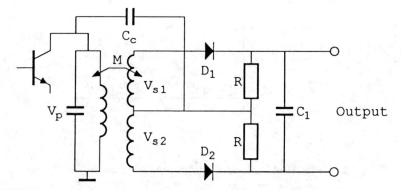

Fig. 12.41 *Foster–Seely discriminator for FM-demodulation.*

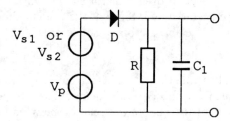

Fig. 12.42 *Rectifier circuit of one diode.*

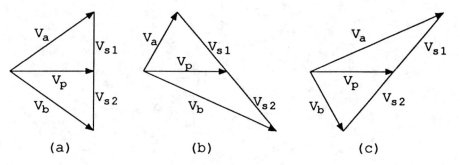

Fig. 12.43 *Phasor diagram showing the formation of the output signal* $|V_a| - |V_b|$; (a) $f = f_0$; (b) $f > f_0$; (c) $f < f_0$.

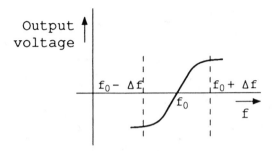

Fig. 12.44 *Static discriminator characteristic.*

relation is larger, the more the applied frequency deviates from the resonant frequency. Hence, the rectified voltage $|V_a| - |V_b|$ depends on the momentary frequency. It appears to be possible to choose the quality factors of the tuned circuits such that the relation between the output voltage of the circuit and the deviation from the carrier frequency is approximately linear. Figure 12.44 depicts the characteristic of this type of discriminator circuit. Various modifications of this prototype circuit have been devised. Since at the present time these circuits are of minor interest we will not elaborate on these variants.

φ-detectors

A φ-detector, also called quadrature detector, converts frequency variations into phase variations, which are subsequently demodulated by a phase demodulator. Once again use is made of the quadrature relation between the voltages on weakly coupled resonant circuits. Figure 12.45 gives an example wherein, for the sake of variation, capacitive coupling is applied. At resonance the impedance of the

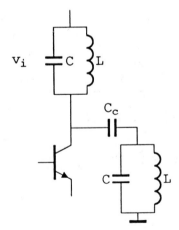

Fig. 12.45 *Conversion of frequency variations into equivalent phase variations.*

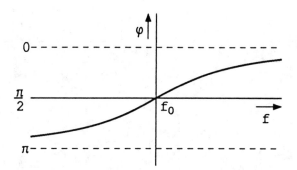

Fig. 12.46 *Phase characteristic of V_o/V_i.*

secondary circuit is $R = Q/\omega_0 C$. If $C_c \ll C/Q$, implying weak coupling, then the current in the secondary circuit is approximately determined by the capacitive coupling through C_c, so that V_o is in quadrature with V_i. When the applied frequency deviates from the carrier frequency, the character of the resonant circuit becomes capacitive ($f > f_0$) or inductive ($f < f_0$). The phase difference between V_o and V_i depends on frequency, as indicated in Figure 12.46. Within a certain region around $f = f_0$ proportionality between φ and $f - f_0$ holds. The demodulation of the phase-modulated signal obtained in this way is performed with a phase demodulator, usually implemented as a simple multiplying circuit with V_o and V_i as input signals.

Counting demodulators
In a frequency-modulated signal the momentary frequency is a direct representation of the modulating signal $S(t)$. Therefore, the most direct demodulation method is determination of the momentary frequency, which involves determination of the number of zero crossings per given time interval. This is exactly what a counting demodulator does. The central element is a circuit that forms a pulse with fixed width at each zero crossing. The average value of this pulse train over a given time interval is proportional to the momentary frequency of the FM-signal, and therefore also to $S(t)$.

The principle of this modulator is simple and direct. No conversion of FM into PM or AM is involved. However, the electronic implementation of an accurate counting demodulator is not so simple. As in any FM-demodulation arrangement, $M(t)$ is first stripped of any amplitude variations with the aid of a limiter. By using hard limiting, $M(t)$ is converted into a square-wave signal (Figure 12.47a). By differentiating this signal, the positive pulse edges are extracted (Figure 12.47b). The signal obtained in this way is a representation of the locations of the zero crossings associated with the positive edges of the squared version of $M(t)$. Subsequently, these edge signals are used to form pulses of standard width. By passing this signal through a low-pass filter the modulating signal $S(t)$ is reconstructed.

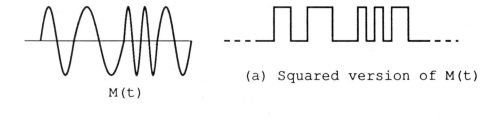

(a) Squared version of M(t)

M(t)

(b) Positive edge signal

Fig. 12.47(a) *Squared version of M(t) obtained by hard limiting.* (b) *Positive edge signal derived from the squared version of M(t).*

The weak point in this process is the formation of the standard-width pulses. The best-known method for this is the use of a monostable multivibrator. Figure 12.48 shows a simple implementation. The circuit has one stable state and one metastable state. In the stable state $T2$ is conducting. When $T1$ is off, the voltage on the base of $T2$ is $V_{B2} = V_{BB} - I_0 R_2/\beta_F$, where β_F is the d.c. current gain I_C/I_B of $T2$. As long as $V_1 < V_{B2}$, $T1$ remains in the off-state. If a short trigger pulse is fed to the base of $T1$, this transistor becomes momentarily conducting. Due to the presence of positive feedback through C and the emitter coupling of $T1$ and $T2$, a regenerative action starts, driving $T1$ into full conduction, irrespective of whether the trigger pulse is taken away or not. In the resulting state $T1$ conducts the full current I_0 and $T2$ is off. However, this state is metastable, i.e. it cannot persist indefinitely. The capacitor starts to discharge through $T1$. Consequently, the voltage drop across R_2 decreases. As long as $V_{BB} - i_c R_2 < V_1$, $T2$ remains in the off-state. When i_c has decreased so far that $V_{B2} = V_1$, $T2$ takes over. This state is stable and it can only be interrupted by a new trigger pulse on $T1$. If the stable state lasts long enough, then the capacitor voltage V_C reaches its steady-state value $V_C = V_{BB} - I_0 R_2/\beta_F$. However, upon the onset of the stable state, the capacitor must adapt its charge to this state. In the circuit according to Figure 12.48, this can only be accomplished through the base current of $T2$. As a consequence, during the transient time the trigger voltage that is needed for activating $T1$ differs slightly from the value needed in the fully established stable state. And, even more deleterious, the duration of the metastable state, and henceforth the width of the output pulse, is erroneous. This implies inaccuracy of the demodulation.

The severity of the errors due to the occurrence of a 'dead time', as described above, depends, of course, on the frequency of the trigger signal. If the time

MODULATION, DEMODULATION AND FREQUENCY CONVERSION 409

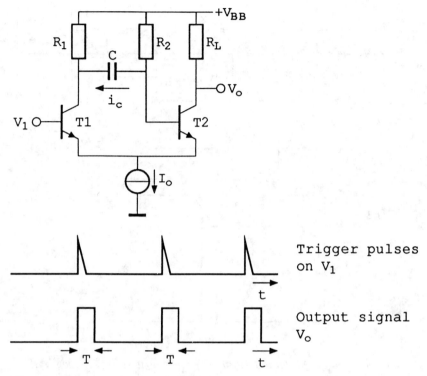

Fig. 12.48 *Monostable multivibrator.*

interval between subsequent trigger pulses is large in comparison with the 'dead time', no problems are encountered.

At high frequencies the use of a monostable circuit is hardly feasible. For high-frequency implementations an attractive alternative is available. Figure 12.49(a) indicates the principle of the so-called delay-line FM-demodulator. The input signal is fed to a mixer, together with a delayed and inverted version of this signal. The mixer responds essentially to the coincidence of the two signals. Figure 12.49(b) depicts the operation in the idealized case of exact square-wave signals. The width of the output pulses obtained is solely determined by the delay time τ_d. Essentially, the mixer operates as an AND-gate. Practice has shown that on the basis of this principle very accurate FM demodulators can be made, even for frequencies of the order of several hundreds of MHz. It goes without saying that to achieve high performance, very careful design is compulsory. In practice balanced versions will always be used. The waveforms need not be ideal square waves, but precise matching is of the essence. The main problem is in mastering the effects of parasitic capacitances. In view of this aspect, monolithic implementation is very attractive because of the predictability of circuit behaviour in a well-modelled fabrication process.

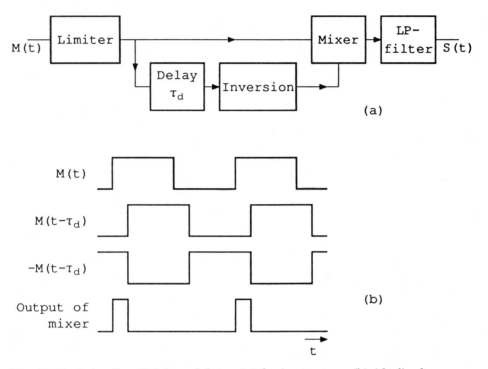

Fig. 12.49 *Delay line FM-demodulator:* (a) *basic structure;* (b) *idealized waveforms.*

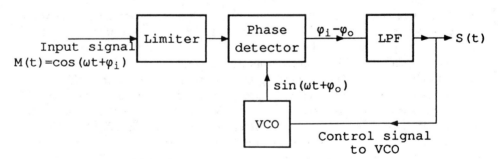

Fig. 12.50 *Basic structure of a phaselock loop.*

PLL-demodulator
A phaselock loop (PLL) is a circuit configuration in which a local oscillator is brought into synchronism with an externally applied signal. Figure 12.50 shows the block diagram of a PLL. A phase detector produces a signal that is proportional to the difference of the arguments of the local oscillator signal and the external signal. This signal is low-pass filtered and is subsequently used to control the frequency of the VCO. The PLL has found widespread application in electronic engineering (a.o. telemetry, television, robotics, mechatronics). The PLL can also be used for frequency demodulation. Any change in frequency of the input signal gives rise to a change in the control signal for the VCO. Obviously, this control signal is a measure of the momentary frequency of the input signal, and consequently of the modulating signal. An analysis of PLL behaviour is beyond the scope of this book. The reader is referred to the specialized textbooks covering the theory and the design of PLLs (Gardner 1979, and Blanchard 1976).

References

Blanchard, A. (1976) *Phase-locked Loops: Application to coherent receiver design*, Wiley, New York.
Eikenbroek, J. W. Th. (1989) Development of an Integrated AM Shortwave Upconversion Receiver Front-end, PhD thesis, Delft University of Technology, Delft.
Gardner, F. M. (1979) *Phaselock Techniques*, Wiley, New York.
Nauta, H. C. (1986) Fundamental Aspects and Design of Monolithically Integrated AM Radio Receivers, PhD thesis, Delft University of Technology, Delft.

Further reading

Several books exist that deal with the theoretical aspects of modulation and demodulation. A well-known specimen is:

Schwarz, M. (1980) *Information Transmission, Modulation and Noise*, McGraw-Hill, New York.

Index

active devices, 9
active filters, 352
active loads, 181
   in MOS-amplifier, 186
   in NMOS, 189
active region of BJT, 18
actuator, 1
additive frequency conversion, 379
aliasing of spectral bands, 358
all-pole function, 275
   of second order, 280
amplifier, 84
   bandpass, 331
   distributed, 291
   for TV-camera, 285
   logarithmic, 209
   low-pass, 84
   multistage, 273
   operational, 190
   operational transconductance, 190
   power, 210
   split-band, 292
   wideband, 266
amplitude response,
   asymptotic representation, 73
   relation with poles and zeros, 71
amplitude modulation, 369
amplitude stabilization in oscillator, 306
AM-radio, 331
analogue signal, 2
   processing, 3
analogue-to-digital (A/D) conversion, 5
angle modulation, 392
angular frequency, 6
anode, 10
a-parameters, 99, 242
   for transformation of noise sources, 242
apodized transducers, 333
architecture,
   of down conversion receiver, 373
   of OPAMP, 190

   of receivers, 371
   of upconversion receiver, 374
ASIC, 65
asymptotic gain model, 101
autocorrelation, 231
   function, 232
automatic gain control,
   in oscillator, 307
   in receiver, 371
   in two-integrator oscillator, 325
autotransformer, 345
auto-zero method of offset compensation, 193
available input power, 246
available noise input power, 233, 246
available power gain, 246
avalanche breakdown, 15
   in transistor, 27
   voltage as reference, 171

back gate, 47
   loading effect of, 189
   transconductance of, 47
balanced output stage, 217
balancing,
   for linearization of conductance, 354
   in mixer, 380
bandgap, 10, 12
   as reference in CMOS, 180
   as reference voltage, 175
   temperature dependence of reference, 176
bandpass amplifier, 331
   with coupled resonators, 347
   multistage with MFM-response, 336
bandwidth narrowing, 267
Barkhausen criterion, 301
base current, 19, 22
base efficiency, 23
base electrode, 10, 18
base resistance, 25
   effect on noise figure, 251
   lowering of, 26

basewidth, 22
  modulation, 23
basic amplifier configurations, 116
biasing, 142
  circuits, 143
  of CE-stage, 146
  of depletion-FET, 146
  of series stage, 146
  of shunt stage, 146
  point, 16
  requirements for, 142
  self-biasing in LC-oscillator, 304
  temperature dependence, 146
  through feedback, 149
BICMOS-technology, 64
binary representation, 2
bipolar transistor (BJT), 18
  active region, 18
  base current, 19
  CE-configuration, 33
  characteristics, 20
  extrinsic, 31
  hybrid-π model, 32
  in CMOS-process, 179
  injection model, 20
  in power amplifier, 211, 216
  intrinsic, 31, 243
  inverse region, 18
  lateral pnp, 56
  monolithic, 54
  noise in, 237
  reciprocity relation, 20
  saturation voltage, 22
  sign conventions, 19
  static behaviour, 21
  substrate pnp, 56
biquad structure, 357
Blakesley transform, 240
Bode plot, 73
body effect, 46
bouncing, 307
breakdown voltage, 15
bridged-T network, 283, 343
built-in field in MOST, 40
buried layer, 55
Butterworth transfer function, 275
bypass-capacitor, 146, 152

camera preamplifier, 285
capacitance coupling of resonators, 338
  pole-zero plot of, 341
capacitive feedback,
  for bandwidth improvement, 271
capacitive signal source noise matching, 252
capacitive tapping, 345
capacitors in monolithic technology, 58
capacity of communication channel, 4
carrier, 368, 392

carrier flow,
  by diffusion, 11
  by electric field, 12
carrier-to-noise ratio, 296
Carson bandwidth, 395
cascading,
  of CS stages and source followers, 289
  of noisy twoports, 247
cascode circuit, 128, 135
  as input stage in integrated amplifier, 183
  MOS-differential pair, 187
cathode, 10
CB-stage, 118, 120, 129
  noise in, 257
  as buffer in wideband amplifiers, 284
CC-stage, see emitter follower
CE-configuration, 30, 33, 116
  as power stage, 214, 221
  current driven, 34, 123
  dominant pole in, 122
  high-frequency behaviour, 120
  transfer parameters, 117
chain parameters, 99
channel,
  communication, 4
  conductance of MOST, 353
  in FET, 35
  length modulation, 42
  selectivity, 372
  stop, 62
characteristic impedance, 88
  amplifier, 95, 254
  in distributed amplifier, 293
characteristics,
  BJT, 20
  discriminator, 406
  FET, 38
  MOST, 41
charge-coupled devices, 361
  charge transport, 362
  noise, 363
charge traps in MOST, 40
Chebyshev response, 273
chip, 51
classes of operation, 217
  class-A, 217
  class-AB, 220
  class-B, 218
    average current in, 219
    efficiency, 219
collector, 10, 18
  dissipation, 143
  maximum junction voltage, 28
Colpitts oscillator, 302, 306, 319
common drain stage, see source follower,
common-gate (CG) circuit, 131
common-mode signal, 155
  external, 158

INDEX 415

common-mode signal (*continued*)
  half circuit, 158
  rejection ratio, 157
compensation methods, 113
  by balancing, 155
  in doping, 53
complementary bipolar process, 56
complementary transistors in output stage, 222
complex frequency, 69
compound active components, 139
compound pnp-transistor, 141, 225
conduction band, 10
controlled transconductance mixer, 381
  with MOSTs, 382
conversion of frequency, 372
  additive, 379
  gain, 385
  methods, 379
  multiplicative, 381
  temperature-to-frequency, 402
conversion of series damping, 297
correlated noise sources, 231
correlation, 231
counting FM-demodulator, 407
coupled resonant circuits, 337
  amplitude and phase response, 342
  capacitive, 338
  inductive, 337, 338
  pole-zero plot, 340
  response far from resonance, 341
coupling factor, 337
  critical, 340
  frequency-dependent, 341
  overcritical, 340, 341
critical coupling factor, 340
cross-modulation, 377
crossover distortion, 220
crystal oscillator, 317
CS-configuration, 45, 124
  cascaded with source follower, 289
  non-unilateral, 126
current amplifier, 92, 96
  with phantom zero, 280
current-controlled oscillator, 395
current driving in power amplifier, 214
current follower, 118, 120
current gain factor,
  dependency on I_C, 213
current mirror, 161
  as active load, 181
  cascoded, 165
  cascoded MOST, 167
  current transforming, 165
  MOST, 165
  noise in, 262
  pole-zero plot of, 168
  signal transfer in, 167
  with external supply of base current, 164

  Wilson type, 162
current source, 159
  cascoded, 161
  noise in, 261
  one-transistor, 160
  with depletion MOST, 188

Darlington pair, 140
  as emitter follower, 141
  in output stage, 225
delay line demodulator, 409
delay time, 79
demodulation,
  AM-signals, 391
  FM-signals, 403
  phase, 393
  synchronous, 392
demodulator,
  counting, 407
  delay-line, 409
  envelope, 391
  FM-discriminator, 404
  phase, 393
  phi-type, 406
  sample-and-hold, 394
depletion layer, 17
  in FET, 37
depletion load, 189
detailed balance, principle of, 13
detuning, 297
deviation of frequency, 394
differential-mode signal, 155
  half circuit, 157
differential operator,
  s as, 70
differential pair, 156
  as limiter, 404
  cascoded, 187
  MOST, 187
  NMOST, 188
  noise in, 259
diffusion,
  buried-n, 55
  capacitance, 28
  constant, 12
  deep-p, 55
  isolation, 55
  length, 14
  of carriers, 11
  technology, 51
  shallow-n, 55
  shallow-p, 55
digital filter, 359
digital signal, 2
discrimination factor, 157
discriminator, 404
dissipation in collector, 143
  in power transistor, 210

distortion, 88
  crossover, 220
  harmonic, 89
  in feedback amplifier, 111
  reduction by error feedforward, 112
  reduction in MOS-channel, 354
distributed amplifier, 291
dominant pole in CE-configuration, 15
doping, 11
  by compensation, 53
  levels in BJT, 24
  profile by diffusion, 52
  reverse profile, 53
double-balanced mixer, 387
double-sideband modulator, 370
down-conversion, 373
drain electrode, 10, 35
drift effects, 149, 152, 154
droop, 79
dual-loop feedback, 96
dynamic range, 4, 88
  in filter networks, 352
  intermodulation-free, 89
  in receivers, 378
  of medium-wave receiver, 372
  of SC-filter, 366

Early-effect, 23
Early-voltage, 24
  in current mirror, 162
  in FET, 38
  in MOST, 42
Ebers-Moll model of BJT, 22
efficiency,
  of class-B stage, 219
  of power amplifier, 211
electrocardiography, 158
electron, specific charge of, 2, 12
electron tube, 10
  in power amplifier, 216
electronic components, 9
emitter, 10, 18
  compensation, 137
  crowding, 27, 213
  efficiency, 23
  scaling, 163
emitter-coupled multivibrator, 398
emitter follower, 118, 132
  as power stage, 215
  complex poles in transfer, 134
  Darlington pair as, 141
  noise in, 257
  with capacitive load, 134
enhancement MOST, 40
envelope demodulator, 391
epitaxial growth, 54
equi-ripple function, 273

equivalent circuit,
  of BJT, 31
  of FET, 44
  of noise sources in BJT, 238
  of noise sources in FET, 239
equivalent noise resistance, 244
error-feedforward, 112
excess noise, 236
excess noise figure, 246
extrinsic transistor, 32

feedback, 91
  capacitance, 125
  capacitive for bandwidth improvement, 217
  local, 111, 118, 289
  noise in feedback amplifiers, 253
  overall, 111
  poles in feedback system, 81
  positive, 112, 301
  series-shunt, 108
  suppression of spurious signals, 111
  two-stage voltage amplifier, 107
field-effect transistor, 35
  capacitances in, 45
  comparison with BJT, 44, 46
  depletion type, 40
  enhancement type, 40
  MOS, 39
  normally-off type, 40
  normally-on type, 40
  saturation region, 39, 41
  small-signal model, 43
  thin film, 36
  threshold voltage, 40
  triode region, 39, 41, 353
filter networks, 332
  digital, 359
  on-chip, 351
  recursive, 361
  surface-acoustic wave, 332
  switched-capacitor, 363
  transversal, 359
  with charge-coupled devices, 361
first-order oscillator, 321
flicker noise, 236
floating ports, 94
FM-modulator, 395
FM-radio, 331
1/f-noise, 236
  in BJT, 238
  in FET, 239
folding of spectral bands, 358
forward-biased junction voltage reference, 174
Foster-Seely discriminator, 405
Fourier transform, 5, 232
frequency,
  carrier, 368
  image, 373

frequency (*continued*)
  intermediate, 373
  momentary, 392
frequency conversion, 372, 379
  additive, 379
  multiplicative, 381
frequency deviation, 394
frequency domain, 5
frequency modulation, 394
Friis, formula of, 247
full-power bandwidth, 198

gain-bandwidth product (GB), 267
  of cascaded stages, 267
  of CB-stage, 130
  of CE-stage, 123
  of CG-stage, 132
gain margin, 195
gamma correction, 87, 114
gate electrode, 10, 35
generation of electron-hole pairs, 11
gradual channel approximation, 38
grid electrode, 10
Gummel number, 22
gyrator as feedback element, 93, 254

harmonic distortion, 89
harmonic oscillator, 296, 302
  high performance, 314
  with series resonator, 316
heterodyne receiver, 372
high-level injection, 27, 213
Hilbert transform, 389
homodyne receiver, 374

image frequency, 373
  rejection of, 390
incremental model, *see* small-signal parameters
inductive coupling of resonators, 337, 338
inductive shunt peaking, 268
inductive tapping, 346
information processing system, 1
injection model of bipolar transistor, 20
input impedance of feedback amplifiers, 104
instability due to positive feedback, 112
integral noise factor, 247
integrating amplifier, 203, 354, 356, 364
interdigital transducers, 332
  apodized, 333
interface, 1
  front, 1
  rear, 1
intermediate frequency, 373
intermodulation, 88
  in receivers, 376
intermodulation-free dynamic range, 89, 377, 386

internal resistance,
  of BJT, 30
  of FET, 44
interstage coupling with resonators, 334
interstage level shifts, 149
intrinsic electron concentration, 10
intrinsic hole concentration, 10
intrinsic semiconductor, 10
intrinsic transistor, 31, 243
inverse region of bipolar transistor, 18
inversion, 40
  moderate, 49
  weak, 48
inversion layer, 40
inverting amplifier, 199
  non-ideal, 201
ion channeling, 53
ion implantation, 53
isolation diffusion, 55

junction breakdown, 15
junction-FET, 35
  depletion layer in, 37
  in bipolar process, 58
  pinch-off in, 37
  planar, 36

knee-voltage, 210

Laplace operator, 70
Laplace transform, 75
lateral pnp-transistor, 156
level shift, 149
  in input stage of integrated amplifier, 183
  in MOST output stage, 226
  in output amplifier, 224
  with capacitor, 150
  with pnp-transistor, 150
  with Zener diode, 150
lifetime of minority carriers, 11, 14
limiting,
  in FM-receiver, 403
  in oscillator, 307
limiting amplifier, 313
limiting transresistance, 316
linear circuits, 68
linearization of MOS-channel conductance, 355
lithium niobate, 332
lithographic process, 51
local feedback, 11, 118, 289
local oscillator, 374
LOCOS (local oxidation of silicon), 61
logarithmic amplifier, 209
long-tailed pair, 128, 134
long-term stability of oscillator, 295
loop-gain, 103
  calculation in two-stage amplifier, 108

418   INDEX

loss compensation in oscillator, 299
   by negative resistance, 301, 328
lossy integrator, 364
low-pass-to-bandpass transformation, 334

majority carriers, 11
mass of electron, 12
master-slave method for filter adjustment, 353
maximally flat magnitude response, 273
   all-pole, 275
   bandpass, 336
   second-order all-pole, 280
medium-wave receiver, 372
metal deposition, 54
Miller capacitance, 126
minimum phase system, 71
minority carriers, 11
   lifetime of, 11
mixer, 372
   controlled conductance, 381
   double-balanced, 387
   noise in, 385
   single-balanced, 387
mixing, reciprocal, 387
mobility of charge carriers, 12
moderate inversion, 49
modulation, 368
   amplitude, 369
   angle, 392
   cross-, 377
   double-sideband, 370
   frequency, 394
   methods for, 379
   phase, 393
   single-sideband, 389
   suppressed-carrier, 370
modulation depth, 369
modulation index, 393
modulator,
   FM, 395
   ring, 388
   single-sideband, 389
   switching, 383
momentary frequency, 392
monolithic technology, 51
   capacitors in, 58
   diodes in, 59
   resistors in, 56
   transistors, 54
monostable multivibrator, 409
MOS-current mirror, 166
   signal transfer in, 168
MOS reference source, 178
MOS-technology, 61
MOS-transistor, 39
   built-in field in, 40
   channel conductance in triode region, 353
   as controlled resistor, 354

planar, 39
   in power amplifier, 212, 216, 226
   subthreshold behaviour, 48
   see also field-effect transistor,
moving coil cartridge, 86
multiple bias currents, formation of, 264
multiple poles, 76
multiplication of signals, 371
multistage amplifier, 273
multivibrator,
   emitter-coupled, 398
   monostable, 409

negative resistance, 301, 308, 310
   realization of, 311, 328
neutralization, 350
neutrodynization, 350
NMOS-amplifier, 188
noise, 3, 230
   bandwidth, 235
   current, 236
   due to base resistance, 251
   excess, 236
   excess noise figure, 246
   factor, 246
   figure, 246, 310
   flicker, 236
   floor in oscillator, 296, 310
   in bipolar transistor, 237, 244
   in CCD, 363
   in current mirrors, 262
   in differential pair, 259
   induced, 236
   in feedback amplifier, 253
   in FET, 238, 245
   in mixer, 385
   in oscillator, 309
   in pn-junction, 237
   in radio receiver, 372
   in RC-network, 234
   matching, 249
   matching with capacitive signal source, 252
   matching in characteristic impedance
      amplifier, 254
   $1/f$, 236
   partition, 236
   physical aspects, 233
   shot, 235
   thermal, 233
   threshold, 88, 378
noisy twoport, 249
   cascading of, 247
non-energic elements, 93
non-ideal inverting amplifier, 201
non-inverting amplifier, 204
non-linearity in receiver, 375
non-minimum phase system, 70

normally-off FET, 40
normally-on FET, 40
Norton representation of signal source, 85
notch-filter, 344
n-type silicon, 11
nullor, 92, 101
n-well process, 63

off-region in bipolar transistors, 18
offset, 191
   autozero correction, 193
   compensation, 192
   compensation through auto-zeroing, 193
   effects in OPAMP, 206
   static and dynamic, 191
on-chip filters, 351
operating point, 142
operational amplifier (OPAMP), 190
   as current-to-voltage converter, 200
   as integrating amplifier, 203
   as inverting amplifier, 200
   as non-inverting amplifier, 204
   as summing amplifier, 202
   dedicated, 198
   in CMOS, 199
   phase compensation in, 196
   slew rate, 195
operational transconductance amplifier, 190
order of linear system, 71
oscillations, conditions for, 300
oscillators, 295
   amplitude stabilization, 306
   Colpitts, 302, 306, 319
   crystal, 317
   current-controlled, 395
   first-order, 321
   harmonic, 296, 302
   local, 374
   noise figure of, 310
   noise sources in, 309
   one-pin, 321
   phase shift, 305
   relaxation, 321, 397
   stabilized by controlled limiting, 307
   two-integrator, 323
   voltage-controlled, 395
   Wien-bridge, 304
   with controlled limiting, 307
   with series resonator, 316
   with transformer coupling, 303
output characteristics of transistor, 142
output impedance of feedback amplifier, 104
output stages, *see* power amplifiers,
overall-feedback, 111
overcritical coupling, 340, 341
overshoot, 79
   in shunt-peaked circuit, 269

parallel LC-resonator, 296
parasitic transistors, 60
partition noise, 236
   in BJT, 236
passive devices, 9
phantom zero, 277
   complex, 283
   in current amplifier, 280
   in input circuit, 282
   in voltage amplifier, 280
phase compensation in OPAMP, 196
phase demodulator, 393
phaselock loop, 411
   as frequency demodulator, 411
phase modulation, 393
   index, 393
phase response,
   relation with poles and zeros, 71
phase shift oscillator, 305
phi-detector, 406
photoresist, 51
piezoelectric effect, 318, 332
pilot signal, 371
pinch-off voltage, 37
pn-junction, 13
   capacitance, 17, 31
   characteristic, 14
   noise in, 237
   potential barrier in, 13
   space charge in, 13
Poisson distribution, 235
Poisson equation, 13
poles, 68
   in feedback system, 81
   multiple, 76
   of biquad structure, 357
   of cascoded MOS-current mirror, 170
   of current mirror, 168
   of multistage wideband amplifier, 273
   of resonant circuit, 299
   of shunt-peaked circuit, 269
   transformation low-pass to bandpass, 335
pole splitting, 128, 196, 277, 279
pole-zero plot, 70
positive feedback, 112
   in oscillator, 301
power amplifiers, 210
   complementary, 222
   current capability, 225
   current driving in, 214
   MOSTs in, 212, 227
power density in spectrum, 231
power efficiency, 212
power matching, 211
power supply rejection ratio (PSRR), 354
power transistor, dissipation in, 210
processing of information, 1
PTAT current, controllable, 402

PTAT-voltage, 173, 402
  by MOSTs in subthreshold, 180
p-type silicon, 11
punch-through, 28
p-well process, 63

quadrature signal, 389
  in two-integrator oscillator, 324
quality factor of resonant circuit, 297, 332
  of quartz-crystal, 318
quartz-crystal, 318
  as series resonant circuit, 320
  in filters, 332
quiescent point, 16

reactance circuit, 396
  capacitive, 397
  inductive, 397
receiver, medium wave, 372
receiver architectures, 371
  down conversion, 373
  heterodyne, 373
  homodyne, 374
  tuned radio frequency, 371
  upconversion, 374
reciprocal mixing, 387
reciprocity relation in bipolar transistor, 20
recombination, 11
  in emitter-base junction, 23, 27
recursive filter, 361
reference current source, 170, 177
reference signal, 371
reference variable, 103
reference voltage source, 170
  avalanche breakdown voltage, 171
  bandgap voltage, 175
  MOS, 178
  thermal voltage, 172
  V_{BE} 174
relaxation oscillator, 321, 397
residue, 76
resistance, realization of negative, 311, 328
resistive broadbanding, 284
resistivity, 57
resistors,
  bulk, 58
  monolithic, 56
  noise model, 233
  temperature coefficient, 57
resonators, 297, 318
  coupled, 337
  in interstage coupling, 334
  LC, 296
  quartz, 318
  tuning fork, 318
ringing, 77, 79
  in shunt-peaked circuit, 269
ring modulator, 388

rise time, 79
  relation with bandwidth, 79
  in switching mixer, 386
  of shunt-peaked amplifier, 270
root loci in second-order system, 276
root locus method, 81
root mean square (rms), 231

safe operating area, 142, 210
sag, 79
sample-and-hold circuit, 394
sampled-data signal, 2, 358
  spectrum of, 358
saturation current in pn-junction, 14
saturation region,
  in bipolar transistor, 18
  in FET, 39, 42
saturation voltage of bipolar transistor, 22
Schmitt-trigger, 397
second-order loop, 276
selectivity of receiver, 372
self biasing,
  in LC-oscillator, 304
  in reference source, 172
semiconductors, 10
semicustom technology, 65
sensitivity of parameters, 90
sensor, 1, 85
series damping, conversion, 297, 346
series resonant circuit, 298
series-stage, 118, 136
  biasing, 144
  cascoded, 138
  emitter compensation in, 137
  in wideband amplifiers, 285, 291
shape factor of FET, 42
sheet resistance, 57
short-term stability of oscillator, 295
shot noise, 225
shunt peaking, 268
shunt stage, 118, 138
  biasing, 144
  in wideband amplifiers, 291
sideband, 369
  due to noise in oscillators, 296
sidewall formation, 53
signal-to-noise ratio, 3, 246
sign conventions, 6
silicon, 11
  breadboard, 66
  nitride, 39, 61
  oxide, 39, 51
single-balanced mixer, 387
single-sideband modulator, 389
slew rate, 195
  modelling of, 196
small-signal model,
  for BJT, 32

small-signal model (*continued*)
    for FET, 43
small-signal parameters,
    conductance of pn-junction, 17
    of bipolar transistor, 29
    of FET, 43
solid solubility of dopants, 52
source electrode, 10, 35
source follower, 133
    noise in, 257
source peaking, 272
space charge in pn-junction, 13
specific resistivity, 57
specific transconductance, 49
spectral distribution of oscillator power, 296
spectral domain, 6
spectral power density, 231, 232
spectral purity of oscillator signal, 295
spectrum,
    of AM-signal, 370
    of local oscillator signal, 386
    of switching modulator, 385
split-band amplifier, 292
spot noise figure, 247
stability of first-order oscillator, 323
stabilization of oscillator amplitude, 307
standard deviation, 33
step function, 74
step response, *see* transient response
stereo information, 371
strong inversion, 48
substrate doping in MOST, 46
substrate pnp-transistor, 56, 224
subthreshold behaviour of MOST, 48
    in formation of PTAT-current, 180
subtracting amplifier, 205
summing amplifier, 202
sum-of-moduli amplitude detector, 327
superfet, 141
suppressed-carrier modulation, 370
suppression filter, 344
surface-acoustic waves, 332
surface mobility, 42
switched-capacitor filters, 365
    dynamic range of, 366
switching function, 384
switching modulator, 383
    with MOSTs, 388
symbols in diagrams, 6
synchronous demodulation, 392

tapped inductance, 345
tapping, capacitive, 346
Thévenin representation of signal source, 85
thermal feedback, 149
thermal noise, 233
    in FET-channel, 239

thermal voltage, 16
    as reference, 171, 172
thin film transistor, 36
threshold, noise, 88
threshold voltage, 40
    as reference, 179
tracking, frequency, 373
transadmittance amplifier, 92, 96
    noise in, 256
transconductance,
    of back gate, 47
    of bipolar transistor, 30
    of FET, 43
    of mixer, 381
    specific, 49
transducer, 1, 85
    input, 1
    output, 1
    piezoelectric, 86
transfer function of linear system, 69
transformation of low-pass to bandpass, 334
transformer,
    as coupling element in oscillator, 303
    as feedback element, 93, 99
    in noise matching, 250
    in power matching, 212
    shifting of noise sources in, 241
transient response, 74
    calculation by Laplace transform, 75
    of circuit with source peaking, 272
    of resonant circuit, 78
transimpedance amplifier, 92, 96, 105, 138
    noise in, 254
transit frequency, 124, 267
    of FET, 125
transmission parameters, 99
transport factor, 23
transport model of BJT, 20
transversal filter, 359
travelling-wave amplifier, 292
triode region of FET, 39, 41
T-section, 293
tuned-radio-frequency receiver, 371
TV-display tube, 87
twin-well process, 63
two-integrator oscillator, 323
twoport network, 95, 99, 242
    noise in, 243

unilateralization, 125
upconversion, 374

valence band, 10
variance, 231
varicap, 395
vestigial-sideband signal, 390
voltage amplification capability of transistors, 45

voltage amplifier, 92, 96, 107
  with phantom zero, 280
voltage-controlled oscillator, 395
voltage follower, 95, 285

weak inversion, 48
white noise, 233
wideband amplifiers, 266
Wien-bridge oscillator, 304
Wilson current mirror, 162
Wingtrap, 343

y-parameters, 99, 344
Y-Δ transformation, 339

Zener breakdown, 15
Zener diode, 16
zeros, 68
  effect on transient response, 271
zinc oxide, 333
z-transform, 361